Frontiers in Mathematics

Advisory Editors

Laurent Saloff-Coste, Cornell University, Ithaca, NY, USA

Igor Shparlinski, The University of New South Wales, Sydney, NSW, Australia

Wolfgang Sprößig, TU Bergakademie Freiberg, Freiberg, Germany

This series is designed to be a repository for up-to-date research results which have been prepared for a wider audience. Graduates and postgraduates as well as scientists will benefit from the latest developments at the research frontiers in mathematics and at the "frontiers" between mathematics and other fields like computer science, physics, biology, economics, finance, etc. All volumes are online available at SpringerLink.

José Antonio Ezquerro Fernández •
Miguel Ángel Hernández Verón

Convexity in Newton's Method

José Antonio Ezquerro Fernández
Department of Mathematics and Computation
University of La Rioja
Logroño, La Rioja, Spain

Miguel Ángel Hernández Verón
Department of Mathematics and Computation
University of La Rioja
Logroño, La Rioja, Spain

ISSN 1660-8046   ISSN 1660-8054   (electronic)
Frontiers in Mathematics
ISBN 978-3-031-85753-9   ISBN 978-3-031-85754-6   (eBook)
https://doi.org/10.1007/978-3-031-85754-6

© The Editor(s) (if applicable) and The Author(s), under exclusive license to Springer Nature Switzerland AG 2025

This work is subject to copyright. All rights are solely and exclusively licensed by the Publisher, whether the whole or part of the material is concerned, specifically the rights of translation, reprinting, reuse of illustrations, recitation, broadcasting, reproduction on microfilms or in any other physical way, and transmission or information storage and retrieval, electronic adaptation, computer software, or by similar or dissimilar methodology now known or hereafter developed.
The use of general descriptive names, registered names, trademarks, service marks, etc. in this publication does not imply, even in the absence of a specific statement, that such names are exempt from the relevant protective laws and regulations and therefore free for general use.
The publisher, the authors and the editors are safe to assume that the advice and information in this book are believed to be true and accurate at the date of publication. Neither the publisher nor the authors or the editors give a warranty, expressed or implied, with respect to the material contained herein or for any errors or omissions that may have been made. The publisher remains neutral with regard to jurisdictional claims in published maps and institutional affiliations.

This book is published under the imprint Birkhäuser, www.birkhauser-science.com by the registered company Springer Nature Switzerland AG
The registered company address is: Gewerbestrasse 11, 6330 Cham, Switzerland

If disposing of this product, please recycle the paper.

To Emma, the light that illuminates the way.
—JAE
To my family, the best gift that life has given me.
—MAHV

Preface

Many problems in mathematics, engineering, and other disciplines can be presented as a nonlinear equation using mathematical models. Generally, it is not possible to determine the solutions of such equation exactly, so we usually use approximation methods, which are normally iterative. In this context, the Newton method, which was proposed by Newton in 1669 and later by Raphson in 1690 [93], stands out especially. Hence it is also occasionally called the Newton-Raphson method. The method remains a central technique, probably the best known and most widely used iterative method for finding a solution of an equation, and is often the basis of the most widely used techniques for solving nonlinear equations.

Initially, the Newton method was used to solve polynomial equations. It was later extended to more complicated equations involving transcendent functions, complex functions, and systems of nonlinear equations [45]. The Russian mathematician L. V. Kantorovich was the first to extend the Newton method to general spaces and his research focused on the study of the Newton method in Banach spaces. Hence, in this context, it is also known as *the Newton-Kantorovich method*. This generalization allows combining functional analysis and numerical analysis techniques to address nonlinear problems as diverse as the solution of integral equations, ordinary differential and partial differential equations, variational calculus problems, etc.

With Kantorovich the "modern" study of the Newton method began with numerous interesting publications, both from a theoretical and practical point of view. Kantorovich developed a technique around the 1950s, in various works [64–67], called *the majorant principle*, and gave a powerful result, known today as *the theorem of Newton-Kantorovich*, which is the basis of the well-known *Kantorovich theory* for the Newton method. A notable feature of Kantorovich's result is that it does not assume the existence of a solution of an equation; it is not only a result on the convergence of the Newton method in Banach spaces but also a result on the existence and uniqueness of the solution of the equation to solve. Since then, many other researchers, including ourselves, have been polishing, expanding, and extending Kantorovich's ideas, in one way or another, in several contexts.

This monograph completes the trilogy that we have done on the Newton method in Banach spaces. In the first monograph, *Newton's Method: An Updated Approach of Kantorovich's Theory* [42], we analyze Kantorovich's theory based on the well-known majorant principle of Kantorovich [65] and the method of majorizing sequences of Ortega [74]. There we present an adapted approach of this theory that includes old results, refines old results, proves the most relevant results, and gives alternative approaches that lead to new sufficient semilocal convergence criteria for the Newton method.

In the second monograph, *Mild Differentiability Conditions for Newton's Method in Banach Spaces* [45], we analyze the semilocal convergence of the Newton method under mild differentiability conditions on the first derivative of the operator involved by using a technique based on recurrence relations. The main conclusion is that the domain of parameters associated with the Newton-Kantorovich theorem is improved.

The general aim of this monograph is the study of different iterative methods in Banach spaces and, in particular, those that, in one way or another, are obtained from the best known, the Newton method. We emphasize that, based on the influence of the convexity of the function involved, we can obtain, from certain iterative methods, other iterative methods that improve the previous ones in some sense. Thus, throughout our trilogy on the Newton method, we have considerably extended Kantorovich's initial theory for the Newton method. All of this would not have been possible without the meritorious contribution of many other researchers throughout all these years.

This monograph is addressed to researchers interested in the theory of the Newton method and other iterative methods in Banach spaces. Each chapter contains several theoretical results that are illustrated with interesting applications to nonlinear equations (scalar equations, system of equations, integral equations, boundary value problems, or elliptic problems are included), which clarify the ideas presented previously.

Taking into account the geometric interpretation of the Newton method in the scalar case, we observe that the lower the convexity of the curve $y = f(t)$, the sequence given by the method,

$$t_{n+1} = t_n - \frac{f(t_n)}{f'(t_n)}, \quad n \geq 0, \quad \text{with } t_0 \text{ given,}$$

approaches the solution of the equation $f(t) = 0$ more quickly. To measure the convexity of the function f, we can use the degree of logarithmic convexity [57],

$$L_f(t) = \frac{f(t)f''(t)}{f'(t)^2},$$

which is a point-wise measure of convexity. This measure was introduced by Garay and Hernández-Verón in [49], and is based on the Bohr-Mollerup Theorem [17], which devises a measure of the convexity of a curve when defining the Gamma function. In Chap. 1, we present a detailed study of the degree of logarithmic convexity.

Starting from the degree of logarithmic convexity, we analyze in Chap. 2 the influence that the convexity of the function involved has on the speed of convergence of the Newton method. Next, we extend the degree of logarithmic convexity to Banach spaces, study its existence, and, from it, analyze the convergence of the Newton method in Banach spaces under classical Kantorovich conditions.

From the influence that logarithmic convexity has on the speed of convergence of the Newton method in the scalar case, we obtain in Chap. 3 three acceleration procedures of the Newton method, which lead us to the well-known third-order iterative methods of Chebyshev, Halley, and Super-Halley. We study these three methods in the scalar case and in Banach spaces, and analyze the semilocal convergence of the methods in Banach spaces based on the degree of logarithmic convexity and the well-known method of majorizing sequences introduced by Ortega in [74] for the Newton method.

Notice the importance of the degree of logarithmic convexity, since third-order one-point iterative methods can be written in terms of it.

In Chap. 4 we present a family of third-order one-point Newton-type iterative methods, which includes the three methods previously obtained as accelerations of the Newton method and in addition to many others. We analyze the semilocal convergence of the family in Banach spaces using the degree of logarithmic convexity under Kantorovich type conditions and the well-known technique of recurrence relations [45]. Next, by the already mentioned technique of majorizing sequences, we study the semilocal convergence of the family under milder convergence conditions than those imposed previously. We finish the chapter by presenting a family of iterative methods, included in the initial family, that has order of convergence four when it is applied to solve quadratic equations, and analyze its convergence by the method of majorizing sequences.

We finish the monograph with Chap. 5, where we do an additional study of the Chebyshev method in Banach spaces. In this chapter, taking the Chebyshev method as an example, we try to solve the two main problems of the third-order one-point iterative methods: the evaluation of the second derivative of the operator involved in the algorithm and, as a consequence, the strong convergence conditions usually required. For this, from the Chebyshev method, we construct some families of iterative methods free of second derivatives, but with the same order of convergence three as the Chebyshev method. In particular, we present four modifications of the Chebyshev method that improve it in some sense. The first two modifications improve the efficiency of the Chebyshev method in certain cases. The third modification improves the accessibility of the Chebyshev method and we see that it has the same accessibility as the Newton method. The fourth and last modification consists of constructing derivative-free iterative methods that improve the efficiency of the secant method, which is the best-known derivative-free iterative method. We carry out a convergence study of each modification using the technique of recurrence relations in all of them.

We have developed in detail the proofs of all the theoretical results presented with the aim of facilitating the reading of the monograph. We also highlight that all the ideas presented have been developed by the authors over recent years. The monograph is full of citations to our works, as well as those of other authors who have contributed to an in-depth study of iterative methods in Banach spaces.

Logroño, La Rioja, Spain
March, 2024

José Antonio Ezquerro Fernández
Miguel Ángel Hernández Verón

Contents

1 **The Degree of Logarithmic Convexity** 1
   1.1 Logarithmically Convex Functions 2
   1.2 The Degree of Logarithmic Convexity 6

2 **The Newton Method and Convexity** 15
   2.1 Scalar Equations 16
       2.1.1 Convergence of the Newton Method 16
       2.1.2 Influence of Convexity in the Newton Method 19
   2.2 Operator Equations in Banach Spaces 23
       2.2.1 The Operator Degree of Logarithmic Convexity 23
       2.2.2 Convergence Conditions on the Degree of Logarithmic Convexity . 28
       2.2.3 Kantorovich-Type Convergence Conditions 61

3 **Accelerations of the Newton Method** 89
   3.1 Acceleration Procedures of the Newton Method 89
       3.1.1 Local Convex Acceleration of the Newton Method 90
       3.1.2 Global Convex Acceleration of the Newton Method 92
       3.1.3 Direct Reduction of the Degree of Logarithmic Convexity of a Function 94
       3.1.4 Iterative Methods Obtained by Means of Convex Accelerations of the Newton Method 95
   3.2 The Chebyshev Method 99
       3.2.1 The Chebyshev Method and Convexity 99
       3.2.2 The Chebyshev Method in Banach Spaces 104
   3.3 The Super-Halley Method 114
       3.3.1 The Super-Halley Method and Convexity 115
       3.3.2 The Super-Halley Method in Banach Spaces 119
   3.4 The Halley Method 123
       3.4.1 The Halley Method and Convexity 124
       3.4.2 The Halley Method in Banach Spaces 127

4	**Newton-Like Methods with High Order of Convergence**		131
	4.1	Kantorovich-Type Convergence Conditions	134
		4.1.1 Semilocal Convergence	134
		4.1.2 R-Order of Convergence	141
		4.1.3 Particular Cases	143
		4.1.4 Applications	145
	4.2	Mild Convergence Conditions	147
		4.2.1 Semilocal Convergence	147
		4.2.2 Application	158
	4.3	Quadratic Equations	162
		4.3.1 A Family of Iterations with R-Order of Convergence At Least Four	163
		4.3.2 Semilocal Convergence	164
		4.3.3 R-Order of Convergence	168
		4.3.4 Applications	170
5	**Optimization of the Chebyshev Method**		179
	5.1	Preliminary Analysis	180
	5.2	Optimization 1	185
		5.2.1 Construction of the Method	185
		5.2.2 Analysis of the Method	186
		5.2.3 Convergence Analysis	187
		5.2.4 Relaxing Convergence Conditions	203
	5.3	Optimization 2	210
		5.3.1 Construction of the Method	211
		5.3.2 Analysis of the Method	211
		5.3.3 Convergence Analysis	213
	5.4	Optimization 3	219
		5.4.1 Semilocal Convergence	221
		5.4.2 Application	224
	5.5	Optimization 4	226
		5.5.1 Construction of Two Methods	228
		5.5.2 Analysis of the Methods	229
		5.5.3 Convergence Analysis	233
Bibliography			239

The Degree of Logarithmic Convexity 1

The study of the concavity and convexity of a scalar function is an old problem studied by the mathematicians. It is perfectly known when a function is concave or convex. However, it is not so developed how to measure this concavity or convexity. The degrees of convexity introduced by Jensen and Popoviciu [23] are interesting from the theoretical standpoint, but their practical application is too much difficult. Another measure of the convexity is suggested by the Bohr-Mollerup Theorem [17], where the concept of *log*-convex function appears [83], that is, a function whose logarithm is a convex function. These degrees of convexity are global measures of convexity over an interval. If we think of a pointwise measure of convexity, as the curvature of a scalar function [73], the degree of logarithmic convexity introduced in [59] (see also [55–57]) is a measure of this kind of convexity. This concept is based on the fact that if we apply a concave scalar function to a convex scalar function, the latter reduces its convexity. Thus, taking into account that the logarithm is a concave scalar function, by a successive application of the logarithm to a convex scalar function f at a point x_0, we can measure the resistance that f has to stop being convex at the point x_0. That is, the number of times that the logarithm must be applied to the scalar function f, so that f stops being convex at the point x_0 and becomes concave at x_0, gives an index of pointwise measurement of the convexity of f at the point x_0. The degree of logarithmic convexity is defined in this way.

We begin by introducing the concept of logarithmically convex function, which is a pointwise measure of the convexity of a scalar function, and by defining a few different types of convexity of the function at a point. We then extend the above concept and define an index of pointwise measurement of the convexity that we call degree of logarithmic convexity.

© The Author(s), under exclusive license to Springer Nature Switzerland AG 2025
J. A. Ezquerro Fernández, M. Á. Hernández Verón, *Convexity in Newton's Method*, Frontiers in Mathematics, https://doi.org/10.1007/978-3-031-85754-6_1

1.1 Logarithmically Convex Functions

As we have indicated in the introduction, the Bohr-Mollerup Theorem introduces the concept of *log*-convex function and uses the fact that the gamma function, $\Gamma(x) = \int_0^\infty t^{x-1} e^{-t} \, dt$, is *log*-convex in $\mathbb{R}_+$, as we can see in Fig. 1.1. It is clear that not every convex function is *log*-convex. As we see in Fig. 1.2, the parabola $x^2 + 1$ is a convex function in the interval $[1, 4]$, but it is not *log*-convex in such an interval. As we have also indicated in the introduction, we consider a pointwise measure of convexity, which is given by the concept of *log*-convex function.

It is known that the function $\log : \mathbb{R}_+ \to \mathbb{R}_+$ is concave and infinitely differentiable. Then, for a function f that is convex at x_0 and at least twice differentiable on a neighborhood $\mathcal{V}$ of x_0, the curvature of f at x_0 is given by O'Neill [73]

$$K(f)(x_0) = \frac{f''(x_0)}{\left(1 + f'(x_0)^2\right)^{\frac{3}{2}}}.$$

As we apply the logarithm to f, we first normalize $f(x_0) = 1$, and it then follows that $K(\log(f))(x_0) < K(f)(x_0)$, since

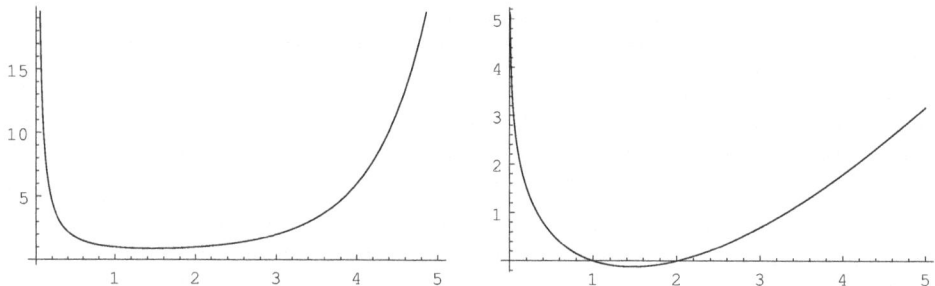

Fig. 1.1 On the left, the Γ function, and on the right, the function $\log \Gamma$

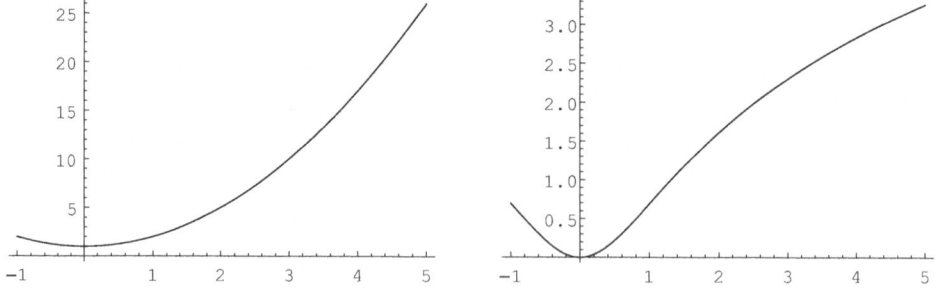

Fig. 1.2 On the left, the function $x^2 + 1$, and on the right, the function $\log(x^2 + 1)$

1.1 Logarithmically Convex Functions

$$(\log(f))'(x_0) = f'(x_0),$$
$$(\log(f))''(x_0) = f''(x_0) - f'(x_0)^2.$$

So, if we compose a convex function with a concave function, we obtain a function "less convex." Therefore, it seems clear that the geometric idea of construction of the degree of logarithmic convexity is verified.

Next, we study the concept of *log*-convex function. For this, throughout this section, we consider a neighborhood $\mathcal{V}$ of x_0 and a positive convex function $f \in \mathcal{C}^m(\mathcal{V})$, $m \geq 2$, in $\mathcal{V}$ and give the following definition.

Definition 1.1 The function f is *log*-convex at x_0 if $\log(f)$ is a convex function at x_0.

If the function f is *log*-convex at x_0, then there exists a function h that is at least twice differentiable and convex, so that $f(x) = \exp(h(x))$ and $h(x) = \log(f(x))$. Then, we have

$$f'(x) = h'(x)\exp(h(x)) = h'(x)f(x),$$
$$f''(x) = h''(x)f(x) + h'(x)f'(x) = (h''(x) + h'(x)^2)f(x),$$

so that

$$h''(x) = \frac{f(x)f''(x) - f'(x)^2}{f(x)^2}, \tag{1.1}$$

which allows studying the convexity of h and $\log(f(x))$.

Now, we remember some different types of convexity of a function at a point.

Definition 1.2 Let $f \in \mathcal{C}^m(\mathcal{V})$, $m \geq 2$. Then:

(A) The function f is strictly convex at x_0 if $f''(x_0) > 0$.
(B) The function f is non-strictly convex at x_0 if there exists an even number $k \in \mathbb{N}$ such that $f''(x_0) = \cdots = f^{(k-1)}(x_0) = 0$ and $f^{(k)}(x_0) > 0$.
(C) The function f has a strong minimum at x_0 if $f'(x_0) = 0$ and $f''(x_0) > 0$.
(D) The function f has a non-strong minimum at x_0 if there exists an even number $k \in \mathbb{N}$ such that $f'(x_0) = f''(x_0) = \cdots = f^{(k-1)}(x_0) = 0$ and $f^{(k)}(x_0) > 0$.

Taking into account the different types of convexity, we obtain the following results. The first one is for cases (A) and (B) of Definition 1.2.

Theorem 1.3 *Suppose that x_0 is not a critical point, i.e., $f'(x_0) > 0$.*

(a) *If the function f is strictly convex at x_0, then we have:*

(a.1) *f is strictly log-convex at x_0, i.e., $\log(f(x))$ is a strictly convex function at x_0 if and only if*

$$\frac{f(x_0)f''(x_0)}{f'(x_0)^2} > 1. \tag{1.2}$$

(a.2) *f is non-strictly log-convex at x_0, i.e., $\log(f(x))$ is a non-strictly convex function at x_0 if and only if there exists an even number $k \in \mathbb{N}$, $k \leq m$, and such that*

$$\frac{f(x_0)^{j-1}f^{(j)}(x_0)}{f'(x_0)^j} = 1, \quad \text{for } 2 \leq j \leq k-1, \quad \text{and}$$

$$\frac{f(x_0)^{k-1}f^{(k)}(x_0)}{f'(x_0)^k} > 1. \tag{1.3}$$

(b) *If f is a non-strictly convex function at x_0, then f is not a log-convex function.*

Proof First, item (a.1) follows immediately from (1.1). Second, since $h(x) = \log(f(x))$ is convex in a non-strict sense at x_0, there exists an even number $k \in \mathbb{N}$ such that $h''(x_0) = \cdots = h^{(k-1)}(x_0) = 0$ and $h^{(k)}(x_0) > 0$. In addition, as $f'(x_0) = h'(x_0)f(x_0)$, we have

$$f^{(n+1)}(x_0) = \sum_{j=0}^{n} \binom{n}{j} h^{(j+1)}(x_0) f^{(n-j)}(x_0). \tag{1.4}$$

Then, for $n = 1$, it follows $f''(x_0) = \frac{f'(x_0)^2}{f(x_0)}$. If we now consider $n = 2$, we then obtain $f'''(x_0) = \frac{f'(x_0)^3}{f(x_0)^2}$. Hence, by recurrence and choosing $n = k-2$, it is easy to prove that $f^{(k-1)}(x_0) = \frac{f'(x_0)^{k-1}}{f(x_0)^{k-2}}$, so that the first expression in (1.3) is proved.

Besides, taking $n = k-1$, we obtain $f^{(k)}(x_0) = \frac{f'(x_0)^k}{f(x_0)^{k-1}} + h^{(k)}(x_0)f(x_0)$. So, as k is an even number, $f(x_0) > 0$ and $h^{(k)}(x_0) > 0$, and then the proof of (1.3) is complete.

To prove that the conditions given in (1.3) are sufficient, just carry out the process described above, but clearing $h^{(j)}(x_0)$ at each step, so that it is easy to check that $h(x) = \log(f(x))$ is convex in a non-strict sense at x_0.

Finally, we observe that $h''(x_0) \leq 0$ follows from (1.1). Thus, $h(x) = \log(f(x))$ is not convex, and therefore $f(x)$ is not logarithmic convex. Item (b) is then complete. ∎

As we can see in the following example, conditions (1.3) with $j = 2$ are not sufficient for the function $f(x)$ to be non-strictly *log*-convex, since we cannot guarantee that $h(x) = \log(f(x))$ is convex if $\frac{f(x_0)f''(x_0)}{f'(x_0)^2} = 1$, as we can see in the following example.

1.1 Logarithmically Convex Functions

Example 1.4 The functions

$$f_1(x) = \exp\left(x + \frac{x^4}{4}\right), \qquad f_2(x) = \exp\left(x - \frac{x^4}{4}\right), \qquad \text{and}$$

$$f_3(x) = \exp\left(x + \frac{x^3}{3}\right)$$

satisfy the conditions at $x_0 = 0$ (1.3) with $j = 2$, since $f_i(0) f_i''(0) = f_i'(0) = 2$ with $i = 1, 2, 3$. However, the function $F_1(x) = \log(f_1(x))$ is convex at 0, since $F_1''(0) > 0$, $F_2(x) = \log(f_2(x))$ is concave at 0, since $F_2''(0) < 0$, and $F_3(x) = \log(f_3(x))$ has an inflection point at 0. See Fig. 1.3.

Second, we consider situations (C) and (D) of Definition 1.2, so that if x_0 is a critical point of f, it is a minimum, and we have the following result.

Theorem 1.5 *If x_0 is a minimum of $f(x)$, then*

$$\lim_{x \to x_0} \frac{f(x) f''(x)}{f'(x)^2} = +\infty. \qquad (1.5)$$

Moreover, $f(x)$ has a strong minimum at x_0 if and only if $\log(f(x))$ has a strong minimum at x_0 and $f(x)$ has a non-strong minimum at x_0 if and only if $\log(f(x))$ has a non-strong minimum at x_0.

Proof If $f'(x_0) = 0$, then two situations are possible:

- $f''(x_0) > 0$.
- $f'(x_0) = f''(x_0) = \cdots = f^{(k-1)}(x_0) = 0$ and $f^{(k)}(x_0) > 0$, where k is an even number.

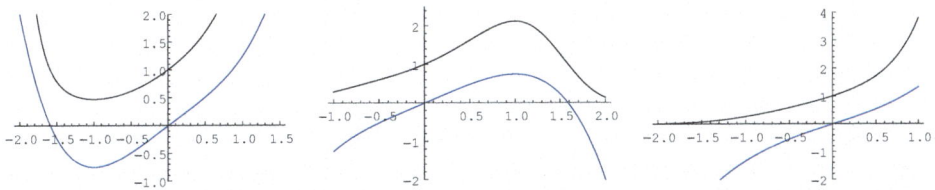

Fig. 1.3 Functions from Example 1.4. On the left, the functions $f_1(x)$ and $F_1(x)$, respectively, black and blue lines. In the middle, the functions $f_2(x)$ and $F_2(x)$, respectively, black and blue lines. On the right, the functions $f_3(x)$ and $F_3(x)$, respectively, black and blue lines

First, if $f''(x_0) > 0$, then (1.5) follows immediately, since $f(x_0) > 0$. Moreover, from (1.1), we have $h''(x_0) > 0$, where $h'(x_0) = 0$, and $h(x) = \log(f(x))$ has a strong minimum at x_0. Again, by (1.1), the reciprocal is clear.

Second, if the second situation is satisfied, we obtain (1.5) by applying L'Hôpital's rule in the limit. In addition, from (1.4), it follows that $f(x)$ has a non-strong minimum at x_0 if and only if $h(x) = \log(f(x))$ has a non-strong minimum at x_0. ∎

Some interesting properties of logarithmically convex functions can be seen in Roberts' book [83].

1.2 The Degree of Logarithmic Convexity

From the concept of logarithmically convex function seen in the previous section, we can consider the extension of this concept by reiterating the successive application of the logarithm function to a function f. Thus, the degree of logarithmic convexity considered in this section measures the number of times that f must be composed with the logarithm to obtain a concave function.

In what follows, we consider a neighborhood $\mathcal{V}$ of the point x_0 and a function $f \in \mathcal{C}^m(\mathcal{V})$, $m \geq 2$, which is positive and convex in $\mathcal{V}$. In addition, to simplify the calculations, we assume that $f(x_0) = 1$. Later, we indicate what to do in another case.

If we denote $\mathcal{C}^m_{(x_0,r)}(\mathcal{V}) = \{g \in \mathcal{C}^m(\mathcal{V}) : g(x_0) = r\}$ and define the operator that normalizes to 1 at the point x_0 as

$$N : \mathcal{C}^m_{(x_0,0)}(\mathcal{V}) \to \mathcal{C}^m_{(x_0,1)}(\mathcal{V}), \quad \text{such that} \quad N(g)(x) = g(x) + 1,$$

we can define the following two sequences:

$$F_1(x) = \log f(x), \qquad G_1(x) = N(F_1)(x),$$
$$F_2(x) = \log G_1(x), \qquad G_2(x) = N(F_2)(x),$$
$$\vdots \qquad\qquad\qquad \vdots$$
$$F_n(x) = \log G_{n-1}(x), \qquad G_n(x) = N(F_n)(x).$$

Notice that $\{F_n(x_0)\} \in \mathcal{C}^m_{(x_0,0)}(\mathcal{V})$ and $\{G_n(x_0)\} \in \mathcal{C}^m_{(x_0,1)}(\mathcal{V})$, for each $n \in \mathbb{N}$.

To measure the resistance of f to be concave by the logarithm function, we study the convexity of the functions F_n at x_0. In the following example, this idea is shown.

Example 1.6 If we consider $f(x) = x - 3\cos(x-1) + 3$ and $x_0 = 1$, we obtain $F_1''(x_0) = 2$, $F_2''(x_0) = 1$, $F_3''(x_0) = 0$, and $F_4''(x_0) = -1$, so that the function $f(x)$ remains convex

1.2 The Degree of Logarithmic Convexity

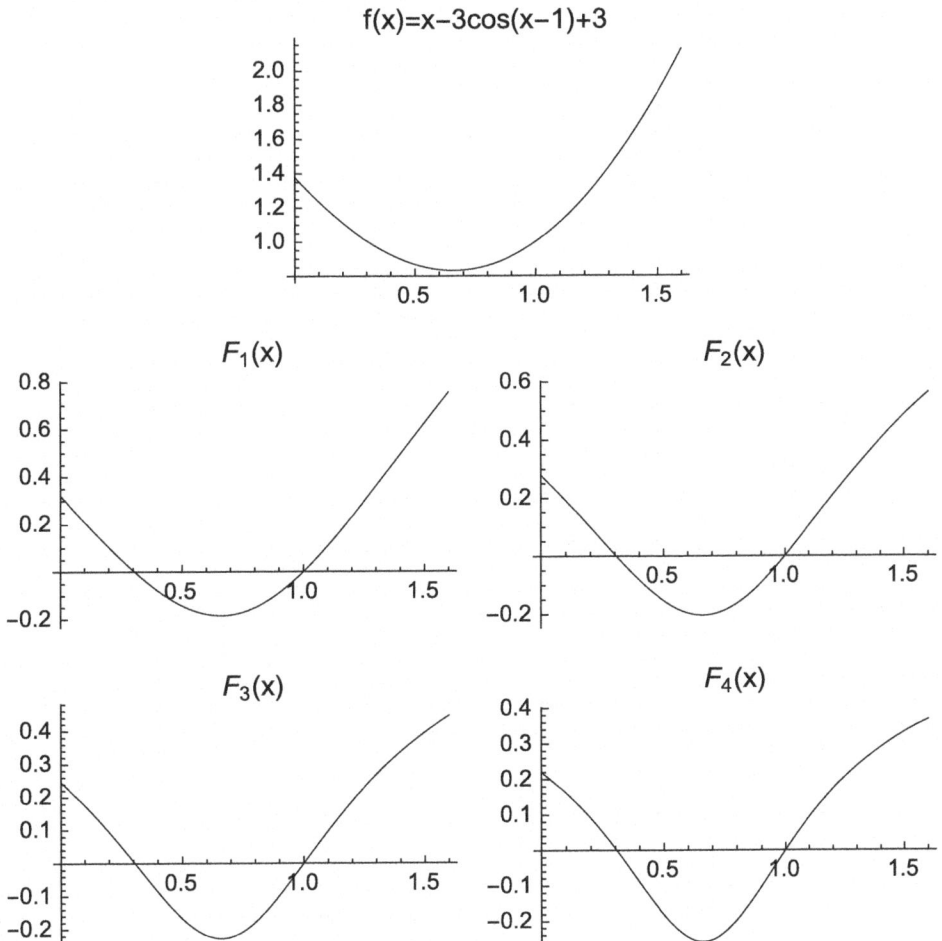

Fig. 1.4 The functions $f(x)$, $F_1(x)$, $F_2(x)$, $F_3(x)$, and $F_4(x)$ from Example 1.6

after two applications of the logarithm function at x_0, we have an inflection point at x_0 after three applications, and we already obtain a concave function at x_0 after four applications. See Fig. 1.4.

After that, we introduce the concept of n-th log-convex function, which is a generalization of the log-convex function analyzed previously.

Definition 1.7 The function $f(x)$ is n-th log-convex at x_0 if $F_n(x)$ is convex at x_0. If $f(x)$ is n-th log-convex at x_0, for all $n \in \mathbb{N}$, we say that $f(x)$ is infinitely log-convex.

Next, we give an analytic characterization of these concepts and obtain an expression for the successive derivatives of each function F_n, $n \in \mathbb{N}$.

Lemma 1.8 *For each $n \in \mathbb{N}$, we have:*

(a) $F_n'(x_0) = f'(x_0)$.
(b) $F_n''(x_0) = f''(x_0) - nf'(x_0)^2$.

Proof Both items (a) and (b) are proved by mathematical induction. Item (a) is clear for $n = 1$. Suppose then $F_p'(x_0) = f'(x_0)$, for all $p \in \mathbb{N}$, with $p < n$. As $F_n(x) = \log G_{n-1}(x)$, it follows $F_n'(x_0) = G_{n-1}'(x_0)$, since $G_{n-1} \in \mathcal{C}_{(x_0,1)}^m(V)$. From the construction of the sequences $\{F_n\}_{n \in \mathbb{N}}$ and $\{G_n\}_{n \in \mathbb{N}}$, it is clear that $G_{n-1}'(x_0) = F_{n-1}'(x_0)$, and, by induction hypothesis, we obtain $F_n'(x_0) = G_{n-1}'(x_0) = f'(x_0)$, and the proof of item (a) is complete.

From $F_1(x) = \log f(x)$ and (1.1), item (b) is satisfied for $n = 1$. Next, if we suppose $F_p''(x_0) = f''(x_0) - pf'(x_0)^2$, for all $p \in \mathbb{N}$, with $p < n$, then

$$F_n'(x) = \frac{F_{n-1}'(x)}{F_{n-1}(x) + 1} \quad \text{and} \quad F_n''(x) = \frac{F_{n-1}''(x)(F_{n-1}(x) + 1) - F_{n-1}'(x)^2}{(F_{n-1}(x) + 1)^2},$$

since $F_n(x) = \log G_{n-1}(x) = \log(F_{n-1}(x) + 1)$. Therefore, taking into account item (a) and $F_{n-1} \in \mathcal{C}_{(x_0,0)}^m(V)$ and applying induction hypothesis, we obtain item (b). ∎

Notice that if the function f is n-th log-convex at x_0, then f is a p-th log-convex function at x_0 with $p \in \mathbb{N}$ and $1 \leq p \leq n$.

Now, taking into account the different types of convexity that can be given in x_0, we analytically characterize the concept of n-th log-convex function at x_0.

As in the previous section, we begin analyzing cases (A) and (B) of Definition 1.2. We then start this study when x_0 is not a critical point of f. So, there are two situations: $f(x)$ is strictly convex at x_0 or $f(x)$ is non-strictly convex. Denote the integer part of a real number r by $[r]$.

Theorem 1.9 *Let $f(x)$ be a strictly convex function at x_0 with $f'(x_0) \neq 0$, and denote $r = \frac{f''(x_0)}{f'(x_0)^2}$. Then:*

(a) *If $r \notin \mathbb{N}$, then:*
   (a.1) *For $p = [r] \in \mathbb{N}$, $F_p(x)$ is strictly convex at x_0, and $F_{p+1}(x)$ is strictly concave at x_0, so that $f(x)$ is p-th log-convex at x_0 and is not $(p + 1)$-th log-convex at x_0,*
   (a.2) *For $n \leq p$, $f(x)$ is n-th log-convex at x_0.*

1.2 The Degree of Logarithmic Convexity

(b) *If $r \in \mathbb{N}$, then:*

(b.1) $f(x)$ is $(r-1)$-th log-convex at x_0, $F_{r-1}(x)$ strictly convex at x_0, and $F_{r+1}(x)$ is strictly concave at x_0, so that $f(x)$ is not $(r+1)$-th log-convex at x_0.

(b.2) $f(x)$ is r-th log-convex at x_0 if and only if there exists an even number $k \in \mathbb{N}$ such that

$$F_{r-1}^{(j)}(x_0) = f'(x_0)^j, \quad \text{for} \quad 2 \leq j \leq k-1, \quad \text{and} \quad F_{r-1}^{(k)}(x_0) > f'(x_0)^k.$$

Proof Item (a.1) follows easily from item (b) of Lemma 1.8, since $f''(x_0) > pf'(x_0)^2$ and $f''(x_0) < (p+1)f'(x_0)^2$ with $p = [r]$. Item (a.2) is clear from item (b) of Lemma 1.8.

Item (b.1) is immediate from item (b) of Lemma 1.8. To prove item (b.2), we take into account that $F_r(x) = \log G_{r-1}(x)$ and $G_{r-1}(x) = F_{r-1}(x) + 1$ and have

$$F_{r-1}^{(i+1)}(x_0) = F_r^{(i+1)}(x_0) + \sum_{j=0}^{i-1} \binom{i}{j} F_r^{(j+1)}(x_0) F_{r-1}^{(i-j)}(x_0). \tag{1.6}$$

Moreover, $F_r''(x_0) = f''(x_0) - rf'(x_0)^2 = 0$, so that f is r-th *log*-convex at x_0 if and only if there exists $k \in \mathbb{N}$, even, such that $F_r''(x_0) = F_r'''(x_0) = \cdots = F_r^{(k-1)}(x_0) = 0$, $F_r^{(k)}(x_0) > 0$, and $F_r'(x_0) = f'(x_0) \neq 0$. Therefore, $F_r(x)$ is a convex function in a non-strict sense at x_0.

Furthermore, from (1.6), we have

$$F_{r-1}''(x_0) = F_r'(x_0) F_{r-1}'(x_0) = f'(x_0)^2,$$

$$F_{r-1}'''(x_0) = F_r'(x_0) F_{r-1}''(x_0) + F_r''(x_0) F_{r-1}'(x_0)$$

$$= F_r'(x_0) F_{r-1}''(x_0)$$

$$= f'(x_0)^3.$$

In addition, by recurrence, if we suppose $F_{r-1}^{(k-2)}(x_0) = f'(x_0)^{k-2}$, then

$$F_{r-1}^{(k-1)}(x_0) = F_r^{(k-1)}(x_0) + \sum_{j=0}^{k-3} \binom{k-2}{j} F_r^{(j+1)}(x_0) F_{r-1}^{(k-2-j)}(x_0)$$

$$= \binom{k-2}{0} F_r'(x_0) F_{r-1}^{(k-2)}(x_0)$$

$$= f'(x_0)^{k-1}.$$

Finally,

$$F_{r-1}^{(k)}(x_0) = F_r^{(k)}(x_0) + \sum_{j=0}^{k-2} \binom{k-1}{j} F_r^{(j+1)}(x_0) F_{r-1}^{(k-1-j)}(x_0)$$

$$= F_r^{(k)}(x_0) + F_r'(x_0) F_{r-1}^{(k-1)}(x_0)$$

$$= F_r^{(k)}(x_0) + f'(x_0)^k$$

$$> f'(x_0)^k,$$

and item (b.2) is proved. ∎

Next, we see an example where item (b) of the last result is illustrated.

Example 1.10 For the functions

$$f_1(x) = \exp\left(\exp\left(x + \frac{x^4}{4}\right) - 1\right),$$

$$f_2(x) = \exp\left(\exp\left(x - \frac{x^4}{4}\right) - 1\right),$$

$$f_3(x) = \exp\left(\exp\left(x + \frac{x^3}{3}\right) - 1\right),$$

and $x_0 = 0$, we have $\dfrac{f_i''(x_0)}{f_i'(x_0)^2} = 2$, with $i = 1, 2, 3$. If we denote by $\{F_n^i\}$ the sequence $\{F_n\}$ corresponding to f_i with $i = 1, 2, 3$, we can check that $F_2^{1''}(x_0) > 0$, $F_2^{2''}(x_0) < 0$, and $F_2^3(x)$ has an inflection point at the origin. See Fig. 1.5.

The situation that remains to be analyzed is when the function $f(x)$ is convex, in a non-strict sense, at x_0 it is clear, since we know that in this case there exists an even $k \in \mathbb{N}$

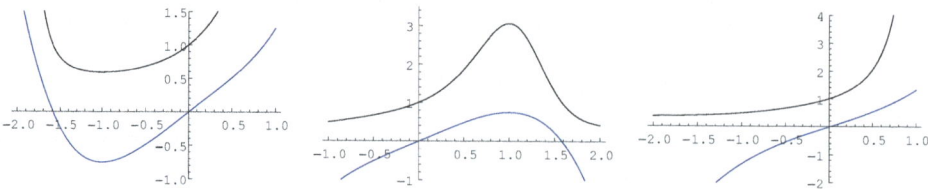

Fig. 1.5 Functions from Example 1.10. On the left, the functions $f_1(x)$ and $F_2^1(x)$, respectively, black and blue lines. In the middle, the functions $f_2(x)$ and $F_2^2(x)$, respectively, black and blue lines. On the right, the functions $f_3(x)$ and $F_2^3(x)$, respectively, black and blue lines

1.2 The Degree of Logarithmic Convexity

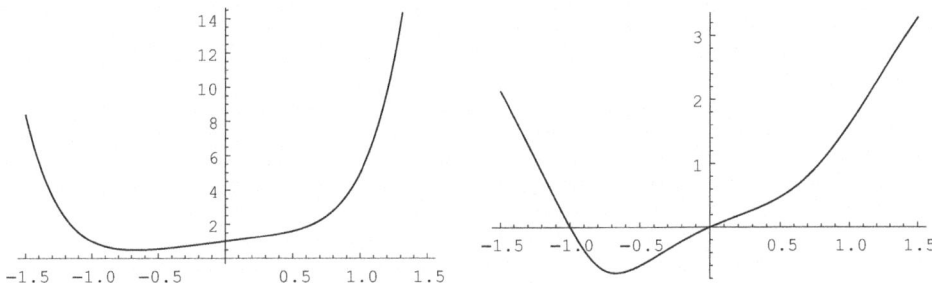

Fig. 1.6 Functions $f(x) = x^6 + x^5 + x^4 + x + 1$ and $F_1(x) = \log f(x)$ and $x_0 = 0$

such that $f''(x_0) = f'''(x_0) = \cdots = f^{(k-1)}(x_0) = 0$ and $f^{(k)}(x_0) > 0$. In addition, as $f'(x_0) \neq 0$, taking into account item (b) of Lemma 1.8, it follows immediately that $F_n(x)$ is a strictly concave function at x_0, for all $n \in \mathbb{N}$. Figure 1.6 illustrates this situation.

After that, we characterize the concept of n-th *log*-convex function at x_0 when x_0 is a minimum of f.

Theorem 1.11 *With the above notation, we have:*

(a) *If x_0 is a minimum of $f(x)$, then x_0 is a minimum of the same type of $F_n(x)$, for all $n \in \mathbb{N}$.*
(b) *x_0 is a minimum of $f(x)$ if and only if $f(x)$ is infinitely log-convex at x_0.*

Proof We start with item (a). We suppose that x_0 is a strict minimum of $f(x)$, so that $f'(x_0) = 0$ and $f''(x_0) > 0$. From items (a) and (b) of Lemma 1.8, it follows that $F_n'(x_0) = 0$ and $F_n''(x_0) > 0$ for all $n \in \mathbb{N}$. Therefore, x_0 is a strict minimum of $F_n(x)$, for all $n \in \mathbb{N}$.

Next, if x_0 is a non-strict minimum of $f(x)$, then there exists an even $k \in \mathbb{N}$ such that $f'(x_0) = f''(x_0) = \cdots = f^{(k-1)}(x_0) = 0$ and $f^{(k)}(x_0) > 0$. Now, by invoking mathematical induction, we see that x_0 is a non-strict minimum of any $F_n(x)$.

From $F_1(x) = \log f(x)$ and (1.4), we obtain

$$f^{(n+1)}(x_0) = \sum_{j=0}^{n} \binom{n}{j} F_1^{(j+1)}(x_0) f^{(n-j)}(x_0).$$

As a consequence, $f'(x_0) = F_1'(x_0) f(x_0)$ and $F_1'(x_0) = 0$. If we assume that $F_1^{(i)}(x_0) = 0$, for $i = 1, 2, \ldots, k-1$, it follows in the same way that

$$f^{(k-1)}(x_0) = \sum_{j=0}^{k-2} \binom{k-2}{j} F_1^{(j+1)}(x_0) f^{(k-2-j)}(x_0) = F_1^{(k-1)}(x_0) f(x_0),$$

and $F_1^{(k-1)}(x_0) = 0$. Moreover,

$$f^{(k)}(x_0) = \sum_{j=0}^{k-2} \binom{k-1}{j} F_1^{(j+1)}(x_0) f^{(k-1-j)}(x_0) = F_1^{(k)}(x_0) f(x_0) = F_1^{(k)}(x_0).$$

Therefore, $F_1'(x_0) = F_1''(x_0) = \cdots = F_1^{(k-1)}(x_0) = 0$ and $F_1^{(k)}(x_0) = f^{(k)}(x_0) > 0$, so that $F_1(x)$ has a non-strict minimum at x_0.

Then, we assume that x_0 is a non-strict minimum of $F_1(x), F_2(x), \ldots, F_{n-1}(x)$. Since $F_n(x) = \log G_{n-1}(x)$, it follows from (1.4) that

$$G_{n-1}^{(i+1)}(x_0) = \sum_{j=0}^{i} \binom{i}{j} F_n^{(j+1)}(x_0) G_{n-1}^{(i-j)}(x_0).$$

Taking into account $G_{n-1}(x) = F_{n-1}(x) + 1$ and $G_{n-1}(x_0) = 1$, we have

$$F_{n-1}^{(i+1)}(x_0) = F_n^{(i+1)}(x_0) + \sum_{j=0}^{i-1} \binom{i}{j} F_n^{(j+1)}(x_0) F_{n-1}^{(i-j)}(x_0).$$

Proceeding as above for $n = 1$ and applying that x_0 is a non-strict minimum of $F_{n-1}(x)$, we obtain that x_0 is a non-strict minimum of $F_n(x)$, for all $n \in \mathbb{N}$.

For item (b), if x_0 is a minimum of $f(x)$, then, from item (a), x_0 is a minimum of $F_n(x)$, for all $n \in \mathbb{N}$. Thus, $F_n(x)$ is convex at x_0, for all $n \in \mathbb{N}$, and therefore $f(x)$ is infinitely log-convex at x_0.

Reciprocally, if $f(x)$ is infinitely log-convex at x_0, then $F_n(x)$ is convex at x_0, for all $n \in \mathbb{N}$. From item (b) of Lemma 1.8, it follows that $f''(x_0) - nf'(x_0)^2 \geq 0$, for all $n \in \mathbb{N}$, and, from item (a) of Lemma 1.8, we have $f'(x_0) = F_n'(x_0) = 0$. As a consequence,

$$f'(x_0) = 0 \quad \text{and} \quad f''(x_0) > 0 \qquad \text{or} \qquad f'(x_0) = f''(x_0) = 0.$$

The first case means that x_0 is a strict minimum of $f(x)$. The second case leads us to the fact that x_0 is a non-strict minimum of $f(x)$, since $f(x)$ is convex. ∎

Once the different possibilities of convexity that can occur at the point x_0 have been studied, it is easy to observe that these are perfectly determined by analyzing the value of $\frac{f''(x_0)}{f'(x_0)^2}$, which allows studying the concavity or convexity of the function $F_n(x)$ at x_0, for all $n \in \mathbb{N}$. This is the origin of the definition of *degree of logarithmic convexity*.

Definition 1.12 The degree of logarithmic convexity of a function $f(x)$ at the point x_0 is the real number

1.2 The Degree of Logarithmic Convexity

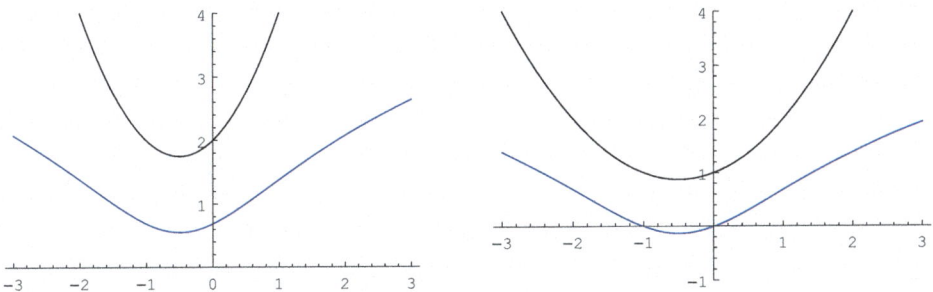

Fig. 1.7 On the left, the functions $f(x) = x^2 + x + 2$ and $\log f(x)$, respectively, black and blue lines. On the right, the functions $f^*(x)$ and $\log f^*(x)$, respectively, black and blue lines. Point $x_0 = 0$

$$\widehat{L}_f(x_0) = \frac{f''(x_0)}{f'(x_0)^2},$$

provided that x_0 is not a minimum of $f(x)$, in which case $\widehat{L}_f(x_0) = +\infty$.

Notice that $\widehat{L}_f : \mathcal{C}^m_{(x_0,1)}(\mathcal{V}) \to \mathbb{R}$, and, if we consider convex functions, the image of $\widehat{L}_f$ is in $[0, +\infty]$.

In the case where the function $f(x)$ takes an arbitrary positive value at the point x_0, we consider the function $f^*(x) = \dfrac{f(x)}{f(x_0)}$, since $\log f(x)$ and $\log f^*(x)$ have "the same convexity" at the point x_0, since they differ in a constant (see Fig. 1.7). Then, we obtain that the degree of logarithmic convexity of $f(x)$ at x_0 is the real number

$$L_f(x_0) = \widehat{L}_{f^*}(x_0) = \frac{f(x_0) f''(x_0)}{f'(x_0)^2},$$

provided that x_0 is not a minimum of $f(x)$, in which case $L_f(x_0) = +\infty$.

Note that if we consider a positive function in a neighborhood of the point x_0, it is clear that the degree of logarithmic convexity of a convex function is a positive real number and is "more convex" with respect to the logarithm the larger this number is. On the other hand, although we lose the geometric meaning of the concept, it is easy to understand that we can formally extend the definition of L_f to the set $\mathcal{C}^2(\mathcal{V})$.

Besides, it is interesting to say that, in the case that the function $f(x)$ does not take the value 1 at the point x_0, the normalization considered is not the usual one, since it is normally considered $N(f)(x) = f(x) - f(x_0) + 1$. This is because the translation of a function influences the concavity or convexity of its logarithm. So, for example, if we take the function $f(x) = \lambda + \exp(x)$, $x_0 = 0$, and $\lambda \in (-1, +\infty)$ and consider the function $F(x) = \log f(x)$, we obtain $\text{sgn}(F''(x_0)) = \text{sgn}(\lambda)$. Therefore, if the graph

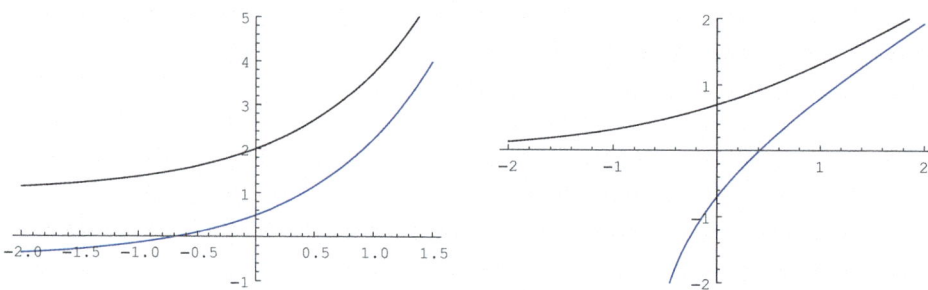

Fig. 1.8 On the left, the functions $f(x) = 1 + \exp(x)$ and $g(x) = -\frac{1}{2} + \exp(x)$, respectively, black and blue lines. On the right, the functions $\log f(x)$ and $\log g(x)$, respectively, black and blue lines. Point $x_0 = 0$

of the exponential function is translated, so that it takes a value between -1 and 0 at the origin, the logarithm of this function is a concave function, where the logarithm of any translation is convex with $\lambda \in \mathbb{R}_+$. Then, it is clear that the degree of logarithmic convexity is not invariant by translations. If, for example, we choose $f(x) = 1 + \exp(x)$ and $g(x) = -\frac{1}{2} + \exp(x)$, we observe in Fig. 1.8 that its logarithms have different convexities at $x_0 = 0$.

Finally, some properties of the degree of logarithmic convexity are given in [57].

The Newton Method and Convexity

It is well known that the most widely used iterative method for solving nonlinear equations is the Newton method. From the geometric interpretation of the method in the scalar case, we see that the method runs faster the lower the convexity of the function involved in the equation to be solved. To measure the convexity of the function, we can use the degree of logarithmic convexity defined in the previous chapter, which, as we have seen, is a point measure of convexity. We begin the chapter studying the convergence of the Newton method in $\mathbb{R}$ and continue looking at the influence that the convexity of the function involved has on the speed of convergence of the method [31]. Next, we do two analyses of the convergence of the method in Banach spaces. For this, we first extend the degree of logarithmic convexity to Banach spaces and, although the geometric sense that it has in the scalar case is lost, we analyze its existence based on the classical Kantorovich conditions. Some ideas collected here are developed in [30, 52], where we find results of semilocal, global, and local convergence. In the first study, the convergence of the method is analyzed based on the type of conditions required to the degree of logarithmic convexity operator, emphasizing the majorant principle introduced by Kantorovich to study the semilocal convergence of the Newton method [67], the fixed point techniques for the study of the global convergence [47], which guarantee the extension of the domain of starting points, together with the use of auxiliary points, and the ideas contributed by Dennis and Schnabel in [28] to obtain the local convergence of the method. In the second study, we use the degree of logarithmic convexity operator under Kantorovich-type conditions [46]. In semilocal convergence studies, we use the theoretical significance of the method to draw conclusions about the existence and uniqueness of solution in some cases and to give a priori error estimates based on the Ostrowski technique [76].

2.1 Scalar Equations

To approximate a solution of a scalar equation $f(x) = 0$, the Newton method is constructed from an initial approximation x_0, so that the following approximation x_1 to the solution of the equation is obtained as the cutoff point of the tangent line to the function f at the point x_0 and the axis of abscissas. Obviously, if f represents a line, we then obtain that x_1 is a solution. Therefore, as we will see, the convexity of the function f plays an important role in the studies of the convergence of the Newton method and its speed of convergence. We use the degree of logarithmic convexity to measure the convexity of f at each point and to establish conditions on the convergence of the method.

2.1.1 Convergence of the Newton Method

The iterative method

$$x_{n+1} = N_f(x_n) = x_n - \frac{f(x_n)}{f'(x_n)}, \quad n \geq 0, \quad \text{with } x_0 \text{ given,} \qquad (2.1)$$

for solving a nonlinear real equation $f(x) = 0$ is known as the Newton method, and sometimes it is alluded to as the Newton-Raphson method [93]. The sequence (2.1) is generated under sufficient conditions. For this, we consider that f is defined on an interval $[a, b]$ and assume that f satisfies the so-called *Fourier conditions* [75] that guarantee the convergence of (2.1):

Let $f : [a, b] \to \mathbb{R}$ be a function such that $f \in C^2([a, b])$ and

- $f(a)f(b) < 0$.
- $f'(x) \neq 0$, for all $x \in [a, b]$.
- $f''(x)$ does not change sign in $[a, b]$.

Then, if $x_0 \in [a, b]$ such that $f(x_0)f''(x_0) > 0$, the Newton method (2.1) converges to the unique solution of $f(x) = 0$ in $[a, b]$.

From the last, in order to guarantee the convergence of the method, the second derivative f'' must not change sign in the interval $[a, b]$. We then wonder what happens if f'' changes sign in $[a, b]$. If, for example, f is nondecreasing in $[a, b]$ and there exists $\xi \in [a, b]$ such that $f''(\xi) = 0$, with $f''(\xi - \epsilon) > 0$, $f''(\xi + \epsilon) < 0$, and $\epsilon > 0$, two situations can be given. First, if the solution of the equation $f(x) = 0$ is an inflection point of f in $[a, b]$ (as we can see in the graph on the left of Fig. 2.1), then, whether starting point x_0 is to the right or to the left of the solution, it must be guaranteed that sequence $\{x_n\}$ lies in $[a, b]$. Second, if the solution of equation $f(x) = 0$ is not an inflection point of f in $[a, b]$ (as we can see in the graph on the right of Fig. 2.1), there can be more than

2.1 Scalar Equations

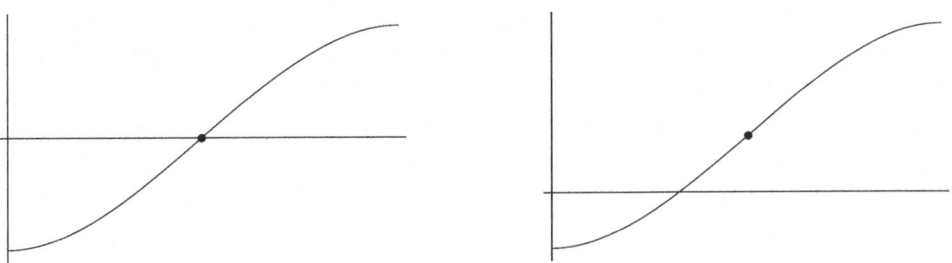

Fig. 2.1 Graphs of functions with inflection points in $[a, b]$

one inflection point in the interval $[a, b]$. Suppose, for example, that there are n inflection points $\xi_1, \xi_2, \ldots, \xi_n \in [a, b]$. We can then order them, for example, in increasing order, $a \leq \xi_1 \leq \xi_2 \leq \cdots \leq \xi_n \leq b$, so that the solution x^* is such that $x^* \in (\xi_i, \xi_{i+1})$, with $i = 1, 2, \ldots, n-1$, or $x^* \in (a, \xi_1)$ or $x^* \in (\xi_n, b)$. In all these cases, we are already in the Fourier conditions in the interval in which the solution x^* is.

The first situation is reflected in the following result, which solves the problem that the Fourier conditions are insufficient to guarantee the convergence of the Newton method when f'' changes sign. First, we remember that the degree of logarithmic convexity of f is given by

$$L_f(x) = \frac{f(x) f''(x)}{f'(x)^2} \tag{2.2}$$

and observe then that $L_f(x) = N_f'(x)$ with N_f defined in (2.1).

Theorem 2.1 *Let $f \in C^2([a,b])$ such that $f(a)f(b) < 0$, $f'(x) \neq 0$, and $|L_f(x)| \leq K < 1$ in $[a, b]$. Then, the Newton method (2.1) starting at any $x_0 \in [a, b]$ converges to a solution x^* of $f(x) = 0$ with $a \leq N_f(x_0) \leq b$.*

Proof First, we prove that $\{x_n\} \subset [a, b]$. Obviously, $x_1 = N_f(x_0) \in [a, b]$, and by the Mean Value Theorem, we obtain

$$x_2 - x^* = N_f(x_1) - N_f(x^*) = N_f'(\theta_1)(x_1 - x^*), \quad \theta_1 \in (x^*, x_1).$$

In addition, since $|N_f'(x)| = |L_f(x)| \leq K < 1$, we have $|x_2 - x^*| < K|x_1 - x^*| < |x_1 - x^*|$. Then, by mathematical induction on n, it is easy to follow that $|x_n - x^*| < K^{n-1}|x_1 - x^*| < |x_1 - x^*|$. Therefore, $x_n \in [a, b]$, for all $n \in \mathbb{N}$.

Second, to prove the convergence of (2.1), it is enough to take into account that $|x_n - x^*| < K^n|x_0 - x^*|$ in $[a, b]$, for all $n \in \mathbb{N}$. Then, $\lim_n |x_n - x^*| = 0$, and the sequence (2.1) converges to x^*. We see that x^* is a solution of $f(x) = 0$ without more than going to the limit in (2.1), since N_f is continuous. ∎

Note that the last result allows improving the commonly used conditions of Fourier. In the next example, the function has an inflection point in a zero, so that the function does not satisfy the Fourier conditions, and as a consequence, we cannot guarantee the convergence of the Newton method from these conditions.

Example 2.2 Consider $f(x) = -x^3 + 3x^2 - 2$ in the interval $\left[\frac{1}{10}, \frac{19}{10}\right]$. Note this function has an inflection point in the zero $x^* = 1$, so that f'' changes the sign in $\left[\frac{1}{10}, \frac{19}{10}\right]$ and the Fourier conditions are then not satisfied. On the other hand, as $L_f(x) = \frac{6(-x^3+3x^2-2)(1-x)}{-3x^2+6x}$ satisfies $|L_f(x)| < 1$ in $\left[\frac{1}{10}, \frac{19}{10}\right]$, it is enough to prove that $x_1 \in \left[\frac{1}{10}, \frac{19}{10}\right]$ to obtain that the Newton sequence (2.1) lies in $\left[\frac{1}{10}, \frac{19}{10}\right]$. If we now choose $x_0 = 1.6$, then $x_1 = 0.775 \in \left[\frac{1}{10}, \frac{19}{10}\right]$, and by Theorem 2.1, the Newton sequence (2.1) converges to $x^* = 1$ in $\left[\frac{1}{10}, \frac{19}{10}\right]$. See Table 2.1.

Moreover, from Theorem 2.1, we can also improve the domain of starting points, as we see in the following example.

Example 2.3 Consider $f(x) = \frac{1}{2} + \sin x$. Observe that f has an inflection point in $x = 0$. As $L_f(x) = -\frac{\sin x(1+2\sin x)}{2\cos^2 x}$, we have that $|L_f(x)| < 1$ in $[-1.00297\ldots, 0.634867\ldots]$, and the conditions of Theorem 2.1 are then satisfied. So, if we now choose $x_0 = 0.6$, then $x_1 = -0.6899 \in [-1.00297\ldots, 0.634867\ldots]$, and as a consequence, the Newton sequence (2.1) lies in the interval $[-1.00297\ldots, 0.634867\ldots]$. Then, by Theorem 2.1, the Newton method converges to the solution $x^* = -0.5235987755982988711\ldots$ See Table 2.2.

From the graph of $f(x) = \frac{1}{2} + \sin x$, Fig. 2.2, we notice that we would have to restrict to the interval $[-\frac{\pi}{2}, \varepsilon]$, with $\varepsilon < 0$, in order to apply the Fourier conditions. Therefore, from Theorem 2.1, we enlarge the domain of starting points for the Newton method.

Table 2.1 Newton's sequence (2.1) from Example 2.2

n	x_n
0	1.6000000000000000...
1	0.7750000000000000...
2	1.0079986833443050...
3	0.9999996588133421...
4	1.0000000000000000...

2.1 Scalar Equations

Table 2.2 Newton's sequence (2.1) from Example 2.3

n	x_n
0	0.600000000000000000000...
1	-0.6899509655978506667...
2	-0.5129726247150719697...
3	-0.5235667752006047706...
4	-0.5235667753027045709...
5	-0.5235667755982988737...
6	-0.5235667755982988705...
7	-0.5235667755982988742...
8	-0.5235667755982988711...
9	-0.5235667755982988711...

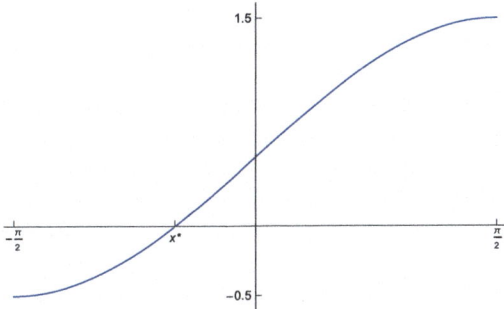

Fig. 2.2 Graph of $f(x) = \frac{1}{2} + \sin x$ in $[-\pi/2, \pi/2]$

2.1.2 Influence of Convexity in the Newton Method

In this section, we analyze the influence of the convexity of the curve $y = f(x)$ in the speed of convergence of the Newton sequence (2.1). To measure the convexity of the curve $y = f(x)$, we use the degree of logarithmic convexity given in (2.2) if $f'(x) \neq 0$. Remember that $N'_f(x) = L_f(x)$ with N_f defined in (2.1).

Taking into account the geometric interpretation of the Newton method, which is based on the approximation of the function $f(x)$ by a tangent line (see Fig. 2.3), we deduce that the lower the convexity of the cure $y = f(x)$, the faster the speed of convergence of sequence (2.1) to a solution x^* of the equation $f(x) = 0$. In particular, if $y = f(x)$ is a line, then $x_1 = x^*$.

Next, we develop the justification of these geometric considerations. For this, we consider a scalar function f defined in an interval $[a, b]$, where the Fourier conditions are satisfied. We can assume without loss of generality that f is convex and increasing. In another case, it is enough to change $f(x)$ by $f(-x)$, $-f(x)$, or $-f(-x)$. Under these conditions, it is clear that f has a unique zero x^* in $[a, b]$.

We consider a function g that satisfies the same conditions as f in $[a, b]$, assume that x^* is also the unique zero of g in $[a, b]$, and think about

Fig. 2.3 Geometric interpretation of the Newton method

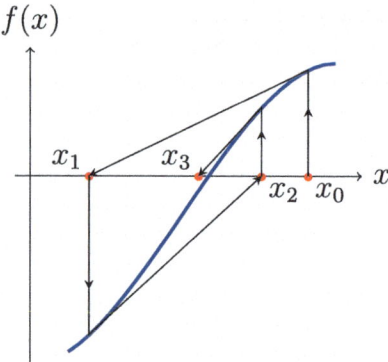

$$y_{n+1} = N_g(y_n) = y_n - \frac{g(y_n)}{g'(y_n)}, \quad n \geq 0, \quad \text{with } y_0 = x_0. \tag{2.3}$$

Then, by the degree of logarithmic convexity of the functions f and g, we compare the speed of convergence of sequences (2.1) and (2.3).

Now, we remember the important concept of order of convergence that allows us to measure how quickly a sequence converges [88].

Definition 2.4 Let $\{v_n\}_{n \in \mathbb{N}}$ be a sequence such that $v_n \to v^*$ and $v_n \neq v^*$ for all $n \in \mathbb{N}$. If there exist a real number p and a nonzero constant α such that

$$\lim_{n \to \infty} \frac{|v_{n+1} - v^*|}{|v_n - v^*|^p} = \alpha,$$

we say that the sequence $\{v_n\}_{n \in \mathbb{N}}$ converges to v^* with order p and asymptotic error constant α.

In general, a sequence converges more quickly the higher its order of convergence and the lower the value of its asymptotic error constant. And we say that an iterative method is of order p if the sequence of generated points converges with order p.

When the speed of convergence of two sequences with the same order of convergence is compared, the problem of accessibility to a solution of an equation appears. So, we first study, for the Newton method, the problem of accessibility to the solution.

Theorem 2.5 *Let f be an increasing convex function in $[a, b]$, x^* the unique zero of f in $[a, b]$, $x_0 \in [a, b]$ with $f(x_0) > 0$, and $\{x_n\}$ the sequence (2.1). If $x_{n-1} \neq x^*$ and $x_n = x^*$, then $f(x) = ax + b$ with $a, b \in \mathbb{R}$, for all $x \in (x^*, x_{n-1})$. Moreover, $x_{n+k} = x^*$ for all $k \in \mathbb{N}$.*

2.1 Scalar Equations

Proof From N_f, defined in (2.1), we deduce that $N_f(x) \leq x^* = N_f(x_{n-1}) = x_n$ in (x^*, x_{n-1}), since N_f is a nondecreasing function in $[x^*, b]$ and $N_f'(x) = L_f(x) \geq 0$. By means of the Mean Value Theorem, we obtain

$$N_f(x) - x^* = N_f(x) - N_f(x^*) = N_f'(\theta)(x - x^*), \quad \theta \in (x^*, x),$$

and then $N_f(x) \geq x^*$. Therefore, $N_f(x) = x^*$, for $x \in (x^*, x_{n-1})$, so that N_f is constant in (x^*, x_{n-1}). In addition, $0 = N_f'(x) = L_f(x)$ in (x^*, x_{n-1}), and since $f(x) \neq 0$, we have $f''(x) = 0$ in (x^*, x_{n-1}). As a consequence, $f(x) = ax + b$ in (x^*, x_{n-1}) with $a, b \in \mathbb{R}$.

Moreover, it is clear that $x_{n+k} = x^*$, for all $k \in \mathbb{N}$, since x^* is a fixed point of N_f. ∎

Now, we compare the speed of sequences defined in (2.1) and (2.3) by the following result, which is an improved version of that given by Hernández in [56].

Theorem 2.6 *Let f and g be two increasing convex functions in $[a, b]$ with the same zero x^* in $[a, b]$ and $x_0 \in [a, b]$ such that $f(x_0) > 0$ and $g(x_0) > 0$. If $|L_f(x)| \geq |L_g(x)| > 0$ in $[a, b] - \{x^*\}$, then the sequence (2.3) starting at $y_0 = x_0$ converges to x^* faster than the sequence (2.1).*

Proof From the hypotheses, it follows that the sequences (2.1) and (2.3) are decreasing to x^*. On the one hand, if $x_n \neq x^*$, for all $n \in \mathbb{N}$, we prove by mathematical induction on n that $y_n \leq x_n$, for all $n \in \mathbb{N}$.

From $N_f(x^*) = x^* = N_g(x^*)$ and $x_0 \in [a, b]$ with $f(x_0) > 0$, we obtain

$$x_1 - y_1 = N_f(x_0) - N_g(x_0) = (N_f - N_g)(x_0) - (N_f - N_g)(x^*) = (N_f - N_g)'(\theta_0)(x_0 - x^*),$$

where $\theta_0 \in (x^*, x_0)$. Since $L_f(x) \geq L_g(x) > 0$ in $(x^*, b]$, then $(N_f - N_g)'(x) = L_f(x) - L_g(x) \geq 0$ in $(x^*, b]$ and $x_1 \geq y_1$.

After that, we assume $x_k \geq y_k$, for $k = 1, 2, \ldots, n - 1$. As $x_n \neq x^*$, then $x_{n-1} \neq x^*$. Moreover,

$$\begin{aligned} x_n - y_n &= N_f(x_{n-1}) - N_g(y_{n-1}) \\ &\geq N_f(x_{n-1}) - N_g(x_{n-1}) \\ &= (N_f - N_g)(x_{n-1}) - (N_f - N_g)(x^*) \\ &= (N_f - N_g)'(\theta_{n-1})(x_{n-1} - x^*), \end{aligned}$$

with $\theta_{n-1} \in (x^*, x_{n-1})$. Then, by proceeding as for $k = 1$, it follows that $x_n \geq y_n$.

On the other hand, if $x_{n-1} \neq x^*$ and $x_n = x^*$, we obtain $y_n = x^*$ with $x_k \geq y_k$, for $k = 1, 2, \ldots, n - 1$. Indeed, for $n = 1$, we have that $y_0 = x_0 = x^*$. Then, $y_k = x_k = x^*$,

$k \geq 0$. After that, we suppose that $x_k \geq y_k$, for $k = 1, 2, \ldots, n-1$. In addition, $x^* \leq y_n$ by construction and $x^* \geq y_n$, since $x_{n-1} \geq y_{n-1}$, N_g is nondecreasing, and

$$\begin{aligned}
x^* - y_n &= x_n - y_n \\
&= N_f(x_{n-1}) - N_g(y_{n-1}) \\
&\geq N_f(x_{n-1}) - N_g(x_{n-1}) \\
&= (N_f - N_g)(x_{n-1}) \\
&= (N_f - N_g)(x_{n-1}) - (N_f - N_g)(x^*) \\
&= (N_f - N_g)'(\theta)(x_{n-1} - x^*) \\
&> 0,
\end{aligned}$$

for some $\theta \in (x^*, x_{n-1})$. As $y_n = x_n = x^*$, it is easy to check that $y_{n+k} = x_{n+k} = x^*$, for all $k \in \mathbb{N}$, since $g(x^*) = x^*$ and $N_g(x^*) = x^*$. ∎

Therefore, Theorem 2.6 states that the speed of convergence of the Newton method is faster the lower the degree of logarithmic convexity.

Notice that if $f(x_0) < 0$, we need to ensure that $x_1 \leq b$ and $y_1 \leq b$ to guarantee that the sequences (2.1) and (2.3) lie in $[a, b]$. Then, both sequences are decreasing from x_1 and y_1 and $L_f(x) < L_g(x)$ in $[x_0, x^*)$, since $-L_f(x) > -L_g(x) > 0$ in $[x_0, x^*)$. Then, with mild modifications, we can obtain an analogous result to Theorem 2.6.

Example 2.7 Let $f(x) = \dfrac{x^3}{216} - 1$ and $g(x) = \dfrac{x^2}{36} - 1$ be two functions with the same zero $x^* = 6$ in $[3, 10]$. Both functions are increasing and convex in $[3, 10]$, see Fig. 2.4. In addition, $L_f(x) = \frac{2}{3} - \frac{144}{x^3}$, $L_g(x) = \frac{1}{2} - \frac{18}{x^2}$, and $|L_f(x)| > |L_g(x)|$ in $[3, 10]$.

Fig. 2.4 Graphs of $f(x) = \frac{x^3}{216} - 1$ and $g(x) = \frac{x^2}{36} - 1$ in $[3, 10]$, respectively, black and blue lines

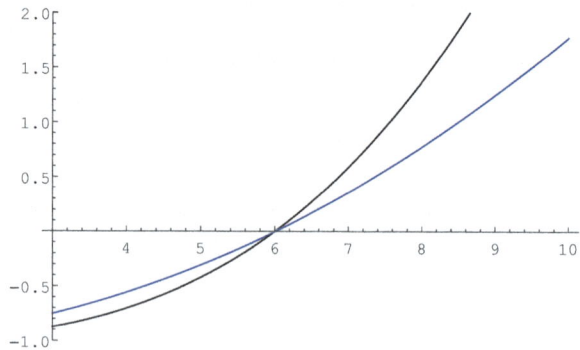

Table 2.3 Sequences (2.1) and (2.3) from Example 2.7

n	x_n	y_n
0	3.0000000000000...	3.0000000000000...
1	10.0000000000000...	7.5000000000000...
2	7.3866666666667...	6.1500000000000...
3	6.2440237430147...	6.0018292682927...
4	6.0094124974239...	6.0000002787669...
5	6.0000147350265...	6.0000000000000...
6	6.0000000000362...	6.0000000000000...

If we choose $y_0 = x_0 \in [3, 10]$ such that $f(x_0) > 0$, then it follows from Theorem 2.6 that the sequence $\{y_n\}$ converges to $x^* = 6$ faster than $\{x_n\}$. In particular, if $y_0 = x_0 = 3$, we see this by the approximations given in Table 2.3.

On the other hand, if we choose as x_0 a point that is to the left of the solution $x^* = 6$, for example, $x_0 = 2$, then $f(2) < 0$, $10 \geq N_f(3) = 10$, and $10 \geq N_g(3) = 7.5$. Therefore, $x_1, y_1 \in [3, 10]$, and $\{x_n\}$ and $\{y_n\}$ lie in $[6, 10]$ for all $n \geq 1$.

2.2 Operator Equations in Banach Spaces

The Russian mathematician L. V. Kantorovich was the first researcher to generalize the Newton method to the Banach spaces. The Newton method has the form

$$x_{n+1} = N_F(x_n) = x_n - [F'(x_n)]^{-1} F(x_n), \quad n \geq 0, \quad \text{with } x_0 \text{ given,} \quad (2.4)$$

to solve the nonlinear equation $F(x) = 0$, where $F : X \to Y$, X and Y are Banach spaces, and F is a continuously Fréchet differentiable operator.

In the middle of the twentieth century, Kantorovich published several works, where the semilocal convergence of the Newton method is analyzed in Banach spaces, among which, we underline [64, 65] and gave the seminal *Newton-Kantorovich theorem* that was the beginning of what we know today as *Kantorovich's theory* for the Newton method [42].

From now on, we denote the set of bounded linear operators from Y to X by $\mathcal{L}(Y, X)$, $B(x, \varrho) = \{z \in X : \|z - x\| < \varrho\}$, and $\overline{B(x, \varrho)} = \{z \in X : \|z - x\| \leq \varrho\}$ and use the infinity norm in X. In addition, we use the infinity norm for $\mathbb{R}^m$ and $\mathcal{C}([a, b])$ in the examples.

2.2.1 The Operator Degree of Logarithmic Convexity

Given a scalar function $f : [a, b] \to \mathbb{R}$, we have seen that the degree of logarithmic convexity of f is the function $L_f : [a, b] \to \mathbb{R}$ defined by

$$L_f(t) = \frac{f(t)f''(t)}{f'(t)^2}.$$

In [42], we extend the degree of logarithmic convexity to Banach spaces. For this, we suppose that $F : X \to Y$ is a twice continuously Fréchet differentiable operator in Ω, and there exists the operator $[F'(x)]^{-1} : Y \to X$. Moreover, since $F''(x) : X \times X \to Y$, it follows $F''(x)[F'(x)]^{-1}F(x) \in \mathcal{L}(X, Y)$ and

$$L_F(x) : \Omega \xrightarrow{F''(x)[F'(x)]^{-1}F(x)} Y \xrightarrow{[F'(x)]^{-1}} \Omega,$$

so that

$$L_F(x) = [F'(x)]^{-1}F''(x)\left([F'(x)]^{-1}F(x)\right) \in \mathcal{L}(X, X), \quad \text{with} \quad [F'(x)]^{-1} \in \mathcal{L}(Y, X).$$

It should be clear that this operator is not a measure of convexity of the operator F, so that the geometric sense of the scalar case is lost. However, the study of this operator allows analyzing the Newton method and other iterative methods with R-order of convergence at least three in Banach spaces by a different approach to that given by other authors.

Then, we first define the R-order of convergence as follows [77]:

Let $\{x_n\}$ be a sequence of points of a Banach space X converging to a point $x^* \in X$, and let $\sigma \geq 1$ and

$$e_n(\sigma) = \begin{cases} n, & \text{if } \sigma = 1, \\ \sigma^n, & \text{if } \sigma > 1, \end{cases} \quad n \geq 0.$$

(a) We say that σ is an R-order of convergence of the sequence $\{x_n\}$ if there are two constants $b \in (0, 1)$ and $B \in (0, +\infty)$ such that

$$\|x_n - x^*\| \leq Bb^{e_n(\sigma)}.$$

(b) We say that σ is the exact R-order of convergence of the sequence $\{x_n\}$ if there are four constants $a, b \in (0, 1)$ and $A, B \in (0, +\infty)$ such that

$$Aa^{e_n(\sigma)} \leq \|x_n - x^*\| \leq Bb^{e_n(\sigma)}, \quad n \geq 0.$$

In general, checkup the two inequalities in (b) is difficult, so normally only upper inequalities like (a) are looked for. Therefore, if we find an R-order of convergence σ of sequence $\{x_n\}$, we then say that sequence $\{x_n\}$ has order of convergence at least σ. This is the argument commonly used to study the order of convergence of an iterative method in Banach spaces.

2.2 Operator Equations in Banach Spaces

Next, given a scalar equation $f(t) = 0$, the Newton sequence is obtained by applying the method of successive approximations $t_{n+1} = N_f(t_n)$, $n \geq 0$, to the function

$$N_f(t) = t - \frac{f(t)}{f'(t)},$$

starting at some fixed point t_0. In this case, it is immediate to see that $N'_f(t) = L_f(t)$. We see below the generalization of this result to Banach spaces. For this, we first need the following lemma.

Lemma 2.8 *Let X and Y be two Banach spaces, and consider two operators*

$$Q : X \to Y \quad \text{and} \quad P : X \to \mathcal{L}(Y, X),$$

which are Fréchet differentiable at a point $x_0 \in X$. If $H : X \to X$ is such that

$$H(x) = (P(x)Q)(x), \quad \text{with} \quad X \xrightarrow{Q} Y \xrightarrow{P(x)} X;$$

then H is differentiable at x_0 and $H'(x_0) : X \to X$ is the linear operator

$$H'(x_0)(z) = \big(P'(x_0)(z)Q\big)(x_0) + \big(P(x_0)Q'(x_0)\big)(z), \quad z \in X.$$

Proof From the definition of a Fréchet derivative, we have

$$
\begin{aligned}
H'(x_0)(z) &= \lim_{s \to 0} \frac{H(x_0 + sz) - H(x_0)}{s} \\
&= \lim_{s \to 0} \frac{(P(x_0 + sz)Q)(x_0 + sz) - (P(x_0)Q)(x_0)}{s} \\
&= \lim_{s \to 0} \frac{((P(x_0 + sz) - P(x_0))Q)(x_0 + sz) + P(x_0)(Q(x_0 + sz) - Q(x_0))}{s} \\
&= (P'(x_0)(z)Q)(x_0) + (P(x_0)Q'(x_0))(z),
\end{aligned}
$$

for all $z \in X$. ∎

The Newton sequence (2.4) is obtained by applying the method of successive approximations $x_{n+1} = N_F(x_n)$, $n \geq 0$, to the operator

$$N_F(x) = x - \Big([F'(x)]^{-1}F\Big)(x), \quad x \in X,$$

starting at some fixed point x_0. In the following result we prove that the derivative of the operator N_F at a point $x_0 \in X$ is precisely the operator defined as the degree of logarithmic convexity $L_F(x_0)$, which has been previously introduced.

Theorem 2.9 *From the above mentioned definition, we have $N'_F(x_0) = L_F(x_0)$, for $x_0 \in X$.*

Proof If $H(x) = (\Gamma(x)F)(x)$, where $\Gamma : X \to \mathcal{L}(X, Y)$ and $\Gamma(x) = [F'(x)]^{-1} \in \mathcal{L}(Y, X)$ with $x \in X$, then $N'_F(x_0) = I - H'(x_0)$. Observe that H is as in Lemma 2.8, so that

$$H'(x_0)(z) = \big(\Gamma'(x_0)(z)F\big)(x_0) + \big(\Gamma(x_0)F'(x_0)\big)(z) = \big(\Gamma'(x_0)(z)F\big)(x_0) + z, \quad z \in X,$$

and then $N'_F(x_0)(z) = -\big(\Gamma'(x_0)(z)F\big)(x_0)$.

Next, if we write Γ as $\Gamma = TF'$, where $T : \mathcal{L}(X, Y) \to \mathcal{L}(Y, X)$ and $T(F) = F^{-1}$, then

$$\Gamma : X \xrightarrow{F'} \mathcal{L}(X, Y) \xrightarrow{T} \mathcal{L}(Y, X),$$

and by the chain rule for differentiating compositions of operators, it follows $\Gamma'(x_0) \in \mathcal{L}(X, \mathcal{L}(Y, X))$ and

$$\Gamma'(x_0)(x) = T'(F'(x_0))F''(x_0)(x), \quad x \in X.$$

On the other hand, from the derivative of an inverse operator [79], we have

$$\Gamma'(x_0)(z) = T'(F(x_0))(F''(x_0)(z)) = -\Gamma(x_0)F''(x_0)(z)\Gamma(x_0) \in \mathcal{L}(Y, X), \quad z \in X.$$

Moreover, as $F''(x_0)$ is symmetric, then

$$N'_F(x_0)(z) = \Gamma(x_0)F''(x_0)(z)\Gamma(x_0)F(x_0) = \Gamma(x_0)F''(x_0)\Gamma(x_0)F(x_0)(z)$$
$$= L_F(x_0)(z), \quad z \in X,$$

and the proof is complete. ∎

The condition of existence of the operator L_F in a certain domain seems very strong, since it implies the existence of the inverse operator $[F'(x)]^{-1}$ at each point in the domain. However, by requiring certain conditions to the second derivative F'', the existence of the operator L_F is guaranteed in a domain. For this, we introduce the following Kantorovich conditions:

2.2 Operator Equations in Banach Spaces

(K1) There exists $\Gamma_0 = [F'(x_0)]^{-1} \in \mathcal{L}(Y, X)$, for some $x_0 \in \Omega$, with $\|\Gamma_0\| \leq \beta$ and $\|\Gamma_0 F(x_0)\| \leq \eta$.

(K2) There exists a constant $M \geq 0$ such that $\|F''(x)\| \leq M$, for $x \in \Omega$.

Theorem 2.10 *Let $F : \Omega \subseteq X \to Y$ be a twice continuously Fréchet differentiable operator defined on a non-empty open convex domain Ω of a Banach space X with values in a Banach space Y. If conditions (K1)–(K2) are satisfied, for all $x \in \Omega$, then the operator $L_F(x)$ exists for all $x \in B\left(x_0, \frac{1}{M\beta}\right) \cap \Omega$ and*

$$\|L_F(x)\| \leq \frac{M\beta \|\Gamma_0 F(x)\|}{(1 - M\beta \|x - x_0\|)^2}. \tag{2.5}$$

Proof Observe that

$$\Gamma_0 \int_{x_0}^{x} F''(u)\, du = \Gamma_0 F'(x) - I, \quad x \in \Omega,$$

and

$$\|\Gamma_0 F'(x) - I\| = \left\| \Gamma_0 \int_{x_0}^{x} F''(u)\, du \right\| \leq M\beta \|x - x_0\| < 1.$$

Then, from the Banach lemma on invertible operators, there exists $[\Gamma_0 F'(x)]^{-1}$, for all $x \in B\left(x_0, \frac{1}{M\beta}\right)$, and is such that

$$\|[\Gamma_0 F'(x)]^{-1}\| \leq \frac{1}{1 - M\beta \|x - x_0\|}.$$

Now, from

$$L_F(x) = [\Gamma_0 F'(x)]^{-1} \Gamma_0 F''(x) [\Gamma_0 F'(x)]^{-1} \Gamma_0 F(x),$$

it follows (2.5). ∎

If we observe conditions (K1)–(K2) required for the existence of the operator L_F in Theorem 2.10, commonly known as *the classic conditions of Kantorovich* [42], we see that they are the same as those required in the Newton-Kantorovich theorem to guarantee the semilocal convergence of the Newton method, as we see below.

Theorem 2.11 (The Newton-Kantorovich Theorem [64]) *Let $F : \Omega \subseteq X \to Y$ be a twice continuously Fréchet differentiable operator defined on a non-empty open convex domain Ω of a Banach space X with values in a Banach space Y. Suppose that conditions*

(K1)–(K2) are satisfied, $h = M\beta\eta \leq \frac{1}{2}$, and $B(x_0, s^*) \subset \Omega$ with $s^* = \frac{1-\sqrt{1-2h}}{h}\eta$. Then, the Newton sequence defined in (2.4) and starting at x_0 converges to a solution x^* of the equation $F(x) = 0$, and the solution x^* and the iterates x_n belong to $\overline{B(x_0, s^*)}$, for all $n = 0, 1, 2, \ldots$ Moreover, if $h = M\beta\eta < \frac{1}{2}$, the solution x^* is unique in $B(x_0, s^{**}) \cap \Omega$, where $s^{**} = \frac{1+\sqrt{1-2h}}{h}\eta$, and, if $h = \frac{1}{2}$, the solution x^* is unique in $\overline{B(x_0, s^*)}$. Furthermore, we have the following a priori error estimates:

$$\|x^* - x_n\| \leq \frac{1}{2^{n-1}}(2h)^{2^n-1}\eta, \quad n = 0, 1, 2, \ldots$$

To guarantee the semilocal convergence of the Newton method by Theorem 2.11 of Newton-Kantorovich, we see that the condition $h = M\beta\eta \leq \frac{1}{2}$ is required. In the next section, conditions on the operator L_F are required to guarantee the convergence of the Newton method.

2.2.2 Convergence Conditions on the Degree of Logarithmic Convexity

As we detail in our monograph [45], it is well known that we can carry out three studies when we study the convergence of the Newton method to a solution of an equation: global, semilocal, and local. The global study of the convergence consists of guaranteeing the convergence of the method from some conditions on the operator involved and independently of the initial approximation. The semilocal study of the convergence consists of guaranteeing the convergence of the method from some conditions on the initial approximation and the operator involved and providing the so-called domain of parameters [41]. The local study of the convergence consists of guaranteeing the convergence of the method from some conditions on the solution of the equation to solve and the operator involved and providing the so-called ball of convergence [28] of the Newton sequence that shows the accessibility to the solution from the initial approximation belonging to the ball. As we can see, conditions on the operator involved are required in the three studies of the convergence. However, conditions required in the solution in the initial approximation or none of these determine the type of study.

The local study of the convergence has the disadvantage of being able to guarantee that the solution, which is unknown, can satisfy certain conditions. In general, the global study of the convergence is very specific as regards the type of operators to consider, as a consequence of absence of conditions on the initial approximation and on the solution. There is a plethora of studies on the weakness and/or extension of the hypothesis made on the underlying operators.

For the semilocal convergence of the Newton method, we require three types of conditions: on the initial approximation, the operator involved, and conditions that connect the previous conditions. The choice of the initial approximation is a basic aspect in the study of semilocal convergence because, together with the theoretical significance,

2.2 Operator Equations in Banach Spaces

it allows drawing conclusions about the existence and uniqueness of the solution of the equation to solve.

2.2.2.1 Semilocal Convergence

Kantorovich proves Theorem 2.11 by means of the so-called *majorant principle* [65, 66]. The idea of this principle is to compare an iterative method given in Banach spaces with a scalar iteration that *majorizes* the previous iterative method and has convergence properties. In particular, Kantorovich proves the convergence of the Newton method (2.4) from the convergence of the method of successive approximations (see [42]). This ingenious procedure devised by Kantorovich has the disadvantage that the convergence conditions required are very restrictive and difficult to satisfy for other iterative methods to analyze. For this, in [42], we develop this Kantorovich's principle from another point of view that is easier to understand and present and is based on the proof of the Newton-Kantorovich theorem given by Ortega in [74]. Moreover, Kantorovich considers an open ball as domain of the operator involved, which is very restrictive, while we consider an open convex region. This point of view rests on the concept of *majorizing sequence* and shows how this is used to prove the convergence of the Newton method. The basic idea is established in the following definition and theorem.

Definition 2.12 If $\{x_n\}$ is a sequence in a Banach space X and $\{t_n\}$ is a real sequence, then $\{t_n\}$ is a *majorizing sequence* of $\{x_n\}$ if $\|x_{n+1} - x_n\| \leq t_{n+1} - t_n$, for all $n \geq 0$.

From the last inequality, we observe that the sequence $\{t_n\}$ is nondecreasing. The interest of the majorizing sequence is that the convergence of the sequence $\{x_n\}$ in the Banach space X is deduced from the convergence of the scalar sequence $\{t_n\}$, as we see in the following result given by Ortega in [74].

Theorem 2.13 *Let $\{x_n\}$ be a sequence in a Banach space X and $\{t_n\}$ a majorizing sequence of $\{x_n\}$. Then, if $\{t_n\}$ converges to $t^* \in \mathbb{R}$, there exists a $x^* \in X$ such that $x^* = \lim_n x_n$ and $\|x^* - x_n\| \leq t^* - t_n$, for all $n \geq 0$.*

From Definition 2.12 and Theorem 2.13, Theorem 2.11 of Newton-Kantorovich is proved by applying the Newton method

$$t_0 = 0, \quad t_{n+1} = N_p(t_n) = t_n - \frac{p(t_n)}{p'(t_n)}, \quad n \geq 0,$$

to the polynomial

$$p(t) = \frac{M}{2}t^2 - \frac{t}{\beta} + \frac{\eta}{\beta}, \tag{2.6}$$

which is usually named as *the Kantorovich polynomial*. This approach rests essentially on the requirement that the majorizing method has the same form as the underlying method and is called the *method of majorizing sequences*. See [42] for more details.

The method of majorizing sequences is also a powerful tool that has been applied to establish convergence results for different variants of the Newton method, so that the concept of a majorizing sequence and estimates of type of those given in Definition 2.12 have been extended to give a convergence theory for a large class of iterative methods [81].

In [42], we see that the scalar function involved in the construction of a majorizing sequence plays a key role, and the studies are based on the construction and existence of scalar functions used to construct majorizing sequences. For this, we define what a majorant function is for us.

Definition 2.14 Let $f \in \mathcal{C}^1([t_0, +\infty))$ be a real function and the Newton sequence

$$t_{n+1} = N_f(t_n) = t_n - \frac{f(t_n)}{f'(t_n)}, \quad n \geq 0, \quad \text{with } t_0 \in \mathbb{R} \text{ given.} \tag{2.7}$$

If (2.7) is a majorizing sequence of the Newton sequence (2.4) defined in the Banach space X, then the real function f is called *majorant function* of the operator F.

To obtain the majorizing sequence given in (2.7), Kantorovich considers a problem of interpolation fitting where the function f is the second-degree polynomial given in (2.6), since it is the simplest scalar function with bounded second derivative and the coefficient of the polynomial is fixed by means of conditions (K1)–(K2). On the other hand, we see in [42] that the polynomial (2.6) can be obtained otherwise, without interpolation fitting, by solving an initial value problem.

In Definition 2.14, we suppose that there exists $f \in \mathcal{C}^1([t_0, +\infty))$, with $t_0 \in \mathbb{R}$. Now, we see that this function exists under the following conditions:

(A1) There exists $\Gamma_0 = [F'(x_0)]^{-1} \in \mathcal{L}(Y, X)$, for some $x_0 \in \Omega$, with $\|\Gamma_0\| \leq \beta$ and $\|\Gamma_0 F(x_0)\| \leq \eta$.
(A2) Let $L_F(x) : \Omega \subseteq X \to X$, where Ω is a non-empty open convex domain, and a constant $K \in [0, 1)$ such that $\|L_F(x)\| \leq K$, for $x \in \Omega$.

Then, we look for such a function by solving an initial value problem and taking into account that

$$\|L_F(x)\| \leq K = L_f(t) \quad \text{for} \quad \|x - x_0\| \leq t - t_0.$$

If $t_0 \geq 0$, we can then find a scalar function f, which allows finding a majorizing sequence of the Newton sequence (2.4), as the unique solution of the initial value problem

2.2 Operator Equations in Banach Spaces

$$\begin{cases} y(t)y''(t) = Ky'(t)^2, \\ y(t_0) = \dfrac{\eta}{\beta}, \quad y'(t_0) = -\dfrac{1}{\beta}. \end{cases}$$

In addition, we can establish the following result that proves the existence of such a function.

Theorem 2.15 *For any nonnegative real numbers $K < 1$, $\beta \neq 0$, and η, there exists only one solution $\ell(t)$ of the last initial value problem in $[t_0, +\infty)$; that is,*

$$\ell(t) = \frac{\eta}{\beta} \left(\frac{1-K}{\eta - (1-K)t_0} \right)^{\frac{1}{1-K}} \left(\frac{\eta}{1-K} - t \right)^{\frac{1}{1-K}}. \tag{2.8}$$

Observe that $t^* = \frac{\eta}{1-K} > 0$ is the unique zero of (2.8) and the scalar sequence $\{t_n\}$, defined by

$$t_{n+1} = t_n - \frac{\ell(t_n)}{\ell'(t_n)} = Kt_n + \eta, \quad n \geq 0, \quad \text{with } t_0 \text{ given}, \tag{2.9}$$

is nondecreasing and converges to t^*.

In addition, we can establish the following semilocal convergence result for the sequence (2.4) in the Banach space X.

Theorem 2.16 *Let $F: \Omega \subseteq X \to Y$ be a twice continuously Fréchet differentiable operator defined on a non-empty open convex domain Ω of a Banach space X with values in a Banach space Y. Suppose that conditions (A1)–(A2) are satisfied and $B(x_0, t^* - t_0) \subset \Omega$, where $t^* = \frac{\eta}{1-K}$. Then, the Newton sequence (2.4) converges to a solution x^* of $F(x) = 0$ and $x_n, x^* \in \overline{B(x_0, t^* - t_0)}$, for all $n = 0, 1, 2, \ldots$ Moreover,*

$$\|x_{n+1} - x_n\| \leq t_{n+1} - t_n \quad \text{and} \quad \|x^* - x_n\| \leq t^* - t_n, \quad n \geq 0,$$

where $\{t_n\}$ is defined in (2.9).

Proof First, we observe that the approximations x_n are well defined if $x_n \in B(x_0, t^* - t_0)$, since the operator $[F'(x_n)]^{-1}$ exists as a consequence of the operator L_F being defined in Ω. Then,

$$\|x_1 - x_0\| = \|\Gamma_0 F(x_0)\| \leq \eta = t_1 - t_0 \leq t^* - t_0,$$

so that $x_1 \in B(x_0, t^* - t_0) \subset \Omega$. Now, we assume that $x_1, x_2, \ldots, x_n \in B(x_0, t^* - t_0) \subset \Omega$ and

$$\|x_{i+1} - x_i\| \leq t_{i+1} - t_i, \quad \text{for all} \quad i = 0, 1, \ldots, n-1. \tag{2.10}$$

Thus,

$$\|x_{n+1} - x_n\| = \left\| \int_{x_{n-1}}^{x_n} L_F(x)\,dx \right\|$$

$$\leq \int_0^1 \|L_F(x_{n-1} + \tau(x_n - x_{n-1}))\| \, \|x_n - x_{n-1}\|\,d\tau$$

$$\leq K \int_0^1 (t_n - t_{n-1})\,d\tau$$

$$= K(t_n - t_{n-1})$$

$$= t_{n+1} - t_n$$

and

$$\|x_{n+1} - x_0\| \leq \|x_{n+1} - x_n\| + \|x_n - x_0\| \leq t_{n+1} - t_n + t_n - t_0 = t_{n+1} - t_0 \leq t^* - t_0,$$

so that $x_{n+1} \in B(x_0, t^* - t_0) \subset \Omega$. We have then proved that $x_i \in B(x_0, t^* - t_0) \subset \Omega$ and (2.10) are true for all positive integers i.

Second, as the sequence (2.9) is convergent, the sequence (2.4) is also convergent. As a consequence, $\lim_n x_n = x^* \in \overline{B(x_0, t^* - t_0)}$, where $t^* = \lim_n t_n$, such that $\|x^* - x_n\| \leq t^* - t_n$, for all $n \geq 0$. Taking then into account that N_F is a continuous operator, we obtain

$$x^* = \lim_{n \to +\infty} x_{n+1} = \lim_{n \to +\infty} N_F(x_n) = N_F\left(\lim_{n \to +\infty} x_n\right) = N_F(x^*),$$

and x^* is then a solution of the equation $F(x) = 0$. ∎

Now, we illustrate the last result with the following example.

Example 2.17 Consider the system of nonlinear equations given by

$$F(x, y) = (xy - 1, xy + x - 2y) = 0.$$

Then,

$$F'(x, y) = \begin{pmatrix} y & x \\ y + 1 & x - 2 \end{pmatrix}.$$

If (x, y) is not on the straight line $x + 2y = 0$, there exists $[F'(x, y)]^{-1}$ and

2.2 Operator Equations in Banach Spaces

$$[F'(x,y)]^{-1} = \frac{1}{x+2y}\begin{pmatrix} 2-x & x \\ y+1 & -y \end{pmatrix}.$$

Besides,

$$F''(x,y) = \begin{pmatrix} 0 & 1 \\ 1 & 0 \\ 0 & 1 \\ 1 & 0 \end{pmatrix},$$

and $L_F(x,y) : \mathbb{R}^2 \to \mathbb{R}^2$ is such that

$$L_F(x,y)(s,t) = [F'(x,y)]^{-1}F''(x,y)\left([F'(x,y)]^{-1}F(x,y),(s,t)\right) = \frac{\Delta}{(x+2y)^2}\begin{pmatrix} 2 \\ 1 \end{pmatrix},$$

where $\Delta = t(x^2+x-2)+s(2y^2-y-1)$.

For the max-norm in $\mathbb{R}^2$, that is, $\|(x,y)\| = \max\{|x|,|y|\}$, we have

$$\|L_F(x,y)(s,t)\| = \max\left\{\frac{2|\Delta|}{(x+2y)^2}, \frac{|\Delta|}{(x+2y)^2}\right\}$$

$$= \frac{2|\Delta|}{(x+2y)^2}$$

$$\leq 2\frac{|x^2+x-2|+|2y^2-y-1|}{(x+2y)^2}\|(s,t)\|$$

and

$$\|L_F(x,y)\| \leq 2\frac{|x^2+x-2|+|2y^2-y-1|}{(x+2y)^2} = h(x,y).$$

In addition, $\|L_F(x,y)\| < 1$ in the gray region of the xy-plane shown in Fig. 2.5.

If we now choose the ball $\Omega = B(x_0,\rho)$, with $x_0 = (1.13, 1.13)$ and $\rho = 0.43$, then $\beta = 0.9616\ldots, \eta = 0.1250\ldots$

$$\|L_F(x,y)\| \leq \sup_{(x,y)\in\Omega} \|L_F(x,y)\| \leq h(0.7,0.7) = 0.7 = K < 1,$$

and $t^* = \frac{\eta}{1-K} = 0.4167\ldots \leq \rho$, so that the conditions of Theorem 2.16 are satisfied. Therefore, the Newton sequence (2.4) starting at $x_0 = (1.13, 1.13)$ is convergent to a solution of $F(x,y) = 0$. Observe in Fig. 2.5 that the black ball Ω is in the gray region where $\|L_F(x,y)\| < 1$.

Fig. 2.5 Region $\|L_F(x,y)\| < 1$ from Example 2.17

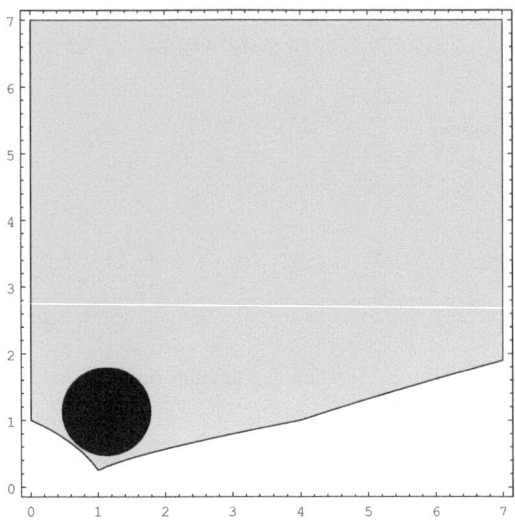

Remark 2.18 At first glance, it may seem that the condition that the operator L_F exists in each approximation of the Newton method is very demanding. However, Theorem 2.10 shows that this is not so, since, under conditions (K1)–(K2) of Theorem 2.11 of Newton-Kantorovich, the operator L_F exists in a ball.

Notice that it is easy to verify that the situation described by Theorem 2.16 is more general than that of Theorem 2.11 of Newton-Kantorovich. To see this, we just consider the following simple example.

Example 2.19 The polynomial

$$F(x) = (1-x)^4$$

satisfies the conditions of Theorem 2.16 if $x_0 = 0$, since $\|[F'(0)]^{-1}\| = \frac{1}{4} = \beta$, $\|[F'(0)]^{-1}F(0)\| = \frac{1}{4} = \eta$, and $L_F(x) = \frac{3}{4} < 1$, for all $x \in \mathbb{R}$, but the hypotheses of Theorem 2.11 of Newton-Kantorovich are not satisfied in the interval $[0,1]$, since $M\beta\eta = \frac{3}{4} > \frac{1}{2}$, where $M = \sup_{x \in [0,1]}\{F''(x)\} = 12$.

We just studied the case that L_F is bounded by a constant less than 1. Now, we study other more general situations. Thus, a general situation is to find a function ω such that

$$\|L_F(x)\| \leq \omega(t), \quad \text{for} \quad \|x - x_0\| \leq t - t_0, \tag{2.11}$$

where $\omega : [0, +\infty) \to \mathbb{R}$ is nondecreasing and continuous and such that $\omega(0) \geq 0$.

2.2 Operator Equations in Banach Spaces

In particular, we construct a majorant function for the operator F under the following conditions:

(B1) There exists $\Gamma_0 = [F'(x_0)]^{-1} \in \mathcal{L}(Y, X)$, for some $x_0 \in \Omega$, with $\|\Gamma_0\| \leq \beta$ and $\|\Gamma_0 F(x_0)\| \leq \eta$.

(B2) Let $L_F(x) : \Omega \subseteq X \to X$, where Ω is a non-empty open convex domain, and there exists a continuous nondecreasing function $\omega : [t_0, +\infty) \to \mathbb{R}$ such that $\|L_F(x)\| \leq \omega(t)$, for $\|x - x_0\| \leq t - t_0$, and $\omega(0) \geq 0$.

For constructing such a scalar function, rely on the fact that the scalar function f must satisfy

$$\|L_F(x)\| \leq L_f(t), \quad \text{for} \quad \|x - x_0\| \leq t - t_0,$$

so that we choose $L_f(t) = \omega(t)$. In addition,

$$\int_{t_0}^{t} L_f(u)\, du = \int_{t_0}^{t} \omega(u)\, du.$$

If

$$u - \frac{f(u)}{f'(u)}\bigg|_{t_0}^{t} = W(u)\big|_{t_0}^{t},$$

where W is a primitive of ω, then

$$t - \frac{f(t)}{f'(t)} - t_0 + \frac{f(t_0)}{f'(t_0)} = t - \frac{f(t)}{f'(t)} - t_0 - \eta = W(t) - W(t_0),$$

so that

$$\frac{f'(t)}{f(t)} = \frac{1}{t - W(t) + W(t_0) - t_0 - \eta}$$

and

$$f(t) = \frac{\eta}{\beta} \exp\left(\int_{t_0}^{t} \frac{du}{u - W(u) + W(t_0) - t_0 - \eta} \right). \tag{2.12}$$

Since the function f has to have at least one zero, it is clear that the zero must appear from the fact that the exponent of the exponential function is of the form $\log \zeta(t)$, where $\zeta(t)$ has at least one zero. One possibility that satisfies this situation is that the expression $u - W(u) + W(t_0) - t_0 - \eta$ is a polynomial with simple real zeros. In this case, applying a process of decomposition in simple fractions to calculate the integral that appears in (2.12), we obtain that f has at least one zero. We observe that it is enough that $W(t)$ is a polynomial to obtain that the expression $u - W(u) + W(t_0) - t_0 - \eta$ is a polynomial, so that the function ω is also a polynomial.

Based on what we just mentioned and taking into account that the simplest case has already been studied ($\omega(t) = K$), we then consider the following simpler case, that is, $\omega(t)$ is a line not parallel to the abscissa axis. Taking into account that the function that defines the majorizing sequence is considered decreasing, convex, and with a zero $\xi > 0$, it is clear that ω is of the form $\omega(t) = \gamma(\xi - t)$ with $\gamma, \xi > 0$. In this case, $W(t) - W(t_0) = \gamma \xi t - \frac{\gamma}{2}t^2 + (\frac{\gamma}{2}t_0^2 - \gamma \xi t_0)$ and $u - W(u) + W(t_0) - t_0 - \eta = \frac{\gamma}{2}u^2 + (1 - \gamma\xi)u - (\frac{\gamma}{2}t_0^2 + (1 - \gamma\xi)t_0 + \eta)$. Then, $u - W(u) + W(t_0) - t_0 - \eta = \frac{\gamma}{2}(u - \xi_1)(u - \xi_2)$, where

$$\xi_1 = \frac{1}{\gamma}\left(\gamma\xi - 1 + \sqrt{1 + 2(\eta + t_0 - \xi)\gamma + (t_0 - \xi)^2\gamma^2}\right) > 0,$$

$$\xi_2 = \frac{1}{\gamma}\left(\gamma\xi - 1 - \sqrt{1 + 2(\eta + t_0 - \xi)\gamma + (t_0 - \xi)^2\gamma^2}\right) < 0.$$

Therefore,

$$\int_{t_0}^{t} \frac{du}{u - W(u) + W(t_0) - t_0 - \eta}$$
$$= \frac{2}{\gamma(\xi_1 - \xi_2)} \log\left(\frac{(t_0 - \xi_2)t + \xi_1(\xi_2 - t_0)}{(t_0 - \xi_1)t + \xi_2(\xi_1 - t_0)}\right), \quad \text{for } t \in [t_0, \xi],$$

since $\xi_1 \geq \xi \geq t$ and $\xi_2 < 0$. And, as a consequence,

$$f(t) = \frac{\eta}{\beta}\left(\frac{(t_0 - \xi_2)t + \xi_1(\xi_2 - t_0)}{(t_0 - \xi_1)t + \xi_2(\xi_1 - t_0)}\right)^{\frac{2}{\gamma(\xi_1 - \xi_2)}}, \quad \text{for } t \in [t_0, \xi], \quad (2.13)$$

so that $L_f(t) = \gamma(\xi - t)$, and the function (2.13) is the unique solution of the initial value problem

$$\begin{cases} y(t)y''(t) = \gamma(\xi - t)y'(t)^2, \\ y(t_0) = \dfrac{\eta}{\beta}, \quad y'(t_0) = -\dfrac{1}{\beta}. \end{cases}$$

Once we have seen what the scalar function must be like to be a majorant function in the sense of Definition 2.14, and under conditions (B1)–(B2), we establish, from the abovementioned definitions, the following result that proves the existence of such a function.

2.2 Operator Equations in Banach Spaces

Theorem 2.20 *For any nonnegative real numbers $\gamma \neq 0$, $\xi \neq 0$, $\beta \neq 0$, and η, the function (2.13) is the unique solution of the last initial value problem in the interval $[t_0, \xi]$.*

After that, we can establish the following convergence result of the Newton sequence (2.4) in the Banach space X.

Theorem 2.21 *Let $F : \Omega \subseteq X \to Y$ be a twice continuously Fréchet differentiable operator defined on a non-empty open convex domain Ω of a Banach space X with values in a Banach space Y. Suppose that the conditions (B1)–(B2), with $\omega(t) = \gamma(\xi - t)$ and $\gamma, \xi > 0$, are satisfied and $B(x_0, t^* - t_0) \subset \Omega$, where $t^* = \xi_1$. Then, the Newton sequence (2.4) converges to a solution x^* of $F(x) = 0$ and $x_n, x^* \in \overline{B(x_0, t^* - t_0)}$, for all $n = 0, 1, 2, \ldots$ Moreover,*

$$\|x_{n+1} - x_n\| \leq t_{n+1} - t_n \quad \text{and} \quad \|x^* - x_n\| \leq t^* - t_n, \quad n \geq 0,$$

where $\{t_n = N_f(t_{n-1}), \, n \in \mathbb{N}\}$, with t_0 given and $f(t)$ defined by (2.13).

Proof We prove first that $\{t_n\}$ is a nondecreasing sequence and convergent to t^*. For this, we note that the sequence $\{t_n\}$ is monotone and bounded.

From $\|\Gamma_0 F(x_0)\| \leq \eta$, we have $0 \leq N_f(t_0) - t_0 = t_1 - t_0$, since $\eta = -\frac{f(t_0)}{f'(t_0)} = t_1 - t_0$, so that $t_1 \geq t_0$. After that, from mathematical induction on n, it follows that $t_n \leq t_{n+1}$, $n \geq 0$, since the function N_f is nondecreasing, as we see from the condition (2.11).

Besides, $t_0 \leq t^* = \xi_1$. In addition, if we suppose that $t_i \leq t^*$, with $i = 0, 1, \ldots, n$, then $t_{n+1} = N_f(t_n) \leq N_f(t^*) = t^*$, since the function N_f is nondecreasing. As a consequence, by mathematical induction on n, it follows easily that $t_n \leq t^*$, for all $n \geq 0$.

Therefore, there exists $v = \lim_n t_n$ such that $v \leq t^*$. Besides, as N_f is a continuous function, it follows

$$v = \lim_{n \to +\infty} t_{n+1} = \lim_{n \to +\infty} N_f(t_n) = N_f\left(\lim_{n \to +\infty} t_n\right) = N_f(v),$$

and then $v = t^* = \xi_1$ is a solution of the equation $N_f(t) = t$, that is, $f(t) = 0$. As a consequence, the sequence $\{t_n\}$ converges to the solution $t^* = \xi_1$ of $f(t) = 0$.

Next, we prove that the approximations x_n are well defined and $\{x_n\}$ is convergent. Obviously, as L_F is defined in Ω, it is clear that there exists $[F'(x)]^{-1}$, for all $x \in \Omega$, and therefore it follows that the operator $[F'(x_{n-1})]^{-1}$ exists, and the iterate x_n is well defined if $x_{n-1} \in B(x_0, t^* - t_0)$. Thus,

$$\|x_1 - x_0\| = \|\Gamma_0 F(x_0)\| \leq \eta = t_1 - t_0 \leq t^* - t_0,$$

so that $x_1 \in B(x_0, t^* - t_0) \subset \Omega$. Now, we assume that $x_1, x_2, \ldots, x_n \in B(x_0, t^* - t_0) \subset \Omega$ and

$$\|x_{i+1} - x_i\| \leq t_{i+1} - t_i, \quad \text{for all} \quad i = 0, 1, \ldots, n-1. \tag{2.14}$$

Then,

$$\|x_{n+1} - x_n\| = \left\| \int_{x_{n-1}}^{x_n} L_F(x) \, dx \right\|$$

$$\leq \int_0^1 \|L_F(x_{n-1} + \tau(x_n - x_{n-1}))\| \, \|x_n - x_{n-1}\| \, d\tau$$

$$\leq \int_0^1 L_f(t_{n-1} + \tau(t_n - t_{n-1}))(t_n - t_{n-1}) \, d\tau$$

$$= \int_{t_{n-1}}^{t_n} L_f(u) \, du$$

$$= t_{n+1} - t_n,$$

since $x = x_{n-1} + \tau(x_n - x_{n-1})$ with $\tau \in [0, 1]$ and

$$\|x - x_0\| \leq \|x_{n-1} - x_0\| + \tau \|x_n - x_{n-1}\| \leq t_{n-1} + \tau(t_n - t_{n-1}) - t_0 = t - t_0,$$

where $t = t_{n-1} + \tau(t_n - t_{n-1})$. In addition,

$$\|x_{n+1} - x_0\| \leq \|x_{n+1} - x_n\| + \|x_n - x_0\| \leq t_{n+1} - t_n + t_n - t_0 = t_{n+1} - t_0 \leq t^* - t_0,$$

and $x_{n+1} \in B(x_0, t^* - t_0) \subset \Omega$.

Then, by mathematical induction, we have seen that $x_i \in B(x_0, t^* - t_0) \subset \Omega$ and (2.14) are true for all positive integers i. As a consequence, $\{t_n\}$ majorizes $\{x_n\}$, and there exists $\lim_n x_n = x^* \in \overline{B(x_0, t^* - t_0)}$, where $t^* = \lim_n t_n$ and such that $\|x^* - x_n\| \leq t^* - t_n$, for all $n \geq 0$. Finally, the fact that x^* is a solution of the equation $F(x) = 0$ such that $\|x^* - x_n\| \leq t^* - t_n$, for all $n \geq 0$, follows as in Theorem 2.16. ∎

Finally, we illustrate the last result with the following simple example.

Example 2.22 Consider the rational function

$$F(x) = \frac{x - 4}{(x + 11)(x - 15)}.$$

In addition, $L_F(x) = \frac{2(x-4)(x^3 - 12x^2 + 543x - 1384)}{(x^2 - 8x + 181)^2}$. If we now choose $\Omega = (0.4876, 4.6210)$, then

$$\|L_F(x)\| \leq \gamma(\xi - t), \quad \text{for} \quad \|x - x_0\| \leq t - t_0,$$

2.2 Operator Equations in Banach Spaces

with $\gamma = 0.05$, $\xi = 5.5$, $x_0 = 2.5$, and $t_0 = 1$. Thus, $t^* = \xi_1 = 2.8432\ldots$ and $B(x_0, t^* - t_0) \subset \Omega$. Therefore, the conditions of Theorem 2.21 are satisfied, and as a consequence, the Newton sequence (2.4) starting at $x_0 = 2.5$ converges to the solution $x^* = 4$ of $F(x) = 0$.

Instead of considering straight lines, as in Theorems 2.16 and 2.21, we can consider other types of functions, so that we can obtain, depending on the degree of logarithmic convexity, new conditions to prove the convergence of the Newton method. However, these results are interesting from the theoretical point of view, but, in practice, it is difficult to find useful bounds for the operator L_F.

2.2.2.2 Global Convergence

In this section, our aim is to locate a solution of the equation to solve and obtain a domain of global convergence, so that any point in the domain can be chosen as starting point to guarantee the convergence of the Newton method to the solution from it. This fact avoids the classic problem of the accessibility of the method: the location of starting points from which the convergence of the Newton method is guaranteed. Our study focuses on the location of restricted global convergence domains for the Newton method which are usually balls.

Conditions on the starting point are required to guarantee the semilocal convergence of the Newton method, as we can see in the Newton-Kantorovich theorem. Now, we try to relax these conditions on the starting point, so that the domain of starting points can be enlarged.

The characterization of domains of starting points for the Newton method is an interesting and difficult problem, even in the scalar case [50]. If an equation has several solutions, we can search the domains of starting points from which the Newton method converges to each of the solutions. This problem is not simple, since close starting points can cause convergent sequences to different solutions or, even, can cause one convergent sequence and another not [79]. We then get into another interesting field, such as chaos and fractals and their relationship with the Newton method. An introduction to these topics can be seen in [14].

In particular, given a nonlinear equation, we give conditions so that a solution of the equation lies in a certain domain and define a set of starting points from which the convergence of the method to the solution is guaranteed.

Notice that the problem of solving an equation of the form $F(x) = 0$ can be transformed into a problem of locating a fixed point of the operator $G(x) = x - F(x)$ and reciprocally. Thus, two equivalent problems result, and therefore the techniques employed in one of the two situations are perfectly extensible to the other.

A procedure commonly used in approximating fixed points is the method of successive approximations. This method arises from the application of the Banach Fixed Point Theorem that allows obtaining the global convergence of the method.

Theorem 2.23 (The Banach Fixed Point Theorem [15]) *Let $T : X \to X$ be a contraction operator in the Banach space X. Then, T admits a unique fixed point x^* in X (i.e., $T(x^*) = x^*$), and it can be approximated by the method of successive approximations, $x_{n+1} = T(x_n)$, $n \geq 0$, for each $x_0 \in X$.*

The application of the Banach Fixed Point Theorem requires quite restrictive hypotheses, and the theorem is usually applied in the full space X, so that only the case in which the solution of the equation is unique in the complete space is considered. A useful modification of the Banach Fixed Point Theorem states a fixed point result on a subset of X instead of the full space X, as we can see in the following.

Theorem 2.24 *If D is a convex and compact set of a Banach space X and the operator $P : D \to D$ is a contraction, then the operator P has a unique fixed point x^* in D, and it can be approximated by the method of successive approximations, $x_{n+1} = P(x_n)$, $n \geq 0$, for each $x_0 \in D$.*

We observe that two conditions are required in Theorem 2.24. First, we must have an operator that transforms a subset of a Banach space X into itself, and second, this operator must be a contraction. We remember that the operator P is a contraction if $\|P(x) - P(y)\| < Q\|x - y\|$ with $Q \in [0, 1)$, for all $x, y \in D$. Moreover, in this case, for all $n \in \mathbb{N}$, we have

$$\|x^* - x_n\| \leq \frac{Q^n}{1 - Q}\|x_1 - x_0\| \quad \text{and} \quad \|x^* - x_n\| \leq Q^n \|x^* - x_0\| \qquad (2.15)$$

for each $x_0 \in D$. If the operator P is differentiable, it is enough that the condition $\|P'(x)\| < 1$, for all $x \in D$, is satisfied to see that P is a contraction.

Note that the Newton method to solve $F(x) = 0$ consists of constructing the sequence (2.4). Therefore, it is clear that this sequence is the same as that obtained by applying the method of successive approximations to the operator N_F, which converges to a fixed point of N_F under the previous conditions. Thus, we can apply the techniques of the Fixed Point Theorem to study the convergence of the Newton method. This idea, which a priori may seem a bit artificial, allows us to provide conditions for the convergence of the Newton method and to analyze in depth the domains of starting points from which the convergence of the method is guaranteed. Regarding this last aspect, we can prove the convergence of the Newton method from starting points where the classical Kantorovich conditions are not satisfied, as we can see later in some examples.

Now, we obtain a first convergence result for the Newton method using the operator degree of logarithmic convexity L_F and arguments similar to those of the Fixed Point Theorem.

2.2 Operator Equations in Banach Spaces

Theorem 2.25 *With the previous notation, we suppose that the operator L_F is defined in a certain closed convex domain Ω of a Banach space X and $\|L_F(x)\| \leq K < 1$, for all $x \in \Omega$. Besides, $x_0 \in \Omega$ and*

$$\rho = \frac{\|x_0 - N_F(x_0)\|}{1 - K} = \frac{\|[F'(x_0)]^{-1} F(x_0)\|}{1 - K}. \tag{2.16}$$

If $\overline{B(x_0, \rho)} \subseteq \Omega$, then

(a) *The sequence $x_{n+1} = N_F(x_n)$, $n \geq 0$, starting at x_0 converges to x^*.*
(b) *There exists a fixed point x^* of N_F in $\overline{B(x_0, \rho)}$.*
(c) *x^* is the unique fixed point of N_F in $\overline{B(x_0, \rho)}$.*

Moreover, $\|x^* - x_n\| \leq K^n \rho$.

Proof First, we see that the sequence $\{x_n\}$ lies in $B(x_0, \rho)$. For this, we use mathematical induction on n. From (2.16), we have

$$\|x_1 - x_0\| = \|N_F(x_0) - x_0\| = (1 - K)\rho < \rho,$$

so that $x_1 \in B(x_0, \rho)$. After that, we suppose that $x_1, x_2, \ldots, x_n \in B(x_0, \rho)$ and

$$\|x_n - x_0\| \leq (1 - K^n)\rho < \rho.$$

Then, from $N'_F(x) = L_F(x)$ (see Theorem 2.9), we obtain

$$\|x_{n+1} - x_n\| = \|N_F(x_n) - N_F(x_{n-1})\| \leq \sup_{z \in [x_{n-1}, x_n]} \|N'_F(z)\| \|x_n - x_{n-1}\| \leq K \|x_n - x_{n-1}\|, \tag{2.17}$$

and, by reiterating the process, it follows

$$\|x_{n+1} - x_n\| \leq K \|x_n - x_{n-1}\| \leq \cdots \leq K^n \|x_1 - x_0\| = K^n (1 - K)\rho, \tag{2.18}$$

$$\|x_{n+1} - x_0\| \leq \|x_{n+1} - x_n\| + \|x_n - x_0\| \leq K^n (1 - K)\rho + (1 - K^n)\rho$$
$$= (1 - K^{n+1})\rho < \rho.$$

Therefore, $x_n \in B(x_0, \rho)$ and $\|x_n - x_0\| \leq (1 - K^n)\rho$, for all $x_n \in B(x_0, \rho)$.

Second, we see that the sequence $\{x_n\}$ is of Cauchy. For $m \geq 1$ and from (2.17), we have

$$\|x_{n+m} - x_n\| \leq \sum_{i=0}^{m-1} \|x_{n+i+1} - x_{n+i}\| \leq \left(\sum_{j=0}^{m-1} K^j\right) \|x_{n+1} - x_n\| = \frac{1 - K^m}{1 - K}\|x_{n+1} - x_n\|,$$

and, taking into account (2.18), we obtain

$$\|x_{n+m} - x_n\| \leq (1 - K^m) K^n \rho, \qquad (2.19)$$

so that $\{x_n\}$ is a Cauchy sequence and therefore convergent, that is, $\lim_{n\to\infty} x_n = x^*$ with $x^* \in \overline{B(x_0, \rho)} \subset \Omega$. As a consequence, items (a) and (b) are true.

Third, as N_F is a differentiable operator, N_F is continuous, and, as x^* is a fixed point of N_F, then

$$x^* = N_F(x^*) = x^* - [F'(x^*)]^{-1} F(x^*)$$

and $[F'(x^*)]^{-1} F(x^*) = 0$. As $x^* \in \overline{B(x_0, \rho)}$, by applying the linear operator $F'(x^*)$ to the last equality, it follows $F(x^*) = 0$, so that x^* is a solution of $F(x) = 0$.

Fourth, we see that x^* is the unique fixed point of N_F in Ω. Suppose that y^* is a fixed point of N_F in Ω and such that $x^* \neq y^*$. Then, from Theorem 2.9 and taking into account that L_F is bounded in $\overline{B(x_0, \rho)}$, we have

$$\|x^* - y^*\| = \|N_F(x^*) - N_F(y^*)\| \leq \sup_{z \in [x^*, y^*]} \|L_F(z)\| \|x^* - y^*\| \leq K \|x^* - y^*\| < \|x^* - y^*\|,$$

so that there is a contradiction, and, as a consequence, item (c) is true.

Finally, from (2.19), it follows $\|x_{n+m} - x_n\| \leq (1 - K^m) K^n \rho$, and, by letting $m \to \infty$, we have

$$\|x^* - x_n\| \leq K^n \rho.$$

The proof is complete. ∎

Observe that the unique fixed point of the operator N_F is the unique solution of the equation $F(x) = 0$.

In the last theorem, we prove the convergence of the Newton method by fixing the starting point as the center of a ball. Next, we obtain a result that guarantees the convergence of the method from a larger domain of starting points. In particular, we can guarantee the convergence of the Newton method starting at any point of a ball.

Theorem 2.26 *Under the conditions of the last theorem, if the sequence $\tilde{x}_{n+1} = N_F(\tilde{x}_n)$, $n \geq 0$, starts at any point $\tilde{x}_0$ of the ball $\overline{B(x_0, \rho)}$, with ρ defined in (2.16), then the sequence*

2.2 Operator Equations in Banach Spaces

$\{\widetilde{x}_n\}$ converges to the unique fixed point of the operator N_F in $\overline{B(x_0, \rho)}$. Moreover, for all $n \in \mathbb{N}$,

$$\|x^* - \widetilde{x}_n\| \leq \frac{K^n}{1-K}\|\widetilde{x}_1 - \widetilde{x}_0\| \quad \text{and} \quad \|x^* - \widetilde{x}_n\| \leq K^n \|x^* - \widetilde{x}_0\|.$$

Proof If x is any point of $\overline{B(x_0, \rho)}$, then

$$\|x_1 - N_F(x)\| = \|N_F(x_0) - N_F(x)\| \leq K\|x_0 - x\| \leq K\rho$$

and $N_F(x) \in \overline{B(x_1, K\rho)}$.

Moreover, if $x \in \overline{B(x_1, K\rho)}$ and we take into account (2.16), then

$$\|x - x_0\| = \|x - x_1\| + \|x_1 - x_0\| \leq K\rho + (1-K)\rho = \rho,$$

so that $\overline{B(x_1, K\rho)} \subseteq \overline{B(x_0, \rho)}$. Then, the operator $N_F : \overline{B(x_0, \rho)} \to \overline{B(x_0, \rho)}$ is a contraction with contractivity factor $K < 1$. The proof is now complete from Theorem 2.24 and (2.15). ∎

After that, we consider the particular case in which the convex domain Ω, where the operator L_F is defined and bounded, is a closed ball $\Omega = \overline{B(w_0, R)}$. We see that Theorem 2.26 is satisfied in relation to the starting points that we can choose for this new situation.

Corollary 2.27 *Suppose the conditions of Theorem 2.25. Let $x_0 \in \Omega = \overline{B(w_0, R)}$ and ρ be given in (2.16). If $\rho > R$ and $\|w_0 - x_0\| \leq R - \rho$, then the Newton sequence (2.4) starting at any point of the ball $\overline{B(x_0, \rho)}$ converges to the unique solution of the equation $F(x) = 0$.*

Proof If $z \in \overline{B(x_0, \rho)}$, then

$$\|z - w_0\| \leq \|z - x_0\| + \|x_0 - w_0\| \leq R,$$

so that $\overline{B(x_0, \rho)} \subseteq \Omega = \overline{B(w_0, R)}$, and the result follows from Theorem 2.26. ∎

2.2.2.3 Extension of the Domain of Starting Points

Next, we pay attention to determine domains of starting points for the Newton method as large as possible.

Theorem 2.26 establishes that the convergence of the Newton method is given from a point and a certain neighborhood of it. Corollary 2.27 establishes how to find a domain of starting points if the domain Ω, where the operator L_F is bounded, is a closed ball $\Omega = \overline{B(w_0, R)}$. Continuing with the fact that the domain Ω is a closed ball, we then suppose that the center of the ball, w_0, is a starting point, that is,

$$\varphi(w_0) \leq R(1-K),$$

where

$$\varphi(w_0) = \|[F'(w_0)]^{-1} F(w_0)\|. \tag{2.20}$$

Then, Corollary 2.27 guarantees that the Newton sequence starting at any point of the ball $\overline{B(w_0, \rho)}$, with

$$\rho = \frac{\varphi(w_0)}{1-K},$$

is convergent. Now, depending on the quantity $\varphi(w_0)$, we enlarge the domain of starting points given by Corollary 2.27.

Theorem 2.28 *Let $\Omega = \overline{B(w_0, R)}$ be a closed ball where the operator $L_F(x)$ exists and is such that $\|L_F(x)\| \leq K < 1$, and $\varphi(w_0)$ is given by (2.20). Then:*

(a) *If $\varphi(w_0) \leq (1-K)^2 R$, the Newton sequence (2.4) starting at any point of Ω converges.*
(b) *If $(1-K)^2 R < \varphi(w_0) \leq \frac{1-K}{1+K} R$, the Newton sequence (2.4) starting at any point of the closed ball*

$$\overline{B\left(w_0, \frac{R}{K} - \frac{\varphi(w_0)}{K(1-K)}\right)}$$

converges.
(c) *If $\frac{1-K}{1+K} R < \varphi(w_0) \leq (1-K)R$, the Newton sequence (2.4) starting at any point of the closed ball*

$$\overline{B\left(w_0, \frac{\varphi(w_0)}{1-K}\right)}$$

converges.

Proof To prove item (a), we suppose that the Newton sequence (2.4) lies in Ω and starts at $x_0 \in \Omega$. So,

$$\|x_{n+1} - x_n\| = \|N_F(x_n) - N_F(x_{n-1})\| \leq \sup_{x \in [x_{n-1}, x_n]} \|L_F(x)\| \|x_n - x_{n-1}\| \leq K \|x_n - x_{n-1}\|$$

and

$$\|x_{n+1} - x_n\| \leq K^n \|x_1 - x_0\|.$$

2.2 Operator Equations in Banach Spaces

Therefore, the sequence $\{x_n\}$ converges to some $x^* \in \Omega$. Moreover, as N_F is continuous and $N_F(x^*) = x^*$, then x^* is a solution of $F(x) = 0$.

As a consequence of the last, we just have to check what conditions must been satisfied by the starting point x_0 so that the Newton sequence lies in the Ω. Thus, we denote $[F'(w_0)]^{-1}$ by $\Gamma(w_0)$ and have

$$\|x_1 - w_0\| \leq \|N_F(x_0) - N_F(w_0) - \Gamma(w_0)F(w_0)\|$$
$$\leq K\|x_0 - w_0\| + \varphi(w_0),$$
$$\|x_2 - w_0\| \leq \|N_F(x_1) - N_F(w_0) - \Gamma(w_0)F(w_0)\|$$
$$\leq K\|x_1 - w_0\| + \varphi(w_0)$$
$$\leq K^2\|x_0 - w_0\| + \varphi(w_0)(1 + K)$$
$$\vdots$$
$$\|x_n - w_0\| \leq K^n\|x_0 - w_0 - \Gamma(w_0)F(w_0)\|(1 + K + \cdots + K^{n-1})$$
$$< K\|x_0 - w_0\| + \frac{1}{1 - K}\varphi(w_0).$$

Hence, $x_n \in \overline{B(w_0, R)}$ if $K\|x_0 - w_0\| + \frac{1}{1-K}\varphi(w_0) \leq R$, that is,

$$\|x_0 - w_0\| \leq \frac{R}{K} - \frac{1}{K(1 - K)}\varphi(w_0). \tag{2.21}$$

Note that the right side of the last inequality is a positive value, since w_0 is a starting point for the Newton method and $\varphi(w_0) \leq R(1 - K)$.

As $\|x_0 - w_0\| \leq R$, (2.21) is satisfied if $R \leq \frac{R}{K} - \frac{\varphi(w_0)}{K(1-K)}$, that is,

$$\frac{1}{K(1 - K)}\varphi(w_0) \leq R\left(\frac{1}{K} - 1\right) \quad \text{if} \quad \varphi(w_0) \leq R(1 - K)^2.$$

Therefore, any point on $\Omega = \overline{B(w_0, R)}$ that we choose as the starting point for the Newton method makes the method being convergent.

To prove item (b), if

$$R(1 - K)^2 < \varphi(w_0) \leq \frac{1 - K}{1 + K}R,$$

then

$$\frac{R}{K} - \frac{1}{K(1 - K)}\varphi(w_0) \geq \frac{\varphi(w_0)}{1 - K}.$$

Hence, if $x \in \overline{B\left(w_0, \frac{R}{K} - \frac{\varphi(w_0)}{K(1-K)}\right)}$, the condition (2.21) is true, and, as a consequence, the Newton sequence starting at any point on the ball is convergent.

Finally, if

$$\frac{1-K}{1+K} R < \varphi(w_0) \leq (1-K)R$$

and we proceed as in items (a) and (b), we obtain item (c). ∎

Remark 2.29 The balls that appear in items (a), (b), and (c) satisfy the following:

$$\overline{B\left(w_0, \frac{\varphi(w_0)}{1-K}\right)} \subseteq \overline{B\left(w_0, \frac{R}{K} - \frac{\varphi(w_0)}{K(1-K)}\right)} \subseteq \Omega.$$

The previous results show that we can enlarge the domain of starting points given by Kantorovich for the Newton method, since we can consider starting points that do not satisfy the conditions of the Newton-Kantorovich theorem.

After that, we see two examples in which we consider starting points for the Newton method that do not satisfy the Kantorovich conditions.

Example 2.30 Consider the polynomial $p(x) = x^2 - 4x + 3$. Then,

$$L_p(x) = \frac{p(x)p''(x)}{p'(x)^2} = \frac{x^2 - 4x + 3}{2(x-2)^2}$$

and $|L_p(x)| < 1$ if $x \in \left(-\infty, 2 - \frac{\sqrt{3}}{3}\right)$.

If we choose $\Omega = [-100, 1.36754]$, then

$$K = \sup_{x \in \Omega} |L_p(x)| = |L_p(1.36754)| = \frac{3}{4}.$$

In addition, we choose $x_0 = 0.88$ and denote

$$\rho = \frac{1}{1-K} \left| \frac{p(x_0)}{p'(x_0)} \right| = -\frac{4p(x_0)}{p'(x_0)} = 0.4542.$$

Since $\overline{B(x_0, \rho)} = [0.4257, 1.3342] \subseteq \Omega$, we follow from Theorem 2.26 that the Newton method starting at any point of the previous interval is convergent. In particular, from the upper end of the interval, $\widetilde{x}_0 = 1.3342$. However, the conditions of Theorem 2.11 of Newton-Kantorovich are not satisfied in this point, since $p(\widetilde{x}_0) = -0.5567\ldots$, $p'(\widetilde{x}_0) = -1.3316\ldots$, $p''(x) = 2$, and $h = 0.6279\ldots > \frac{1}{2}$, so that the convergence of the Newton method is not guaranteed from Theorem 2.11.

2.2 Operator Equations in Banach Spaces

Example 2.31 Consider the nonlinear system given in Example 2.17,

$$F(x, y) = (xy - 1, xy + x - 2y) = 0.$$

If $w_0 = (1.13, 1.13)$ and $R = 0.43$, we see in Example 2.17 that $\sup_{(x,y) \in \Omega} \|L_F(x, y)\| \leq 0.7$, where $\Omega = \overline{B(w_0, R)}$, so that $K = 0.7$. Besides, as

$$\frac{1-K}{1+K} R = 0.0758\ldots < \varphi(w_0) = 0.1250\ldots < (1-K)R = 0.129,$$

item (c) of Theorem 2.28 is satisfied. So,

$$\frac{\varphi(w_0)}{1-K} = 0.4167\ldots,$$

and the Newton method starting at any point of the ball $\overline{B(w_0, 0.4167\ldots)}$ is convergent. However, there are points in the ball where the conditions of Theorem 2.11 of Newton-Kantorovich are not satisfied. For example, if $(x_0, y_0) = (0.73, 0.73)$, then $\beta = 1.1232\ldots$, $\eta = 0.3365\ldots$, $\|F''(x, y)\| = 2$, and $h = 0.7561\ldots > 1/2$, so that we cannot guarantee the convergence of the Newton method by Theorem 2.11.

In Theorem 2.28, the operator L_F is bounded in the closed ball $\Omega = \overline{B(w_0, R)}$, and the convergence of the Newton method is guaranteed starting at any point on a closed ball centered on w_0 and whose radius depends on the values of $\varphi(w_0) = \|[F'(w_0)]^{-1}F(w_0)\|$. Next, by using other technique, we improve considerably this result, since we prove the convergence of the method starting at any point of Ω with no more than requiring that the center of the ball, chosen as starting point, provides a convergent sequence, which is a condition required in Theorem 2.25.

Theorem 2.32 *Let $\Omega = \overline{B(w_0, R)}$ be a closed ball where $\|L_F(x)\| \leq K < 1$ for all $x \in \Omega$. If $\varphi(w_0) \leq (1-K)R$, then the Newton sequence starting at any point of Ω converges to the unique solution x^* of $F(x) = 0$ in Ω.*

Proof From $N_F(x) = x - [F'(x)]^{-1}F(x)$ and $N_F'(x_0) = L_F(x_0)$, it follows that N_F is a contraction operator. Besides, $N_F : \Omega \to \Omega$. Indeed, if x is any point of Ω, then

$$\|x_1 - N_F(x)\| = \|N_F(x_0) - N_F(x)\| \leq K\|x_0 - x\| \leq KR,$$

so that $N_F : \Omega \to \Omega$.

Moreover, $\overline{B(x_1, KR)} \subseteq \overline{B(x_0, R)} = \Omega$. For this, we choose $x \in \overline{B(x_1, KR)}$, and, as

$$\rho = \frac{\varphi(w_0)}{1-K} \leq R,$$

it follows from (2.18) that

$$\|x - x_0\| = \|x - x_1\| + \|x_1 - x_0\| \leq KR + (1-K)\rho \leq KR + (1-K)R = R.$$

And, by Theorem 2.24, the proof is complete. ■

We see below that we can enlarge the domain of starting points obtained in Example 2.31 for the Newton method with the previous result.

Example 2.33 In Example 2.31, we prove that the Newton method starting at any point on the ball $\overline{B(w_0, 0.4167\ldots)}$ with $w_0 = (1.13, 1.13)$ is convergent. As

$$\frac{\varphi(w_0)}{1-K} = 0.4167\ldots \leq R = 0.43,$$

the convergence of the Newton method is guaranteed starting at any point of the ball $\overline{B(w_0, 0.43)}$ from Theorem 2.32, so that the domain of starting points for the Newton method is enlarged.

We conclude with an example where we show that we can use the previous results to guarantee the convergence of the Newton method to the solution of a scalar equation in which there is an inflection point, a situation that is not included under the Fourier conditions and therefore not usually considered.

Example 2.34 The polynomial $p(x) = \frac{(x-1)^3}{6} - (x-1)$ has an inflection point in the zero $x^* = 1$; see Fig. 2.6. With the help of Theorem 2.28, we find an interval of staring points from which the convergence of the Newton sequence is guaranteed.

If we consider $w_0 = 0.8$ and $R = 0.35$, then $\Omega = \overline{B(w_0, R)} = [0.45, 1.15]$ and

$$K = \sup_{x \in \Omega} |L_p(x)| = |L_p(0.45)| = 0.3987.$$

As $\varphi(w_0) = \left|\frac{p(w_0)}{p'(w_0)}\right| = 0.2027\ldots$, then

$$\frac{1-K}{1+K} R = 0.1504\ldots < \varphi(w_0) = 0.2027\ldots < (1-K)R = 0.2104\ldots,$$

so that, by item (c) of Theorem 2.28, the Newton method starting at any point of the ball

$$\overline{B\left(w_0, \frac{\varphi(w_0)}{1-K}\right)} = \overline{B(0.8, 0.3371\ldots)}$$

is convergent. In Table 2.4, we see the Newton sequence starting at $x_0 = 0.6$.

2.2 Operator Equations in Banach Spaces

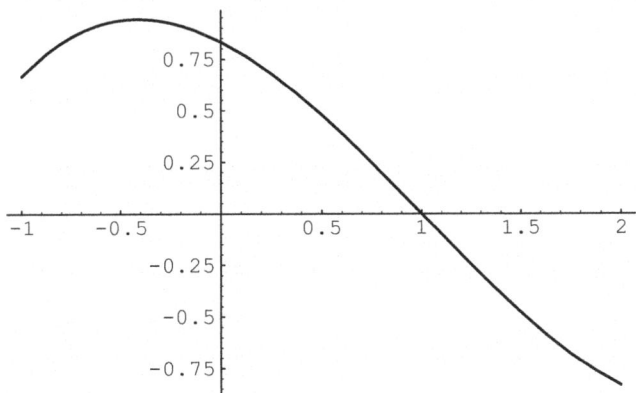

Fig. 2.6 Graph of $p(x) = \frac{(x-1)^3}{6} - (x-1)$ in $[-1, 2]$

Table 2.4 Newton's sequence from Example 2.34

n	x_n
0	0.6000000000000000...
1	1.0231884057971016...
2	0.9999995842730325...
3	1.0000000000000000...
4	1.0000000000000000...

2.2.2.4 Global Convergence with Auxiliary Points

Now, we use the operator N_F, given by (2.4), and a technique based on auxiliary points [43, 44] to obtain results of global convergence restricted to balls for the Newton method under the usual conditions of Kantorovich. For this, we apply Theorem 2.24.

We start by locating a domain that contains a fixed point of the operator N_F. So, we consider a domain $\overline{B(\widetilde{x}, R)} \subseteq \Omega$, where $R \in \mathbb{R}_+$ and $\widetilde{x} \in \Omega$ is an auxiliary point that satisfies the following conditions:

(C1) There exists $\widetilde{\Gamma} = [F'(\widetilde{x})]^{-1} \in \mathcal{L}(Y, X)$, for some $\widetilde{x} \in \Omega$, with $\|\widetilde{\Gamma}\| \leq \beta$ and $\|\widetilde{\Gamma} F(\widetilde{x})\| \leq \eta$.

(C2) There exists a constant $M \geq 0$ such that $\|F''(x)\| \leq M$ for all $x \in \Omega$.

We can then establish the following theorem.

Theorem 2.35 *Let $\widetilde{x} \in \Omega$, and suppose that conditions (C1)–(C2) are satisfied. If $M\beta\eta \leq \frac{1}{6}$, then there exists $R > 0$ such that $R \in [R_-, R_+]$, where $R_- = \frac{1-\sqrt{1-6M\beta\eta}}{3M\beta}$ and $R_+ = \frac{1+\sqrt{1-6M\beta\eta}}{3M\beta}$, and $N_F : \overline{B(\widetilde{x}, R)} \to \overline{B(\widetilde{x}, R)}$.*

Proof First, under the hypotheses of the theorem, we observe that $[R_-, R_+] \neq \emptyset$, so that we can choose R in this interval.

Second, we see that there exists $\Gamma = [F'(x)]^{-1}$ for all $x \in \overline{B(\widetilde{x}, R)}$. Indeed, from

$$\|I - \widetilde{\Gamma} F'(x)\| \leq \|\widetilde{\Gamma}\| \|F'(\widetilde{x}) - F'(x)\| \leq \beta M \|x - \widetilde{x}\| \leq M\beta R < 1,$$

since $R \in [R_-, R_+]$, it follows, by the Banach lemma on invertible operators, that the operator Γ exists with $\|\Gamma\| \leq \frac{\beta}{1-M\beta R}$ and $\|\Gamma F'(\widetilde{x})\| \leq \frac{1}{1-M\beta R}$. As a consequence, the operator N_F is well defined in the domain indicated.

Third, we see that $N_F(x) \in \overline{B(\widetilde{x}, R)}$ for all $x \in \overline{B(\widetilde{x}, R)}$. Indeed, from

$$\|N_F(x) - \widetilde{x}\| = \left\| -\Gamma \left(F(x) + F'(x)(\widetilde{x} - x) \right) \right\|$$

$$= \left\| -\Gamma F(\widetilde{x}) + \Gamma \int_0^1 \left(F'(x + t(\widetilde{x} - x)) - F'(x) \right) dt (\widetilde{x} - x) \right\|$$

$$\leq \|\Gamma F'(\widetilde{x})\| \|\widetilde{\Gamma} F(\widetilde{x})\| + \frac{M}{2} \|\Gamma\| \|\widetilde{x} - x\|^2$$

$$\leq \frac{M\beta R^2 + 2\eta}{2(1 - M\beta R)}$$

$$\leq R,$$

it follows that $N_F(x) \in \overline{B(\widetilde{x}, R)}$, provided that $M\beta\eta \leq \frac{1}{6}$ and $R \in [R_-, R_+]$. ∎

To apply Fixed Point Theorem 2.24 to the operator N_F, we need to guarantee that the operator N_F is a contraction in the domain considered. One way to check that N_F is a contraction is to directly check the condition $\|N_F(x) - N_F(y)\| < Q\|x - y\|$ with $Q \in [0, 1)$, for all $x, y \in \overline{B(\widetilde{x}, R)}$. Other variant that we can consider to prove that N_F is a contraction comes from the fact that N_F is a differentiable operator such that $\|N_F'(x)\| < 1$. For this, we consider the operator degree of logarithmic convexity, which satisfies $L_F(x) = N_F'(x)$, as we see in Sect. 2.2.1.

After that, from conditions (C1)–(C2), we see that the operator N_F is a contraction.

Lemma 2.36 *Let $\widetilde{x} \in \Omega$, and suppose that conditions (C1)–(C2) are satisfied. If $M\beta\eta \leq \frac{1}{6}$, then there exists $R > 0$ such that $R \leq \delta_- = \frac{4 - \sqrt{10 + 6M\beta\eta}}{3M\beta}$, and the operator N_F is a contraction.*

Proof If $x \in \overline{B(\widetilde{x}, R)}$, then

$$\widetilde{\Gamma} F(x) = \widetilde{\Gamma} F(\widetilde{x}) + (x - \widetilde{x}) + \widetilde{\Gamma} \int_0^1 \left(F'(\widetilde{x} + t(x - \widetilde{x})) - F'(\widetilde{x}) \right) (x - \widetilde{x}) dt,$$

so that

2.2 Operator Equations in Banach Spaces

$$\|\widetilde{\Gamma}F(x)\| \leq \|\widetilde{\Gamma}F(\widetilde{x})\| + \|x - \widetilde{x}\| + \frac{M}{2}\|\widetilde{\Gamma}\|\|x - \widetilde{x}\|^2 \leq \frac{M}{2}\beta R^2 + R + \eta.$$

Now, if $x, y \in \overline{B(\widetilde{x}, R)}$ and we denote $\Gamma_x = [F'(x)]^{-1}$ and $\Gamma_y = [F'(y)]^{-1}$, then

$$\begin{aligned}
N_F(x) - N_F(x) &= x - \Gamma_x F(x) - y + \Gamma_y F(y) \\
&= \Gamma_y \left(F'(y)(x - y) - F'(y)\Gamma_x F(x) + F(y) \right) \\
&= \Gamma_y \left(F(x) - \int_y^x (F'(z) - F'(y))\,dz - F'(y)\Gamma_x F(x) \right) \\
&= \Gamma_y \left(\left(\int_0^1 F''(y + t(x - y))\,dt\,(x - y) \right) \Gamma_x F(x) \right. \\
&\quad \left. - \int_y^x (F'(z) - F'(y))\,dz \right),
\end{aligned}$$

$$\begin{aligned}
\|N_F(x) - N_F(y)\| &= \|\Gamma_y\| \left(M\|x - y\|\|\Gamma_x F(x)\| + \frac{M}{2}\|x - y\|^2 \right) \\
&= \|\Gamma_y\|M \left(\|\Gamma_x F(x)\| + \frac{1}{2}\|x - y\| \right) \|x - y\| \\
&\leq \|\Gamma_y\|M \left(\|\Gamma_x F(x)\| + R \right) \|x - y\| \\
&= \alpha \|x - y\|.
\end{aligned}$$

Taking now into account that $\|[F'(x)]^{-1}\| \leq \frac{\beta}{1 - M\beta R}$ for all $x \in \overline{B(\widetilde{x}, R)}$, we have $\|\Gamma_y\| \leq \frac{\beta}{1 - M\beta R}$. Moreover,

$$\|\Gamma_x F(x)\| \leq \|\Gamma_x F'(\widetilde{x})\|\|\widetilde{\Gamma}F(x)\| \leq \frac{\frac{M}{2}\beta R^2 + R + \eta}{1 - M\beta R}.$$

Thus, we can consider $\dfrac{M\beta(2\eta + 4R - M\beta R^2)}{2(1 - M\beta R)^2} = \alpha$. Then, $\alpha < 1$ if $R \leq \delta_-$ and the operator N_F is therefore a contraction. ∎

Next, we locate the value of R.

Lemma 2.37 *Let $\widetilde{x} \in \Omega$, and suppose that conditions (C1)–(C2) are satisfied. If $M\beta\eta \leq \frac{-3 + \sqrt{13}}{4} = 0.1513\ldots$, there always exists $R > 0$ such that $R \in [R_-, \delta_-]$.*

Proof If $M\beta\eta \leq 0.1513\ldots$, we observe that $R_- \leq \delta_-$, since $4(M\beta\eta)^2 + 6(M\beta\eta) - 1 \leq 0$, so that $[R_-, \delta_-] \neq \emptyset$. Besides, as $R_+ > \delta_-$, the proof is complete. ∎

Now, from Fixed Point Theorem 2.24, we obtain the existence of a unique fixed point of N_F in $\overline{B(\widetilde{x}, R)}$ that can be approximated by the Newton method.

Theorem 2.38 *Let $\widetilde{x} \in \Omega$, and suppose that conditions (C1)–(C2) are satisfied. If $M\beta\eta \leq \frac{-3+\sqrt{13}}{4} = 0.1513\ldots$, then there exists $R > 0$ such that $R \in [R_-, \delta_-]$, the operator $N_F : \overline{B(\widetilde{x}, R)} \to \overline{B(\widetilde{x}, R)}$ has a unique fixed point x^*, and the Newton method starting at $x_0 \in \overline{B(\widetilde{x}, R)}$ converges quadratically to x^*.*

Proof As $N_F(x) \in \overline{B(\widetilde{x}, R)}$, for all $x \in \overline{B(\widetilde{x}, R)}$, and the operator $N_F : \overline{B(\widetilde{x}, R)} \to \overline{B(\widetilde{x}, R)}$ is a contraction, then the proof is concluded from the application of Fixed Point Theorem 2.24 to the operator N_F in $\overline{B(\widetilde{x}, R)}$. The fact that the Newton method has quadratic convergence under conditions (C1)–(C2) follows easily from Theorem 2.11 of Newton-Kantorovich. ∎

Since R_- is the smallest value that R can take and δ_- the largest, we can consider $\overline{B(\widetilde{x}, R_-)}$ as the ball of location of the fixed point and $\overline{B(\widetilde{x}, \delta_-)}$ as the ball of uniqueness and thus achieve a greater separation of other possible fixed points of the operator N_F. Besides, we observe that we have global convergence in any ball $\overline{B(\widetilde{x}, R)}$ with $R \in [R_-, \delta_-]$.

On the other hand, we prove that the operator N_F is a contraction from a condition on $N_F' = L_F$ and then has a unique fixed point.

Theorem 2.39 *Let $\widetilde{x} \in \Omega$, and suppose that conditions (C1)–(C2) are satisfied. If $M\beta\eta \leq \frac{1}{6}$ and $\|L_F(x)\| \leq K < 1$, for all $x \in B(\widetilde{x}, R)$, then the operator $N_F : \overline{B(\widetilde{x}, R)} \to \overline{B(\widetilde{x}, R)}$ has a unique fixed point x^* for all $R \in [R_-, R_+]$, and the Newton method starting at $x_0 \in \overline{B(\widetilde{x}, R)}$ converges quadratically to x^*.*

Proof Under the conditions of the theorem, we have seen in Theorem 2.38 that $N_F(x) \in \overline{B(\widetilde{x}, R)}$ for all $x \in \overline{B(\widetilde{x}, R)}$ with $R \in [R_-, R_+]$. Moreover, as $\|N_F'(x)\| \leq \|L_F(x)\| \leq K < 1$, for all $x \in \overline{B(\widetilde{x}, R)}$, then the operator N_F is a contraction. Therefore, from the application of Fixed Point Theorem 2.24 to the operator N_F in $\overline{B(\widetilde{x}, R)}$, the proof is complete. ∎

In this case, we have $\overline{B(\widetilde{x}, R_-)}$ as the ball of location of the fixed point and $\overline{B(\widetilde{x}, R_+)}$ as the ball of uniqueness of the fixed point. We note that the most favorable ball of global convergence in this case is the last.

We illustrate the previous study with a Fredholm integral equation.

Example 2.40 Consider the following integral equation:

$$x(s) = \sin(\pi s) + \frac{1}{4}\int_0^1 \cos(\pi s)\sin(\pi t)x(t)^2\,dt, \quad s \in [0, 1]. \tag{2.22}$$

2.2 Operator Equations in Banach Spaces

Equations of this type have been studied by different authors (see, e.g., [72, 80]).

Solving (2.22) is equivalent to solving the equation $\mathcal{F}(x) = 0$, where $\mathcal{F} : \mathcal{C}([0, 1]) \to \mathcal{C}([0, 1])$ and

$$[\mathcal{F}(x)](s) = x(s) - \sin(\pi s) - \frac{1}{4}\int_0^1 \cos(\pi s)\sin(\pi t)x(t)^2 dt, \quad s \in [0, 1].$$

Then,

$$[\mathcal{F}'(x)y](s) = y(s) - \frac{1}{2}\int_0^1 \cos(\pi s)\sin(\pi t)x(t)y(t)\,dt,$$

$$[\mathcal{F}''(x)yz](s) = -\frac{1}{2}\int_0^1 \cos(\pi s)\sin(\pi t)y(t)z(t)\,dt,$$

and $\|F(x)\| \leq M = \frac{1}{\pi}$.

If we choose $\widetilde{x}(s) = \sin \pi s$, then $\beta = \frac{\pi}{\pi-1}$, $\eta = \frac{1}{2(\pi-1)}$, and $M\beta\eta = \frac{1}{2(\pi-1)^2} = 0.1090\ldots \leq \frac{-3+\sqrt{13}}{4} = 0.1513\ldots$ Therefore, we can apply Theorem 2.38. Thus, we then have that the operator N_F has a fixed point in $\overline{B(\sin(\pi s), R)}$ with $R \in [0.2940\ldots, 0.5253\ldots]$, so that the ball of location of fixed points is $\overline{B(\sin(\pi s), 0.2940\ldots)}$, and the ball of separation of fixed points (the ball of uniqueness) is $\overline{B(\sin(\pi s), 0.5253\ldots)}$.

Although the conditions on the quantity $M\beta\eta$ seem very restrictive, it is sufficient that the Kantorovich condition $M\beta\eta \leq \frac{1}{2}$ for the auxiliary point is satisfied, since the value of η tends to zero in the Newton sequence. So, once the Kantorovich condition has been verified at the auxiliary point, it is enough to take one or two iterations of the Newton method to fix an auxiliary point that satisfies the required condition. In this case, we find, from an auxiliary point $\widetilde{x}$ such that $M\beta\eta \leq \frac{1}{2}$, a new point $\widetilde{y}$ and obtain global convergence balls of the form $\overline{B(\widetilde{y}, R)}$ that locate a solution of the equation $F(x) = 0$ and separate it from other possible solutions, so that the condition $M\beta\eta \leq \frac{1}{6}$ is relaxed to $M\beta\eta \leq \frac{1}{2}$, and the previous study is then enlarged.

Under conditions (C1)–(C2) and $h = M\beta\eta \leq \frac{1}{2}$, it is clear that Theorem 2.11 of Newton-Kantorovich is satisfied with $x_0 = \widetilde{x}$. In addition, in the proof of the theorem, the following estimates, which we use later, are also deduced (see [64]):

$$h_n = M\beta_n\eta_n \leq \frac{1}{2}(2h)^{2^n}, \tag{2.23}$$

where $\|[F'(x_n)]^{-1}\| \leq \beta_n$ and $\|[F'(x_n)]^{-1}F(x_n)\| \leq \eta_n$.

The first step in locating domains that contain possible fixed points is to describe these domains. For this, we consider balls of the form $\overline{B(\widetilde{y}, R)}$. Thus, the first two aims are to locate the point $\widetilde{y} \in \Omega$ and calculate the value of R, so that $N_F : \overline{B(\widetilde{y}, R)} \to \overline{B(\widetilde{y}, R)}$. After that, we will be able to apply Fixed Point Theorem 2.24. So, we establish the following theorem.

Lemma 2.41 *Let $\widetilde{x} \in \Omega$, and suppose that conditions (C1)–(C2) and $h = M\beta\eta \leq \frac{1}{2}$ are satisfied. Then, there exist $\widetilde{y} \in \overline{B(\widetilde{x}, \rho^*)}$, where $\rho^* = \frac{1-\sqrt{1-2h}}{h}\eta$, and $R > 0$ such that $N_F : \overline{B(\widetilde{y}, R)} \to \overline{B(\widetilde{y}, R)}$ with $R \in [\widetilde{R}_-, \widetilde{R}_+]$, where $\widetilde{R}_- = \frac{1-\sqrt{1-6Mba}}{3Mb}$, $\widetilde{R}_+ = \frac{1+\sqrt{1-6Mba}}{3Mb}$, $\|[F'(\widetilde{y})]^{-1}\| \leq b$, and $\|[F'(\widetilde{y})]^{-1}F(\widetilde{y})\| \leq a$.*

Proof If $M\beta\eta \leq \frac{1}{6}$, then we choose $\widetilde{y} = \widetilde{x}$, so that $Mba \leq \frac{1}{6}$. Otherwise, from (2.23), we observe that if we choose $x_0 = \widetilde{x}$ for the Newton method, we can then consider

$$N_0 = 1 + \left[\frac{\log\left(-\frac{\log 3}{\log 2h}\right)}{\log 2}\right], \quad (2.24)$$

where $[z]$ denotes the integer part of any positive real number z, so that the N_0-th iterate of the method is such that $h_{N_0} \leq \frac{1}{6}$. Therefore, if we choose $\widetilde{y} = x_{N_0}$, it is clear that $Mba \leq \frac{1}{6}$. As a consequence, $[\widetilde{R}_-, \widetilde{R}_+] \neq \emptyset$, and we can then consider R in this interval.

Second, we see that the operator $\Gamma = [F'(x)]^{-1}$ exists for all $x \in \overline{B(\widetilde{y}, R)}$. Indeed, from

$$\|I - [F'(\widetilde{y})]^{-1}F'(x)\| \leq \|[F'(\widetilde{y})]^{-1}\|\|F'(\widetilde{y}) - F'(x)\| \leq bM\|x - \widetilde{y}\| \leq MbR < 1, \quad (2.25)$$

since $R \in [\widetilde{R}_-, \widetilde{R}_+]$, it follows, by the Banach lemma on invertible operators, that the operator Γ exists with $\|\Gamma\| \leq \frac{\beta}{1-MbR}$ and $\|\Gamma F'(\widetilde{y})\| \leq \frac{1}{1-MbR}$. As a consequence, the operator N_F is well defined in the domain indicated.

Third, we see that $N_F(x) \in \overline{B(\widetilde{y}, R)}$ for all $x \in \overline{B(\widetilde{y}, R)}$. Indeed, from

$$\|N_F(x) - \widetilde{y}\| = \left\|-\Gamma\left(F(x) + F'(x)(\widetilde{y} - x)\right)\right\|$$

$$= \left\|-\Gamma F(\widetilde{y}) + \Gamma \int_0^1 \left(F'(x + t(\widetilde{y} - x)) - F'(x)\right) dt (\widetilde{y} - x)\right\|$$

$$\leq \|\Gamma F'(\widetilde{y})\|\|[F'(\widetilde{y})]^{-1}F(\widetilde{y})\| + \frac{M}{2}\|\Gamma\|\|\widetilde{y} - x\|^2$$

$$\leq \frac{MbR^2 + 2a}{2(1 - MbR)}$$

$$\leq R,$$

it follows that $N_F(x) \in \overline{B(\widetilde{y}, R)}$, provided that $Mba \leq \frac{1}{6}$ and $R \in [\widetilde{R}_-, \widetilde{R}_+]$. ■

2.2 Operator Equations in Banach Spaces

Now, we see that the operator N_F is a contraction from conditions (C1)–(C2) and $h = M\beta\eta \leq \frac{1}{2}$.

Lemma 2.42 *Let $\widetilde{x} \in \Omega$, and suppose that conditions (C1)–(C2) and $h = M\beta\eta \leq \frac{1}{2}$ are satisfied. Then, there exist $\widetilde{y} \in \overline{B(\widetilde{x}, \rho^*)}$, where $\rho^* = \frac{1-\sqrt{1-2h}}{h}\eta$, and $R > 0$ such that $R \leq \widetilde{\delta}_- = \frac{4-\sqrt{10+6Mba}}{3Mb}$, with $\|[F'(\widetilde{y})]^{-1}\| \leq b$ and $\|[F'(\widetilde{y})]^{-1}F(\widetilde{y})\| \leq a$, and the operator $N_F : \overline{B(\widetilde{y}, R)} \to \overline{B(\widetilde{y}, R)}$ is a contraction.*

Proof Take $\widetilde{y} = x_0 = \widetilde{x}$. Then, $Mba \leq \frac{1}{2}$, $\widetilde{\delta}_- = \frac{4-\sqrt{10+6Mba}}{3Mb} > 0$, and there exists R such that $0 < R \leq \widetilde{\delta}_-$.

If $x \in \overline{B(\widetilde{y}, R)}$, then

$$[F'(\widetilde{y})]^{-1}F(x) = [F'(\widetilde{y})]^{-1}F(\widetilde{y}) + (x - \widetilde{y}) + [F'(\widetilde{y})]^{-1}$$
$$\times \int_0^1 \left(F'(\widetilde{y} + t(x - \widetilde{y})) - F'(\widetilde{y})\right)(x - \widetilde{y})\,dt,$$

so that

$$\|[F'(\widetilde{y})]^{-1}F(x)\| \leq \|[F'(\widetilde{y})]^{-1}F(\widetilde{x})\| + \|x - \widetilde{y}\| + \frac{M}{2}\|[F'(\widetilde{y})]^{-1}\|\|x - \widetilde{y}\|^2$$
$$\leq \frac{M}{2}bR^2 + R + a.$$

Now, if $x, y \in \overline{B(\widetilde{x}, R)}$, as $MbR < 1$, from (2.25), there exist $\Gamma_x = [F'(x)]^{-1}$ and $\Gamma_y = [F'(y)]^{-1}$, and

$$\|N_F(x) - N_F(y)\| = \widetilde{\alpha}\|x - y\|.$$

Taking now into account that $\|[F'(x)]^{-1}\| \leq \frac{b}{1-MbR}$, for all $x \in \overline{B(\widetilde{y}, R)}$, we have $\|\Gamma_y\| \leq \frac{b}{1-MbR}$. Moreover,

$$\|\Gamma_x F(x)\| \leq \|\Gamma_x F'(\widetilde{y})\|\|[F'(\widetilde{y})]^{-1}F(x)\| \leq \frac{\frac{M}{2}bR^2 + R + a}{1 - MbR}.$$

Thus, we consider $\widetilde{\alpha} = \frac{Mb(2a + 4R - MbR^2)}{2(1 - MbR)^2}$. Therefore, $\widetilde{\alpha} < 1$ if $0 < R \leq \widetilde{\delta}_-$, and the operator N_F is a contraction. ∎

Next, we locate the value of R.

Lemma 2.43 *If $Mba \leq \frac{-3+\sqrt{13}}{4} = 0.1513\ldots$, where $\|[F'(\tilde{y})]^{-1}\| \leq b$ and $\|[F'(\tilde{y})]^{-1}F(\tilde{y})\| \leq a$, there always exists $R > 0$ such that $R \in [\tilde{R}_-, \tilde{\delta}_-]$ with $\tilde{R}_- = \frac{1-\sqrt{1-6Mba}}{3Mb}$ and $\tilde{\delta}_- = \frac{4-\sqrt{10+6Mba}}{3Mb}$.*

Proof If $Mba \leq 0.1513\ldots$, we observe that $\tilde{R}_- \leq \tilde{\delta}_-$, since $4(Mba)^2 + 6(Mba) - 1 \leq 0$, so that $[\tilde{R}_-, \tilde{\delta}_-] \neq \emptyset$. Moreover, as $\tilde{R}_+ > \tilde{\delta}_-$, the proof is complete. ∎

After that, from Fixed Point Theorem 2.24, we obtain the existence of a unique fixed point of N_F in $\overline{B(\tilde{y}, R)}$ that can be approximated by the Newton method.

Theorem 2.44 *Let $\tilde{x} \in \Omega$, and suppose that conditions (C1)–(C2) and $h = M\beta\eta \leq \frac{1}{2}$ are satisfied. Then, there exist $\tilde{y} \in \overline{B(\tilde{x}, \rho^*)}$, where $\rho^* = \frac{1-\sqrt{1-2h}}{h}\eta$, and $R > 0$ such that $R \in [\tilde{R}_-, \tilde{\delta}_-]$, where $\tilde{R}_- = \frac{1-\sqrt{1-6Mba}}{3Mb}$, $\tilde{\delta}_- = \frac{4-\sqrt{10+6Mba}}{3Mb}$, $\|[F'(\tilde{y})]^{-1}\| \leq b$, and $\|[F'(\tilde{y})]^{-1}F(\tilde{y})\| \leq a$. Moreover, the operator $N_F : \overline{B(\tilde{y}, R)} \to \overline{B(\tilde{y}, R)}$ has a unique fixed point x^*, and the Newton method starting at any point in $\overline{B(\tilde{y}, R)}$ converges quadratically to x^*.*

Proof If $M\beta\eta \leq 0.1513\ldots$, then we choose $\tilde{y} = \tilde{x}$, so that $Mba \leq 0.1513\ldots$ Otherwise, from (2.23), we observe that if we choose $x_0 = \tilde{x}$ for the Newton method, we can then consider

$$N_0 = 1 + \left\lceil \frac{\log\left(\frac{\log(0.3026\ldots)}{\log 2h}\right)}{\log 2} \right\rceil,$$

so that the N_0-th iterate of the method is such that $h_{N_0} \leq 0.1513\ldots$ Therefore, if we choose $\tilde{y} = x_{N_0}$, it is clear that $Mba \leq 0.1513\ldots$

Now, from Theorem 2.41, it follows that $N_F(x) \in \overline{B(\tilde{y}, R)}$, for all $x \in \overline{B(\tilde{y}, R)}$, and, from Lemmas 2.42 and 2.43, the operator $N_F : \overline{B(\tilde{y}, R)} \to \overline{B(\tilde{y}, R)}$ is a contraction. Then, the proof is concluded from the application of Fixed Point Theorem 2.24 to the operator N_F in $\overline{B(\tilde{y}, R)}$. The fact that the Newton method has quadratic convergence follows easily from Theorem 2.11 of Newton-Kantorovich. ∎

Since $\tilde{R}_- = \frac{1-\sqrt{1-6Mba}}{3Mb}$ is the smallest value that R can take and $\tilde{\delta}_- = \frac{4-\sqrt{10+6Mba}}{3Mb}$ the largest, we can consider $\overline{B(\tilde{y}, \tilde{R}_-)}$ as the ball of location of the fixed point and $\overline{B(\tilde{y}, \tilde{\delta}_-)}$ as the ball of uniqueness and thus achieve a greater separation of other possible fixed points of the operator N_F. Moreover, we observe that we have global convergence in any ball $\overline{B(\tilde{y}, R)}$ with $R \in [\tilde{R}_-, \tilde{\delta}_-]$.

On the other hand, from the condition on $N'_F = L_F$, we prove that the operator N_F is a contraction and then has a unique fixed point.

Theorem 2.45 *Let $\tilde{x} \in \Omega$, and suppose that conditions (C1)–(C2) and $h = M\beta\eta \leq \frac{1}{2}$ are satisfied. Then, there exists $\tilde{y} \in \overline{B(\tilde{x}, \rho^*)}$, where $\rho^* = \frac{1-\sqrt{1-2h}}{h}\eta$, such that, if $\|L_F(x)\| \leq K < 1$, for all $x \in B(\tilde{y}, R)$, the operator $N_F : \overline{B(\tilde{y}, R)} \to \overline{B(\tilde{y}, R)}$ has a unique fixed point x^* for all $R \in [\tilde{R}_-, \tilde{R}_+]$ with $\tilde{R}_- = \frac{1-\sqrt{1-6Mba}}{3Mb}$, $\tilde{R}_+ = \frac{1+\sqrt{1-6Mba}}{3Mb}$, $\|[F'(\tilde{y})]^{-1}\| \leq b$, and $\|[F'(\tilde{y})]^{-1}F(\tilde{y})\| \leq a$, and the Newton method starting at any point in $\overline{B(\tilde{y}, R)}$ converges quadratically to x^*.*

Proof If $M\beta\eta \leq \frac{1}{6}$, then we choose $\tilde{y} = \tilde{x}$, so that $Mba \leq \frac{1}{6}$. Otherwise, we consider $x_0 = \tilde{x}$ for the Newton method and choose $\tilde{y} = x_{N_0}$ with N_0 given in (2.24), so that $Mba \leq \frac{1}{6}$.

Under the conditions of the theorem, we have seen in Theorem 2.41 that $N_F(x) \in \overline{B(\tilde{y}, R)}$ for all $x \in \overline{B(\tilde{y}, R)}$, where $R \in [\tilde{R}_-, \tilde{R}_+]$. Moreover, as $\|N'_F(x)\| \leq \|L_F(x)\| \leq K < 1$, for all $x \in \overline{B(\tilde{y}, R)}$, the operator N_F is a contraction. Therefore, from the application of Theorem 2.39 to the operator N_F in $\overline{B(\tilde{y}, R)}$, the proof is complete. ∎

In this case, we have $\overline{B(\tilde{y}, \tilde{R}_-)}$ as the ball of location of the fixed point and $\overline{B(\tilde{y}, \tilde{R}_+)}$ as the ball of uniqueness. We note that the most favorable ball of global convergence in this case is the last.

We illustrate the previous study with a Davis type integral equation that is used in [44].

Example 2.46 We consider the integral equation

$$x(s) = s + \frac{7}{5}\int_0^1 G(s,t)x(t)^2 dt, \quad s \in [0,1], \tag{2.26}$$

where the kernel $G(s, t)$ is the Green function

$$G(s,t) = \begin{cases} (1-s)t, & t \leq s, \\ s(1-t), & s \leq t. \end{cases}$$

As we cannot apply the Newton method directly, since we do not know the inverse operator that is involved in the algorithm of the method, we transform the integral equation into a finite-dimensional problem by a process of discretization, where the Gauss-Legendre quadrature formula

$$\int_0^1 \phi(t)\, dt \simeq \sum_{j=1}^8 w_j \phi(t_j),$$

with the nodes t_j and weights w_j known, is used to approximate the integral that appears in (2.26).

By denoting the $x(t_i)$ by x_i, for $i = 1, 2, \ldots, 8$, the integral equation is equivalent to the following system of nonlinear equations:

$$x_j = t_j + \frac{7}{5} \sum_{k=1}^{8} a_{jk} x_k^2, \quad k = 1, 2, \ldots, 8,$$

where

$$a_{jk} = \begin{cases} w_k (1 - t_j) t_k, & k \leq j, \\ w_k (1 - t_k) t_j, & k > j. \end{cases}$$

Now, we write the previous nonlinear system in matrix form as

$$F(\mathbf{x}) \equiv \mathbf{x} - \mathbf{v} - \frac{7}{5} A \, \mathbf{y} = 0, \qquad F : \mathbb{R}^8 \to \mathbb{R}^8, \tag{2.27}$$

where

$$\mathbf{x} = (x_1, x_2, \ldots, x_8)^t, \quad \mathbf{v} = (t_1, t_2, \ldots, t_8)^T, \quad A = (a_{jk})_{j,k=1}^{8}, \quad \mathbf{y} = \left(x_1^2, x_2^2, \ldots, x_8^2\right)^T.$$

If we choose $\widetilde{\mathbf{x}} = (1, 1, \ldots, 1)^t$, then $\beta = 1.4912\ldots$, $\eta = 0.9850\ldots$, and $h = M\beta\eta = 0.3629\ldots$, where $M = 0.2471\ldots$ As $h > 0.1513\ldots$, we cannot apply Theorem 2.38, and, as a consequence, we cannot locate and separate a solution of the integral equation. However, we can apply Theorem 2.44.

For Theorem 2.44, we have $N_0 = 2$ and $h_{N_0} = 0.000192\ldots < 0.1513\ldots$, so that $\widetilde{\mathbf{y}} = \mathbf{x}_2 = (x_1^2, x_2^2, \ldots, x_8^2)$ (the vector shown in Table 2.5), $R \in [0.0006\ldots, 0.9076\ldots]$, the best ball of location of solution is $\overline{B(\mathbf{x}_2, 0.0006\ldots)}$, and the largest ball of convergence is $\overline{B(\mathbf{x}_2, 0.9076\ldots)}$.

If the starting point for the Newton method is $\mathbf{x}_0 = \mathbf{v}$, the method converges to the solution $\mathbf{x}^* = (x_1^*, x_2^*, \ldots, x_8^*)^t$ of system (2.27), which is shown in Table 2.6, after four iterations with stopping criterion $\|\mathbf{x}_n - \mathbf{x}_{n-1}\|_\infty < 10^{-18}$, $n \in \mathbb{N}$.

Table 2.5 The iterate $\mathbf{x}_2$ of the Newton method for (2.27)

i	x_i^2	i	x_i^2
1	0.022629…	5	0.658303…
2	0.115872…	6	0.822537…
3	0.270122…	7	0.932596…
4	0.462132…	8	0.987941…

2.2 Operator Equations in Banach Spaces

Table 2.6 The numerical solution $\mathbf{x}^*$ of (2.27)

i	x_i^*	i	x_i^*
1	0.02267000...	5	0.65888692...
2	0.11607746...	6	0.82291926...
3	0.27057507...	7	0.93276524...
4	0.46275932...	8	0.98797444...

Table 2.7 Absolute errors and $\{\|F(\mathbf{x}_n)\|\}$ from Example 2.46

n	$\|\mathbf{x}^* - \mathbf{x}_n\|$	$\|F(\mathbf{x}_n)\|$
0	$6.7169\ldots \times 10^{-2}$	$9.6624\ldots \times 10^{-1}$
1	$6.4734\ldots \times 10^{-4}$	$5.3956\ldots \times 10^{-4}$
2	$6.0570\ldots \times 10^{-8}$	$5.0516\ldots \times 10^{-8}$
3	$5.4063\ldots \times 10^{-16}$	$4.5077\ldots \times 10^{-16}$

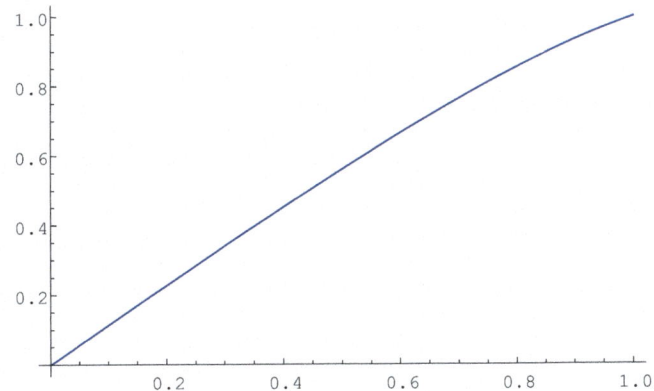

Fig. 2.7 The approximate solution $\mathbf{x}^*$ of system (2.27)

Moreover, the errors $\|\mathbf{x}^* - \mathbf{x}_n\|$ and the sequence $\{\|F(\mathbf{x}_n)\|\}$ are shown in Table 2.7. Observe then that the vector shown in Table 2.6 is a good approximation of a solution of (2.27).

Furthermore, as a solution of (2.26) satisfies $x(0) = 0$ and $x(1) = 1$, if the values of Table 2.6 are interpolated, the approximate solution shown in Fig. 2.7 is obtained. Notice that this approximate solution lies in the domain of location of solution $\overline{B(\mathbf{x}_2, 0.0006\ldots)}$ which is obtained from Theorem 2.44.

2.2.2.5 Local Convergence

While the semilocal convergence results require conditions on the operator F and the starting point x_0, the local convergence results require conditions on the operator F and a solution x^* of $F(x) = 0$. An interesting local result, given by Dennis and Schnabel in [28], for the Newton method requires the following conditions:

(G1) Let x^* be a solution of $F(x) = 0$ such that the operator $[F'(x^*)]^{-1}$ exists and $\|[F'(x^*)]^{-1}\| \leq \gamma$.
(G2) $\|F''(x)\| \leq M$ for all $x \in \Omega$.

Under conditions (G1)–(G2) and $B(x^*, R) \subset \Omega$, Dennis and Schnabel prove that the Newton method starting at any point in $B(x^*, \epsilon)$, where $\epsilon = \min\{R, r\}$ and $r = \dfrac{1}{2M\gamma}$, converges to x^*. The local result provides what we call a ball of convergence, $B(x^*, \epsilon)$. From the value ϵ, this ball of convergence gives information about the accessibility of the solution x^* of the equation to approximate x^* by the Newton method.

In this section, we prove the convergence of the Newton method under the condition $\|L_F(x)\| \leq K < 1$, for all $x \in \Omega$, instead of conditions (G1)–(G2) and compare later both results with an example.

Theorem 2.47 *Let $F : \Omega \subseteq X \to Y$ be a nonlinear twice continuously Fréchet differentiable operator on a non-empty open convex domain Ω of a Banach space X with values in a Banach space Y. Let x^* be a solution of $F(x) = 0$ such that $B(x^*, R) \subset \Omega$ with $R > 0$. If $\|L_F(x)\| \leq K < 1$, for all $x \in \Omega$, then the Newton sequence $\{x_n\}$ is well defined and starting at any $x_0 \in B(x^*, R)$ converges to x^*.*

Proof From the existence of $L_F(x)$, for all $x \in \Omega$, it follows that the operator $[F'(x^*)]^{-1}$ exists. So, if $x_0 \in B(x^*, R)$, there exists $[F'(x_0)]^{-1}$, and $x_1 = x_0 - [F'(x_0)]^{-1} F(x_0)$ is then well defined. Moreover,

$$x_1 - x^* = N_F(x_0) - N_F(x^*) = \int_{x^*}^{x_0} L_F(v)\, dv,$$

so that $\|x_1 - x^*\| \leq K \|x_0 - x^*\| < R$ and $x_1 \in B(x^*, R)$. By reiterating this process, we obtain

$$\|x_n - x^*\| \leq K \|x_{n-1} - x^*\| \leq \cdots \leq K^n \|x_0 - x^*\| < R.$$

Thus, $x_n \in B(x^*, R)$, for all $n \geq 0$, and the Newton sequence $\{x_n\}$ converges to x^*. ∎

Next, we illustrate the previous results with the following example given by Dennis and Schnabel in [28].

Example 2.48 Let $F(x, y, z) = 0$ be a nonlinear system, where

$$F : \Omega \subseteq \mathbb{R}^3 \to \mathbb{R}^3 \quad \text{and} \quad F(x, y, z) = (x, y^2 + y, e^z - 1). \tag{2.28}$$

It is obvious that $(0, 0, 0) = \mathbf{x}^*$ is a solution of the system.
From F, we deduce

2.2 Operator Equations in Banach Spaces

$$F'(\mathbf{x}) = \begin{pmatrix} 1 & 0 & 0 \\ 0 & 2y+1 & 0 \\ 0 & 0 & e^z \end{pmatrix} \quad \text{and} \quad F'(\mathbf{x}^*) = \text{diag}\{1, 1, 1\},$$

where $\mathbf{x} = (x, y, z)$. Hence, $[F'(\mathbf{x})]^{-1}$ exists, provided that $y \neq \frac{1}{2}$, and $\|[F'(x^*)]^{-1}\| = \gamma = 1$.

Moreover, as

$$F''(\mathbf{x}) = \begin{pmatrix} 0\,0\,0 & 0\,0\,0 & 0\,0\,0 \\ 0\,0\,0 & 0\,2\,0 & 0\,0\,0 \\ 0\,0\,0 & 0\,0\,0 & 0\,0\,e^z \end{pmatrix} \quad \text{and} \quad L_F(\mathbf{x}) = \begin{pmatrix} 0 & 0 & 0 \\ 0 & \frac{2y(y+1)}{(2y+1)^2} & 0 \\ 0 & 0 & \frac{e^z-1}{e^z} \end{pmatrix},$$

we have that $\|L_F(\mathbf{x})\| = \max\left\{\left|\frac{2y(y+1)}{(2y+1)^2}\right|, \left|\frac{e^z-1}{e^z}\right|\right\} < 1$ if $\left|\frac{2y(y+1)}{(2y+1)^2}\right| < 1$ and $\left|\frac{e^z-1}{e^z}\right| < 1$. Then, $\|L_F(\mathbf{x})\| < 1$, for all $\mathbf{x} \in B(\mathbf{x}^*, R)$, with $R = \frac{3-\sqrt{3}}{6} = 0.2113\ldots$ Therefore, from Theorem 2.47, the Newton method starting at any $\mathbf{x}_0 \in B(\mathbf{x}^*, 0.2113\ldots)$ is convergent.

If we compare Theorem 2.47 with the aforementioned result of Dennis and Schnabel, we can emphasize three things. The first and most important is that Theorem 2.47 is independent of the value r, since we can choose $\Omega = \mathbb{R}^3$, while Dennis and Schnabel cannot. The second is that Dennis and Schnabel obtain $R = \frac{1}{4}$ if $r < \ln 2$, so that the domain of starting points that we obtain from Theorem 2.47 is improved. The third is that Dennis and Schnabel obtain $R = \frac{1}{2e^r}$ if $r \geq \ln 2$, so that we can improve the domain of starting points that Dennis and Schnabel obtain if $r \geq \ln\left(\frac{3+\sqrt{3}}{2}\right) = 0.8612\ldots$, since $R = \frac{1}{2e^r}$ decreases exponentially depending on r, which does not occur with the value of $R = 0.2113\ldots$ that is obtained from Theorem 2.47.

Notice that the main advantage of Theorem 2.47 is that it does not depend on any bound calculated on the domain $B(x^*, r)$.

2.2.3 Kantorovich-Type Convergence Conditions

While in Sect. 2.2.2 we obtain results of semilocal, global, and local convergence for the Newton method from conditions on the operator L_F, in this section we obtain other results of semilocal, global, and local convergence under Kantorovich-type conditions and using the operator L_F.

2.2.3.1 Semilocal Convergence

Theorem 2.11 of Newton-Kantorovich is based on the classical conditions of Kantorovich (K1)–(K2) and $M\beta\eta \leq \frac{1}{2}$. Under these hypotheses, the scalar function that Kantorovich fixes to prove the convergence of the Newton method and that provides the a priori error estimates in Theorem 2.11 is the Kantorovich polynomial (2.6)

$$p(t) = \frac{M}{2}t^2 - \frac{t}{\beta} + \frac{\eta}{\beta}.$$

Next, by using the operator degree of logarithmic convexity L_F, we see a new proof of the convergence of the Newton sequence under conditions (K1)–(K2).

What we do is to see that under the Kantorovich conditions and the starting point x_0, we can define the following step x_1 of the Newton sequence. In addition, for every x belonging to the segment that joins x_0 and x_1, the existence of the operator $L_F(x)$ is proven and is bounded by the degree of logarithmic convexity of a scalar function. The results obtained for $x \in [x_0, x_1]$ are extended to $x \in [x_1, x_2]$ and generalized to $x \in [x_n, x_{n+1}]$, where $\{x_n\}$ is the Newton sequence (2.4). This scalar function is precisely the polynomial defined in (2.6).

Theorem 2.49 *Let $F : \Omega \subseteq X \to Y$ be a twice continuously Fréchet differentiable operator defined on a non-empty open convex domain Ω of a Banach space X with values in a Banach space Y. Suppose that conditions (K1)–(K2) are satisfied, $M\beta\eta \leq \frac{1}{2}$, and $B\left(x_0, \frac{1}{M\beta}\right) \subset \Omega$. Then, the sequence*

$$t_{n+1} = t_n - \frac{p(t_n)}{p'(t_n)}, \quad n \geq 0, \quad \text{with } t_0 = 0, \tag{2.29}$$

where $p(t)$ is polynomial (2.6), is a majorizing sequence of the Newton method (2.4).

Proof First, from the hypotheses, it follows that x_1 is well defined and $x_1 \in B\left(x_0, \frac{1}{M\beta}\right)$. Moreover, $\|x_1 - x_0\| \leq t_1 - t_0$, since

$$\|x_1 - x_0\| \leq \eta = t_1 = t_0 - \frac{p(t_0)}{p'(t_0)}.$$

Next, from Theorem 2.9, we obtain

$$\int_{x_0}^{x_1} L_F(z)\,dz = N_F(x_1) - N_F(x_0) = x_2 - x_1. \tag{2.30}$$

Taking now into account $x \in [x_0, x_1]$, then $x = x_0 + s(x_1 - x_0) = x_0 - s\Gamma_0 F(x_0)$ with $s \in [0, 1]$, and, from Taylor's formula, it follows

$$F(x) = F(x_0) + F'(x_0)(x-x_0) + \int_{x_0}^{x} F''(z)(x-z)\,dz = (1-s)F(x_0) + \int_{x_0}^{x} F''(z)(x-z)\,dz.$$

2.2 Operator Equations in Banach Spaces

Also,

$$\|\Gamma_0 F(x)\| \leq (1-s)\|\Gamma_0 F(x_0)\| + \left\|\Gamma_0 \int_{x_0}^{x} F''(z)(x-z)\,dz\right\|$$

$$\leq (1-s)\eta + \frac{M\beta}{2}\|x-x_0\|^2$$

$$= (1-s)\eta + \frac{M\beta}{2}s^2\eta^2.$$

Besides, if $t = \|x - x_0\| = s\|x_1 - x_0\| = s\eta$, we have

$$\|\Gamma_0 F(x)\| \leq \eta - t + \frac{M\beta}{2}t^2,$$

and, in particular, for $x = x_1$,

$$\|\Gamma_0 F(x_1)\| \leq -\frac{p(t_1)}{p'(t_0)}. \tag{2.31}$$

As a consequence, for $x \in [x_0, x_1]$ and $t = \|x - x_0\|$, it follows

$$\|L_F(x)\| \leq \frac{M\beta\left(\eta - t + \frac{M\beta}{2}t^2\right)}{(1 - M\beta t)^2}.$$

We then observe that the right side of the last inequality is the degree of logarithmic convexity L_p of the polynomial (2.6). Therefore, if $x \in [x_0, x_1]$ and $t = \|x - x_0\|$, we obtain

$$\|L_F(x)\| \leq L_p(t). \tag{2.32}$$

After that, as

$$\|I - \Gamma_0 F'(x_1)\| \leq \|\Gamma_0\|\|F'(x_0) - F'(x_1)\| \leq M\beta\|x_1 - x_0\| \leq M\beta\eta < 1,$$

it follows, from the Banach lemma on invertible operators, that the operator $\Gamma_1 = [F'(x_1)]^{-1}$ exists, so that we can define x_2. In addition, from (2.30) and (2.32), we have

$$\|x_2 - x_1\| \leq \left\|\int_{x_0}^{x_1} L_F(z)\,dz\right\| \leq \int_{t_0}^{t_1} L_p(t)\,dt = t_2 - t_1,$$

since $\|L_F(z)\| \leq L_p(t)$ and $\|x_1 - x_0\| \leq t_1 - t_0$. In addition,

$$\|x_2 - x_0\| \leq \|x_2 - x_1\| + \|x_1 - x_0\| \leq t_2 - t_1 + t_1 - t_0 = t_2 < \frac{1 - \sqrt{1 - 2M\beta\eta}}{M\beta} < \frac{1}{M\beta},$$

and $x_2 \in B\left(x_0, \frac{1}{M\beta}\right)$. Besides, there exists $\Gamma_2 = [F'(x_2)]^{-1}$, and we can then define x_3. On the other hand, as

$$x_3 - x_2 = \int_{x_1}^{x_2} L_F(z)\, dz,$$

to obtain $\|x_3 - x_2\| \leq t_3 - t_2$, it is enough that $\|L_F(x)\| \leq L_p(t)$, provided that $x \in [x_1, x_2]$ and $\|x - x_1\| = t$. For this, we replace x_0 with x_1 in the previous step.

First, we observe that

$$\|\Gamma_1 F''(x)\| \leq \|\Gamma_1 F'(x_0)\| \|\Gamma_0 F''(x)\|.$$

So, from the Banach lemma on invertible operators, it follows

$$\|\Gamma_1 F'(x_0)\| = \|[\Gamma_0 F'(x_1)]^{-1}\| \leq \frac{1}{1 - M\beta\|x_1 - x_0\|} = \frac{1}{1 - M\beta t_1} = \frac{p'(t_0)}{p'(t_1)}, \quad (2.33)$$

so that

$$\|\Gamma_1 F''(x)\| \leq \frac{M\beta}{1 - M\beta t_1}.$$

Moreover, since

$$\Gamma_1 \int_{x_1}^{x} F''(z)\, dz = \Gamma_1 F'(x) - I,$$

we obtain

$$\|\Gamma_1 F'(x) - I\| \leq \frac{M\beta}{1 - M\beta t_1} \|x - x_1\| < 1$$

if $x = x_1 + s(x_2 - x_1) \in [x_1, x_2]$, with $s \in [0, 1]$, and

$$\|x - x_1\| = s\|x_2 - x_1\| \leq \|x_2 - x_1\| \leq t_2 - t_1 < \frac{1}{M\beta} - t_1 = \frac{1 - M\beta t_1}{M\beta}.$$

Thus, by the Banach lemma on invertible operators, the operator $[\Gamma_1 F'(x)]^{-1}$ exists, and

$$\|[\Gamma_1 F'(x)]^{-1}\| \leq \frac{1}{1 - \frac{M\beta}{1 - M\beta t_1}\|x - x_1\|} = \frac{1 - M\beta t_1}{1 - M\beta(t_1 + \|x - x_1\|)}.$$

2.2 Operator Equations in Banach Spaces

Now, if $x \in [x_1, x_2]$, then $x - x_1 = s(x_2 - x_1) = -s\Gamma_1 F(x_1)$, and, from Taylor's formula, we have

$$\Gamma_1 F(x) = \Gamma_1 F(x_1) + (x - x_1) + \Gamma_1 \int_{x_1}^{x} F''(z)(x - z)\, dz$$

$$= (1 - s)\Gamma_1 F(x_1) + \Gamma_1 \int_{x_1}^{x} F''(z)(x - z)\, dz$$

and

$$\|\Gamma_1 F(x)\| \le (1 - s)\|\Gamma_1 F(x_1)\| + \left\| \int_{x_1}^{x} \Gamma_1 F''(z)(x - z)\, dz \right\|$$

$$\le (1 - s)\|\Gamma_1 F(x_1)\| + \frac{M\beta}{1 - M\beta t_1} \frac{\|x - x_1\|^2}{2}$$

$$= (1 - s)\|\Gamma_1 F(x_1)\| + \frac{M\beta}{1 - M\beta t_1} \frac{s^2 \|\Gamma_1 F(x_1)\|^2}{2}.$$

If we denote $t = \|x - x_1\| = s\|\Gamma_1 F(x_1)\|$, with

$$t \in [0, \|\Gamma_1 F(x_1)\|] = [0, \|x - x_1\|] \subseteq [0, t_2 - t_1],$$

it follows

$$\|\Gamma_1 F(x)\| \le \|\Gamma_1 F(x_1)\| - t + \frac{M\beta}{2(1 - M\beta t_1)} t^2.$$

Besides, from (2.31), we obtain

$$\|\Gamma_1 F(x_1)\| \le \|[\Gamma_0 F'(x_1)]^{-1}\| \|\Gamma_0 F(x_1)\| \le \frac{p(t_1)}{1 - M\beta t_1} = -\frac{p(t_1)}{p'(t_1)}$$

and

$$\|\Gamma_1 F(x)\| \le \frac{1}{1 - M\beta t_1} \left(p(t_1) - t(1 - M\beta t_1) + \frac{M\beta}{2} t^2 \right)$$

$$= \frac{1}{1 - M\beta t_1} \left(\eta - t_1 + \frac{M\beta}{2} t_1^2 - t + M\beta t_1 t + \frac{M\beta}{2} t^2 \right)$$

$$= \frac{p(t_1 + t)}{1 - M\beta t_1}.$$

If $x \in [x_1, x_2]$ and $\|x - x_1\| = t$, then

$$\|L_F(x)\| \leq \|[\Gamma_1 F'(x)]^{-1}\| \|\Gamma_1 F''(x)\| \|[\Gamma_1 F'(x)]^{-1}\| \|\Gamma_1 F(x)\|$$

$$\leq \frac{1 - M\beta t_1}{1 - M\beta(t_1 + t)} \frac{M\beta}{1 - M\beta t_1} \frac{1 - M\beta t_1}{1 - M\beta(t_1 + t)} \frac{p(t_1 + t)}{1 - M\beta t_1}$$

$$= \frac{M\beta p(t_1 + t)}{(1 - M\beta(t_1 + t))^2}$$

$$= \frac{M\beta p(t_1 + t)}{(p'(t_1 + t))^2}$$

$$= L_p(t_1 + t)$$

and

$$\|x - x_1\| \leq s\|x_2 - x_1\| \leq \|x_2 - x_1\| \leq t_2 - t_1.$$

Therefore,

$$\|x_3 - x_2\| = \left\| \int_{x_1}^{x_2} L_F(x)\, dx \right\| \leq \int_{t_1}^{t_2} L_p(z)\, dz = t_3 - t_2.$$

We can prove without difficulty that we can replace x_2 with x_1 in the previous step and, in general, x_{n+1} with x_n, for $n \geq 0$, so that $\|L_F(x)\| \leq L_p(t_{n-1} + t)$ if $x \in [x_{n-1}, x_n]$ and $t = \|x - x_{n-1}\|$. And, as a consequence,

$$\|x_{n+1} - x_n\| = \left\| \int_{x_{n-1}}^{x_n} L_F(x)\, dx \right\| \leq \int_{t_{n-1}}^{t_n} L_p(z)\, dz = t_{n+1} - t_n.$$

The proof is complete. ∎

Since (2.29) is a nondecreasing sequence and convergent, the convergence of the Newton method (2.4) is proved under the Kantorovich conditions (K1)–(K2) and $M\beta\eta \leq \frac{1}{2}$, so that we can establish the following result.

Theorem 2.50 *Under the conditions of Theorem 2.49, the Newton method (2.4) converges to a solution x^* of the equation $F(x) = 0$. Moreover, if t^* and t^{**} are the two zeros of the polynomial (2.6), with $t^* \leq t^{**}$, then $x^* \in \overline{B(x_0, t^*)}$ and x^* is unique in $B(x_0, t^{**}) \cap \Omega$ if $t^* < t^{**}$ or in $\overline{B(x_0, t^*)}$ if $t^* = t^{**}$.*

Proof From Theorem 2.49, it follows $\|x_{n+1} - x_n\| \leq t_{n+1} - t_n$, $n \geq 0$, so that the sequence $\{x_n\}$ converges to a limit $x^* \in \overline{B(x_0, t^*)} \subseteq B\left(x_0, \frac{1}{M\beta}\right)$. Also, $\|\Gamma_n F(x_n)\| \leq -\frac{p(t_n)}{p'(t_n)}$, and, by the convergence of $\{t_n\}$ to a solution of $p(t) = 0$ and the continuity of F, it follows $F(x^*) = 0$.

2.2 Operator Equations in Banach Spaces

Next, we prove the uniqueness of the solution x^*. For this, we suppose that there exists a solution y^* of $F(x) = 0$, such that $y^* \neq x^*$, in $B(x_0, t^{**}) \cap \Omega$ if $t^* < t^{**}$ or in $\overline{B(x_0, t^*)}$ if $t^* = t^{**}$.

First, we suppose that $t^* < t^{**}$. Then,

$$\|y^* - x_0\| = \sigma(t^{**} - t_0) \quad \text{with} \quad \sigma \in (0, 1).$$

In addition, we assume that $\|y^* - x_j\| \leq \sigma^{2^j}(t^{**} - t_j)$, for $j = 0, 1, \ldots, n$, and use mathematical induction on n. For this, we observe that

$$y^* - x_{n+1} = y^* - x_n + \Gamma_n F(x_n)$$
$$= -\Gamma_n(F(y^*) - F(x_n) - F'(x_n)(y^* - x_n))$$
$$= -\Gamma_n \int_{x_n}^{y^*} F''(x)(y^* - x)\, dx$$
$$= [-\Gamma_0 F'(x_n)]^{-1} \int_0^1 \Gamma_0 F''(x_n + \tau(y^* - x_n))(1 - \tau)\, d\tau\, (y^* - x_n)^2,$$

and taking into account $\|\Gamma_0 F''(x)\| \leq M\beta$ and

$$\|[\Gamma_0 F'(x_n)]^{-1}\| \leq \frac{1}{1 - M\beta t_n} = \frac{-1}{p'(t_n)},$$

it follows, from (2.33), that

$$\|y^* - x_{n+1}\| \leq \frac{-1}{p'(t_n)} \int_0^1 M\beta(1-\tau)\, d\tau \|y^* - x_n\|^2 = \frac{-M\beta}{2p'(t_n)} \|y^* - x_n\|^2.$$

By following the same way for the polynomial (2.6), we obtain

$$t^{**} - t_{n+1} = \frac{-M\beta}{2p'(t_n)}(t^{**} - t_n)^2.$$

Now, it is easy to see

$$\|y^* - x_{n+1}\| \leq \frac{-M\beta(t^{**} - t_n)^2}{2p'(t_n)} \left(\frac{\|y^* - x_n\|}{t^{**} - t_n}\right)^2 \leq (t^{**} - t_{n+1})\sigma^{2^{n+1}},$$

which completes the induction. As a consequence, $y^* = x^*$, since $\|y^* - x_n\| \to 0$ as $n \to +\infty$.

Second, if $t^* = t^{**}$ and y^* is a solution of $F(x) = 0$ in $\overline{B(x_0, t^*)}$, then $\|y^* - x_0\| \leq t^* - t_0$. Proceeding similarly to the previous case, we can also prove, by mathematical induction on n, that $\|y^* - x_n\| \leq t^* - t_n$. As $t^* = t^{**}$ and $\lim_n t_n = t^*$, the uniqueness of solution is now easy to conclude. ∎

The next question is to ask whether the commonly considered polynomial (2.6) provides the best error estimate for the Newton method under the Kantorovich conditions [52].

In particular, we consider a function f such that:

(D1) $\quad \|\Gamma_0\| = -\dfrac{1}{p'(t_0)}$

(D2) $\quad \|\Gamma_0 F(x_0)\| = -\dfrac{p(t_0)}{p'(t_0)} < -\dfrac{f(t_0)}{f'(t_0)}$

(D3) $\quad \sup_{x \in \Omega} \|F''(x)\| = p''(t) < f''(t)$, for $t \in [0, t^*)$

where t^* is the smallest zero of the polynomial (2.6). We now wonder which function, p or f, will give us the best error estimate.

For this, we take into account the relationship between the convexity and the speed of convergence of the Newton method seen in Sect. 2.1.2 and thus compare the error estimates obtained from these functions. Then, we consider $t_0 = 0$ and the following result, which is fundamental in the study.

Lemma 2.51 *Let f and g be two decreasing and convex scalar functions in $[0, t^*]$, where t^* is the smallest zero of (2.6), such that $f(t^*) = g(t^*) = 0$, and the following is satisfied:*

(a) $-\dfrac{1}{g'(0)} < -\dfrac{1}{f'(0)}$.

(b) $-\dfrac{g(0)}{g'(0)} < -\dfrac{f(0)}{f'(0)}$.

(c) $g''(t) < f''(t)$, for $t \in [0, t^*)$.

Then, $L_f(t) > L_g(t)$ in $[0, t^*)$.

Proof If we consider $t \in (0, t^*)$, then, by items (a) and (c), we have

$$0 < \int_0^t \left(-\frac{f''(s)}{f'(0)} + \frac{g''(s)}{g'(0)} \right) ds = -\frac{f'(t)}{f'(0)} + \frac{g'(t)}{g'(0)},$$

so that

$$-\frac{f'(t)}{f'(0)} > -\frac{g'(t)}{g'(0)}, \quad t \in (0, t^*). \tag{2.34}$$

2.2 Operator Equations in Banach Spaces

Taking now into account (2.34), it follows

$$0 < \int_0^t \left(-\frac{f'(s)}{f'(0)} + \frac{g'(s)}{g'(0)} \right) ds = \left(-\frac{f(t)}{f'(0)} + \frac{g(t)}{g'(0)} \right) - \left(-\frac{f(0)}{f'(0)} + \frac{g(0)}{g'(0)} \right),$$

and, by (b), we have

$$-\frac{f(t)}{f'(0)} > -\frac{g(t)}{g'(0)}, \quad t \in (0, t^*). \tag{2.35}$$

Next, by items (a) and (c), we obtain

$$f''(t) > \frac{f'(0)}{g'(0)} g''(t), \quad t \in (0, t^*). \tag{2.36}$$

Again, from (2.34), we follow

$$f'(t) > \frac{f'(0)}{g'(0)} g'(t), \quad t \in (0, t^*).$$

In addition, as $f'(t) \leq 0$ in $(0, t^*)$, then

$$\frac{1}{f'(t)^2} > \frac{1}{\left(\frac{f'(0)}{g'(0)} \right)^2 g'(t)^2}, \quad t \in (0, t^*). \tag{2.37}$$

Finally from (2.35), (2.36), and (2.37),

$$L_f(t) = \frac{f(t) f''(t)}{f'(t)^2} > \frac{g(t) g''(t)}{g'(t)^2} = L_g(t), \quad t \in (0, t^*).$$

The proof is complete. ∎

Remark 2.52 The previous result is also valid if the equality is given in any of the three previous conditions (a), (b), and (c).

From Lemma 2.51, we know that $L_p(t) < L_f(t)$, for $t \in [0, t^*)$, under conditions (D1)–(D2)–(D3). As a consequence, from Theorem 2.6, it follows that the Newton method applied to f is slower than that applied to p, so that the error estimate obtained from the polynomial (2.6) is optimal under conditions (D1)–(D2)–(D3). From these observations, we can establish the following result, whose proof follows from the previous comments.

Theorem 2.53 *The polynomial (2.6) provides the best error estimate for the Newton method (2.4) under the Kantorovich conditions and using majorizing sequences.*

After that, we obtain some a priori error estimates for the Newton method by using those given by Ostrowski [76] for a second-degree polynomial. As a consequence, we obtain the quadratic convergence of the Newton method under the Kantorovich conditions (K1)–(K2) and $M\beta\eta \leq \frac{1}{2}$.

Theorem 2.54 *Under the conditions of Theorem 2.49, we obtain the following error estimates:*

$$\|x^* - x_n\| \leq t^* - t_n = \frac{t^{**} - t^*}{1 - \theta^{2^n}} \theta^{2^n}, \quad \text{where } \theta = \frac{t^*}{t^{**}}, \quad \text{if } M\beta\eta < \frac{1}{2},$$

$$\|x^* - x_n\| \leq t^* - t_n = \frac{t^*}{2^n} \quad \text{if } M\beta\eta = \frac{1}{2}.$$

Proof As the sequence (2.29) majorizes the sequence (2.4), then

$$\|x_{n+m} - x_n\| \leq \sum_{i=0}^{m-1} \|x_{n+i+1} - x_{n+i}\| \leq \sum_{i=0}^{m-1} \|t_{n+i+1} - t_{n+i}\| = t_{n+m} - t_n, \quad m \in \mathbb{N},$$

and, by letting $m \to +\infty$, we obtain $\|x^* - x_n\| \leq t^* - t_n$.

Next, we denote $a_n = t^* - t_n$ and $b_n = t^{**} - t_n$, for all $n \geq 0$. Thus,

$$p(t_n) = \frac{M}{2}(t^* - t_n)(t^{**} - t_n) = \frac{M}{2} a_n b_n, \qquad p'(t_n) = -\frac{M}{2}(a_n + b_n),$$

and then

$$a_{n+1} = t^* - t_{n+1} = t^* - t_n + \frac{p(t_n)}{p'(t_n)} = a_n - \frac{a_n b_n}{a_n + b_n} = \frac{a_n^2}{a_n + b_n}, \quad n \geq 0,$$

$$b_{n+1} = t^{**} - t_{n+1} = t^{**} - t_n + \frac{p(t_n)}{p'(t_n)} = b_n - \frac{a_n b_n}{a_n + b_n} = \frac{b_n^2}{a_n + b_n}, \quad n \geq 0.$$

Hence,

$$\frac{a_n}{b_n} = \left(\frac{a_{n-1}}{b_{n-1}}\right)^2 = \left(\frac{a_{n-2}}{b_{n-2}}\right)^{2^2} = \cdots = \left(\frac{a_0}{b_0}\right)^{2^n} = \left(\frac{t^*}{t^{**}}\right)^{2^n} = \theta^{2^n}, \quad n \geq 0.$$

If $M\beta\eta < \frac{1}{2}$, then $t^* < t^{**}$, and there exists $d > 0$ such that $t^{**} = t^* + d$, so that

$$t^* - t_n = (t^{**} - t_n)\theta^{2^n} = (t^* - t_n)\theta^{2^n} + d\theta^{2^n}$$

and

2.2 Operator Equations in Banach Spaces

$$t^* - t_n = \frac{d\theta^{2^n}}{1 - \theta^{2^n}}, \quad n \geq 0.$$

If $M\beta\eta = \frac{1}{2}$, then $t^* = t^{**} = \frac{1}{M\beta}$, $a_n = b_n$, and $a_{n+1} = \frac{a_n}{2}$. Thus, $a_n = \frac{t^*}{2^n}$. ∎

The estimate given is the best possible, since all inequalities are transformed into equalities in the case of working with scalar functions.

Example 2.55 Consider the following nonlinear system of equations:

$$F(x, y) = \left(x^3 - 2y + \frac{1}{3}, y^3 - 4x + \frac{2}{3}\right) = 0. \tag{2.38}$$

In addition,

$$F'(x, y) = \begin{pmatrix} 3x^2 & -2 \\ -4 & 3y^2 \end{pmatrix}, \quad F''(x, y) = \begin{pmatrix} 6x & 0 \\ 0 & 0 \\ 0 & 0 \\ 0 & 6y \end{pmatrix},$$

and $\|F''(x, y)\| \leq 6\|(x, y)\|$ with the max-norm.

If we now choose the starting point $(x_0, y_0) = (0, 0)$, we have

$$[F'(x, y)]^{-1} = \frac{1}{\Delta}\begin{pmatrix} 3y^2 & 2 \\ 4 & 3x^2 \end{pmatrix}, \quad \text{with} \quad \Delta = 9x^2 y^2 - 8,$$

and

$$\Gamma_0 = [F'(x_0, y_0)]^{-1} = \begin{pmatrix} 0 & -1/4 \\ -1/2 & 0 \end{pmatrix}$$

so that $\beta = \|\Gamma_0\| = \frac{1}{2}$ and $\eta = \|\Gamma_0 F(x_0, y_0)\| = \frac{1}{6}$.

Now, we study the solution of (2.38) in a domain. For this, and guaranteeing the existence of $[F'(x, y)]^{-1}$ in a domain, if we choose the ball $\Omega = B((x_0, y_0), 0.97)$, then $\|F''(x, y)\| \leq 5.82 = M$ and $M\beta\eta = 0.485 < \frac{1}{2}$. Therefore, as

$$B\left((x_0, y_0), \frac{1}{M\beta}\right) = B((0, 0), 0.3436\ldots) \subset \Omega = B((0, 0), 0.97),$$

the convergence of the Newton method starting at $(0, 0)$ is guaranteed from Theorem 2.50. In addition, the polynomial (2.6) is reduced to $p(t) = 2.91t^2 - 2t + \frac{1}{3}$. As the two zeros

Table 2.8 A priori error bounds from Example 2.55

n	$t^* - t_n$
0	0.284121965376...
1	0.117455298709...
2	0.038976334308...
3	0.007711681245...
4	0.000442272563...
5	0.000001631050...
6	0.000000000023...
7	0.000000000000...

of the polynomial are $t^* = 0.2841\ldots$ and $t^{**} = 0.4031\ldots$, we write $p(t) = 2.91(t - 0.2841\ldots)(t - 0.4031\ldots)$, and, from Ostrowski's technique given in Theorem 2.54, we can guarantee the quadratic convergence of the Newton method. Moreover, we show in Table 2.8 the a priori error bounds obtained from Theorem 2.54, which allows us to know the number of iterations that we have to do with the Newton method to obtain an approximation of the solution with a predetermined precision.

Hölder Case

By means of the degree of logarithmic convexity L_F, we find in Theorem 2.49 a majorizing sequence under the condition that the second derivative of the operator involved F is bounded. This majorizing sequence is constructed by applying the Newton method to the equation $p(t) = 0$, where p is defined in (2.6). But, sometimes, the condition $M\beta\eta \leq \frac{1}{2}$ of Theorem 2.49 fails, as we can see in the following example, so that we cannot guarantee the convergence of the Newton method from Theorem 2.49. Remember that the conditions required in the Theorem 2.49 are the same as those required in Theorem 2.11 of Newton-Kantorovich, so we cannot guarantee the convergence of the method from Theorem 2.11 either.

Example 2.56 (Zhengda [94]) We consider $\chi : [-1, 1] \to [-1, 1]$ such that

$$\chi(x) = \frac{x^3}{6} + \frac{x^2}{6} - \frac{5}{6}x + \frac{1}{3} \qquad (2.39)$$

and $x_0 = 0$. Then, $\|\chi''(x)\| \leq \frac{4}{3}$, for all $x \in [-1, 1]$, $[\chi'(0)]^{-1} = -\frac{6}{5}$, and $[\chi'(0)]^{-1}\chi 0 = -\frac{2}{5}$. As a consequence, the convergence of the Newton method starting at $x_0 = 0$ cannot be guaranteed based on Theorem 2.49 or Theorem 2.11 of Newton-Kantorovich, since

$$\beta = \frac{6}{5}, \quad \eta = \frac{2}{5}, \quad M = \sup_{t \in [-1,1]} \{\chi''(t)\} = \frac{4}{3} \quad \text{and} \quad M\beta\eta = \frac{16}{25} > \frac{1}{2}.$$

2.2 Operator Equations in Banach Spaces

Later, we prove the convergence of the Newton method starting at $x_0 = 0$ by using new arguments.

Next, we extend the previous study to an operator F such that the second derivative satisfies the following condition [84]:

$$\|F''(x) - F''(x_0)\| \leq M_0 \|x - x_0\|^p, \quad p \in [0, 1], \quad \in \Omega. \tag{2.40}$$

We then say that F'' satisfies a center Hölder condition in Ω.

Observe that $\|F''(x)\| \leq M_0 + \|F''(x_0)\| = M$ for $p = 0$, so that condition (K2) of Theorem 2.49 is satisfied if the condition (2.40) holds.

Conditions of type (2.40) allow guaranteeing the convergence of the Newton sequence in situations in which the Kantorovich conditions do not, as we see later for the function given in Example 2.56. For this, we establish a new semilocal convergence result for the Newton method under the following conditions:

(E1) There exists $\Gamma_0 = [F'(x_0)]^{-1} \in \mathcal{L}(Y, X)$, for some $x_0 \in \Omega$, with $\|\Gamma_0\| \leq \beta$ and $\|\Gamma_0 F(x_0)\| \leq \eta$; moreover, $\|F''(x_0)\| \leq \delta$.

(E2) There exist two constants $M_0 \geq 0$ and $p \in [0, 1]$ such that $\|F''(x) - F''(x_0)\| \leq M_0 \|x - x_0\|^p$, for $x \in \Omega$.

(E3) $\phi(\alpha) \leq 0$, where ϕ is the function

$$\phi(t) = \frac{M_0}{(1+p)(2+p)} t^{2+p} + \frac{\delta}{2} t^2 - \frac{t}{\beta} + \frac{\eta}{\beta} \tag{2.41}$$

and α is the unique positive solution of $\phi'(t) = 0$, and $B(x_0, t^*) \subset \Omega$, where t^* is the smallest positive solution of $\phi(t) = 0$.

Remark 2.57 As α is the unique positive solution of $\phi'(t) = 0$ and $\phi''(\alpha) > 0$, then α is a minimum of $\phi(t)$ such that $\phi(\alpha) \leq 0$, so that (E3) is a necessary and sufficient condition for the existence of two positive solutions t^* and t^{**} of $\phi(t) = 0$ such that $0 < t^* \leq t^{**}$. Moreover, as ϕ is a nonincreasing convex function in $[0, \alpha]$ such that $\phi(\alpha) \leq 0 < \phi(0)$ and $\phi(0)\phi''(0) > 0$, then these conditions are sufficient to guarantee the semilocal convergence of the Newton sequence given by

$$t_0 = 0, \quad t_{n+1} = N_\phi(t_n) = t_n - \frac{\phi(t_n)}{\phi'(t_n)}, \quad n = 0, 1, 2, \ldots, \tag{2.42}$$

to t^*.

Before proving the convergence of the Newton method under conditions (E1)–(E2)–(E3), we give the following technical lemma that is used later.

Lemma 2.58 *Let* $F : \Omega \subseteq X \to Y$ *be a twice continuously Fréchet differentiable operator defined on a non-empty open convex domain* Ω *of a Banach space* X *with values in a Banach space* Y. *Suppose that conditions (E1)–(E2)–(E3) are satisfied. Then, the operator* $L_F(x)$ *exists for* $x \in B(x_0, \alpha)$ *and*

$$\|L_F(x)\| \leq \frac{\beta\left(\delta + M_0 \|x - x_0\|^p\right) \|\Gamma_0 F(x)\|}{\left(1 - \frac{M_0 \beta}{(p+1)} \|x - x_0\|^{p+1} - \beta\delta\|x - x_0\|\right)^2}.$$

Proof From

$$\Gamma_0 \int_{x_0}^{x} (F''(y) - F''(x_0))\, dy = \Gamma_0 F'(x) - I - \Gamma_0 F''(x_0)(x - x_0)$$

and taking into account (D3), it follows

$$\|\Gamma_0 F'(x) - I - \Gamma_0 F''(x_0)(x - x_0)\| = \left\|\Gamma_0 \int_{x_0}^{x} (F''(z) - F''(x_0))\, dz\right\|$$

$$= \left\|\Gamma_0 \int_0^1 (F''(x_0 + \tau(x - x_0))\right.$$

$$\left. - F''(x_0))(x - x_0)\, d\tau\right\|$$

$$\leq M_0 \beta \|x - x_0\|^{p+1} \int_0^1 \tau^p\, d\tau$$

$$= \frac{M_0 \beta \|x - x_0\|^{p+1}}{p+1}$$

and

$$\|\Gamma_0 F'(x) - I\| \leq \frac{M_0 \beta}{p+1} \|x - x_0\|^{p+1} + \beta\delta\|x - x_0\|.$$

Since $\|x - x_0\| \leq \alpha$ and $\phi'(t) < 0$, for $0 \leq t < \alpha$, we obtain, by the Banach lemma on invertible operators, that there exists $[\Gamma_0 F'(x)]^{-1}$, and

$$\|[\Gamma_0 F'(x)]^{-1}\| \leq \frac{1}{1 - \frac{M_0 \beta}{p+1}\|x - x_0\|^2 - \beta\delta\|x - x_0\|}.$$

After that, from (E2), we have

$$\|\Gamma_0 F''(x)\| \leq \beta(\delta + M_0 \|x - x_0\|^p),$$

2.2 Operator Equations in Banach Spaces

and taking into account

$$L_F(x) = [\Gamma_0 F'(x)]^{-1} \Gamma_0 F''(x) [\Gamma_0 F'(x)]^{-1} \Gamma_0 F(x),$$

the thesis follows immediately. ∎

Now, we are ready to establish the semilocal convergence of the Newton method under conditions (E1)–(E2)–(E3).

Theorem 2.59 *Let $F : \Omega \subseteq X \to Y$ be a twice continuously Fréchet differentiable operator defined on a non-empty open convex domain Ω of a Banach space X with values in a Banach space Y. Suppose that conditions (E1)–(E2)–(E3) are satisfied. Then, the Newton method (2.4) starting at x_0 converges to a solution x^* of $F(x) = 0$. Moreover, $x_n, x^* \in \overline{B(x_0, t^*)}$, for all $n \in \mathbb{N}$,*

$$\|x_{n+1} - x_n\| \leq t_{n+1} - t_n \quad \text{and} \quad \|x^* - x_n\| \leq t^* - t_n, \quad n \geq 0,$$

where $t_n = t_{n-1} - \frac{\phi(t_{n-1})}{\phi'(t_{n-1})}$, with $n \in \mathbb{N}$, $t_0 = 0$, and $\phi(t)$ defined in (2.41).

Proof We follow a reasoning analogous to that of the proof of Theorem 2.49. Under the hypotheses, x_1 is well defined and

$$\|x_1 - x_0\| = \|\Gamma_0 F(x_0)\| \leq \eta = t_1 < t^* \leq \alpha.$$

Besides, by Lemma 2.58, there exists $L_F(x_1)$, so that x_2 can be defined and

$$\int_{x_0}^{x_1} L_F(z)\, dz = N_F(x_1) - N_F(x_0) = x_2 - x_1.$$

If $x \in [x_0, x_1]$, then $x = x_0 + s(x_1 - x_0) = x_0 - s\Gamma_0 F(x_0)$ with $s \in [0, 1]$, and, from Taylor's formula, we have

$$F(x) = F(x_0) + F'(x_0)(x - x_0) + \frac{1}{2} F''(x_0)(x - x_0)^2 + \int_{x_0}^{x} (F''(z) - F''(x_0))(x - z)\, dz.$$

Moreover,

$$\|\Gamma_0 F(x)\| \leq (1 - s)\|\Gamma_0 F(x_0)\| + \frac{1}{2}\|\Gamma_0 F''(x_0)\|\|(x - x_0)^2\|$$

$$+ \left\|\Gamma_0 \int_{x_0}^{x} (F''(z) - F''(x_0))(x - z)\, dz\right\|.$$

After that, from

$$\left\| \int_{x_0}^{x} (F''(z) - F''(x_0))(x - z)\, dz \right\| \le M_0 \|x - x_0\|^{2+p} \int_0^1 s^p (1-s)\, ds = \frac{M_0 \|x - x_0\|^{2+p}}{(p+1)(p+2)},$$

it follows

$$\|\Gamma_0 F(x)\| \le (1-s)\eta + \frac{\beta\delta}{2} s^2 \eta^2 + \frac{M_0 \beta}{(p+1)(p+2)} s^{2+p} \eta^{2+p}.$$

If $s\eta = t$, then

$$\|\Gamma_0 F(x)\| \le \eta - t + \frac{\beta\delta}{2} t^2 + \frac{M_0 \beta}{(p+1)(p+2)} t^{2+p},$$

and, by Lemma 2.58, we obtain

$$\|L_F(x)\| \le \frac{\left(\eta - t + \frac{\beta\delta}{2} t^2 + \frac{M_0 \beta}{(p+1)(p+2)} t^{2+p}\right) (\beta\delta + M_0 \beta t^p)}{\left(1 - \beta\delta t - \frac{M_0 \beta}{p+1} t^{1+p}\right)^2} = L_\phi(t),$$

provided that $x \in [x_0, x_1]$ and $t = \|x - x_0\|$. In addition,

$$\|x_2 - x_1\| = \left\| \int_{x_0}^{x_1} L_F(x)\, dx \right\| \le \int_{t_0}^{t_1} L_\phi(t)\, dt = t_2 - t_1.$$

Next, following the proof of Theorem 2.49, we can exchange x_1 for x_0 and x_2 for x_1 in the previous reasoning to obtain that x_3 is defined and such that $\|x_3 - x_2\| \le t_3 - t_2$. In general, by mathematical induction on n, we prove

$$\|x_{n+1} - x_n\| \le t_{n+1} - t_n.$$

As a consequence, (2.42) is a majorizing sequence of the sequence (2.4), and the convergence of (2.42) implies the convergence of (2.4).

From the last inequality follows immediately

$$\|x_n - x_0\| \le t_n - t_0 = t_n < t^*,$$

so that $x_n \in B(x_0, t^* - t_0) \subset \Omega$, for all $n \ge 0$. Moreover, as the sequence $\{t_n\}$ is convergent, $\{x_n\}$ is also convergent. As a consequence, $\lim_n x_n = x^* \in \overline{B(x_0, t^*)}$, where $t^* = \lim_n t_n$. From $\|L_F(x_n)\| \le L_\phi(t_n)$, for all $n \ge 0$, the continuity of the operator F, and letting $n \to +\infty$, we obtain $F(x^*) = 0$.

2.2 Operator Equations in Banach Spaces

In addition, for $i > 0$, we have

$$\|x_{n+i} - x_n\| \leq \sum_{j=1}^{i} \|x_{n+j} - x_{n+j-1}\| \leq \sum_{j=1}^{i} \|t_{n+j} - t_{n+j-1}\| = t_{n+i} - t_n,$$

and, by letting $i \to +\infty$, it follows that $\|x^* - x_n\| \leq t^* - t_n$, for all $n \geq 0$. ∎

Example 2.60 Now, we apply Theorem 2.59 to the function (2.39) of Example 2.56. For (2.39) and $x_0 = 0$, we have $\beta = \frac{5}{6}$, $\eta = \frac{2}{5}$, and $\delta = \frac{1}{3}$. In addition, $M_0 = p = 1$ and

$$\phi(t) = \frac{t^3}{6} + \frac{t^2}{6} - \frac{6}{5}t + \frac{12}{25}$$

so that $\phi'(\alpha) = -0.4340\ldots < 0$, where $\alpha = 1.2513\ldots$ is the unique positive solution of $\phi'(t) = 0$. Finally, as $B(x_0, t^*) = B(0, 0.4383\ldots) \subset \Omega = [-1, 1]$, conditions (E1)–(E2)–(E3) of Theorem 2.59 are satisfied, and, as a consequence, the Newton method starting at $x_0 = 0$ converges to a solution of the equation $\chi(x) = 0$, and the domain of existence of solution is $\{x \in [-1, 1] : |x| \leq t^* = 0.4383\ldots\}$.

After that, we see the quadratic convergence of the Newton method under conditions (E1)–(E2)–(E3). We obtain the following theorem from Ostrowski's technique [76]. Notice first that $\phi(t)$ has two real zeros t^* and t^{**} such that $t_0 < t^* \leq t^{**}$, so that we can then write

$$\phi(t) = (t^* - t)(t^{**} - t)g(t),$$

with $g(t^*) \neq 0$ and $g(t^{**}) \neq 0$.

Theorem 2.61 *Under the hypotheses of Theorem 2.59, we have the following:*

(a) *If $t^* < t^{**}$ and there exist $m_1 > 0$ and $M_1 > 0$ such that $m_1 = \min\{Q_1(t); t \in [t_0, t^*]\}$ and $M_1 = \max\{Q_1(t); t \in [t_0, t^*]\}$, where $Q_1(t) = \frac{(t^{**}-t)g'(t)-g(t)}{(t^*-t)g'(t)-g(t)}$, then*

$$\frac{(t^{**}-t^*)\theta^{2^n}}{m_1 - \theta^{2^n}} \leq t^* - t_n \leq \frac{(t^{**}-t^*)\Delta^{2^n}}{M_1 - \Delta^{2^n}}, \quad n \geq 0,$$

where $\theta = \frac{t^*}{t^{**}}m_1$, $\Delta = \frac{t^*}{t^{**}}M_1$, and provided that $\theta < 1$ and $\Delta < 1$.

(b) *If $t^* = t^{**}$ and there exist $m_2 > 0$ and $M_2 > 0$ such that $m_2 \leq \min\{Q_2(t); t \in [t_0, t^*]\}$ and $M_2 \geq \max\{Q_2(t); t \in [t_0, t^*]\}$, where $Q_2(t) = \frac{(t^*-t)g'(t)-g(t)}{(t^*-t)g'(t)-2g(t)}$, then*

$$m_2^n t^* \leq t^* - t_n \leq M_2^n t^*, \quad n \geq 0,$$

provided that $m_2 < 1$ *and* $M_2 < 1$.

Proof Let $t^* < t^{**}$, and denote $a_n = t^* - t_n$ and $b_n = t^{**} - t_n$, for all $n \geq 0$. Then,

$$\phi(t_n) = a_n b_n g(t_n), \quad \phi'(t_n) = a_n b_n g'(t_n) - (a_n + b_n) g(t_n)$$

and

$$a_{n+1} = t^* - t_{n+1} = t^* - t_n + \frac{\phi(t_n)}{\phi'(t_n)} = \frac{a_n^2 \left(b_n g'(t_n) - g(t_n) \right)}{a_n b_n g'(t_n) - (a_n + b_n) g(t_n)}.$$

From $\dfrac{a_{n+1}}{b_{n+1}} = \dfrac{a_n^2 \left(b_n g'(t_n) - g(t_n) \right)}{b_n^2 \left(a_n g'(t_n) - g(t_n) \right)}$, it follows

$$m_1 \left(\frac{a_n}{b_n} \right)^2 \leq \frac{a_{n+1}}{b_{n+1}} \leq M_1 \left(\frac{a_n}{b_n} \right)^2.$$

In addition,

$$\frac{a_{n+1}}{b_{n+1}} \leq M_1^{2^{n+1}-1} \left(\frac{a_0}{b_0} \right)^{2^{n+1}} = \frac{\Delta^{2^{n+1}}}{M_1} \quad \text{and} \quad \frac{a_{n+1}}{b_{n+1}} \geq m_1^{2^{n+1}-1} \left(\frac{a_0}{b_0} \right)^{2^{n+1}} = \frac{\theta^{2^{n+1}}}{m_1}.$$

Taking then into account that $b_{n+1} = (t^{**} - t^*) + a_{n+1}$, it follows

$$\frac{(t^{**} - t^*) \theta^{2^{n+1}}}{m_1 - \theta^{2^{n+1}}} \leq t^* - t_{n+1} \leq \frac{(t^{**} - t^*) \Delta^{2^{n+1}}}{M_1 - \Delta^{2^{n+1}}}.$$

If $t^* = t^{**}$, then $a_n = b_n$ and

$$a_{n+1} = \frac{a_n \left(a_n g'(t) - g(t) \right)}{a_n g'(t) - 2 g(t_n)}.$$

As a consequence, $m_2 a_n \leq a_{n+1} \leq M_2 a_n$ and

$$m_2^{n+1} t^* \leq t^* - t_{n+1} \leq M_2^{n+1} t^*.$$

The proof is complete. ∎

From the last theorem, it follows that the convergence of the Newton method, under conditions (E1)–(E2)–(E3), is quadratic if $t^* < t^{**}$ and linear if $t^* = t^{**}$.

2.2 Operator Equations in Banach Spaces

Remark 2.62 Provided that the function involved in the semilocal convergence of the Newton method in the scalar case can be factored as $(t^* - t)(t^{**} - t)g(t)$ with $g(t^*) \neq 0$ and $g(t^{**}) \neq 0$, we can obtain a result about the order of convergence of this method as that given in the last theorem. In addition, from the proof of this theorem, we deduce that the order of convergence of the method is independent of the scalar function involved.

One particular case of condition (E2) is when F'' satisfies in Ω the following condition:

$$\|F''(x) - F''(x_0)\| \leq M_0 \|x - x_0\|, \quad \text{for} \quad x \in \Omega. \tag{2.43}$$

Observe that condition (E2) is reduced to the condition (2.43) if $p = 1$. We then say that F'' satisfies a center Lipschitz condition in Ω. In this case, the function (2.41) is reduced to the third-degree polynomial

$$\psi(t) = \frac{M_0}{6} t^3 + \frac{\delta}{2} t^2 - \frac{t}{\beta} + \frac{\eta}{\beta}. \tag{2.44}$$

Example 2.63 Consider the system of nonlinear equations

$$F(x, y) = \left(\frac{x^3}{24} + \frac{y^2}{4} - x + \frac{1}{3}, \frac{y^3}{8} + \frac{x^2}{4} - 3y + 1 \right) = 0, \tag{2.45}$$

where $F : \Omega = [-1, 1] \times [-1, 1] \to \mathbb{R}^2$.

Thus,

$$F'(x, y) = \begin{pmatrix} \frac{x^2}{8} - 1 & \frac{y}{2} \\ \frac{x}{2} & \frac{3}{8} y^2 - 3 \end{pmatrix} \quad \text{and} \quad F''(x, y) = \begin{pmatrix} \frac{x}{4} & 0 \\ 0 & \frac{1}{2} \\ \frac{1}{2} & 0 \\ 0 & \frac{3}{4} y \end{pmatrix}.$$

If we choose the starting point $(x_0, y_0) = (0, 0)$, then

$$\Gamma_0 = [F'(x_0, y_0)]^{-1} = \begin{pmatrix} -1 & 0 \\ 0 & -\frac{1}{3} \end{pmatrix},$$

since

$$[F'(x, y)]^{-1} = \frac{1}{\Delta} \begin{pmatrix} \frac{3}{8} y^2 - 3 & -\frac{y}{2} \\ -\frac{x}{2} & \frac{x^2}{8} - 1 \end{pmatrix} \quad \text{and} \quad \Delta = \frac{3}{64} x^2 y^2 - \frac{3}{8} x^2 - \frac{3}{8} y^2 - \frac{1}{4} xy + 3,$$

so that $\beta = \|\Gamma_0\| = 1$ and $\eta = \|\Gamma_0 F(x_0, y_0)\| = \frac{1}{3}$. Moreover, $\|F''(x, y) - F''(0, 0)\| \leq \frac{3}{4}\|(x, y)\|$. In addition, Eq. (2.41) is reduced to

$$\phi(t) = \frac{t^3}{8} + \frac{t^2}{4} - t + \frac{1}{3},$$

whose smallest positive zero is $t^* = 0.3751\ldots$, and $\phi(\alpha) = -0.2977\ldots \leq 0$, where $\alpha = 1.0971\ldots$, and $B((x_0, y_0), t^*) = B((0, 0), 0.3751\ldots) \subset \Omega$. As a consequence, from Theorem 2.59, the Newton method starting at $(0, 0)$ is convergent, and the sequence $\{t_n\}$ defined from the Newton method applied to ϕ is a majorant sequence of the Newton method applied to F.

Notice that the conditions of Theorem 2.50 are also satisfied. So, $\|F''(x, y)\| \leq \frac{1}{2} + \frac{3}{4}\|(x, y)\| \leq \frac{5}{4} = M$, $p(t) = \frac{5}{8}t^2 - t + \frac{1}{3}$, and $M\beta\eta = \frac{5}{12} < \frac{1}{2}$. Besides, $B\left((x_0, y_0), \frac{1}{M\beta}\right) = B((0, 0), 0.8) \subset \Omega$. Thus, the sequence $\{s_n\}$ defined from the Newton method applied to the Kantorovich polynomial p is a majorant sequence of the Newton method applied to F.

If we denote the solution of $F(x, y) = 0$ by x^*, the Newton sequence to approximate a solution of $F(x, y) = 0$ by $\{x_n\}$, and the smallest positive zero of $p(t)$ by $s^* = 0.4734\ldots$, then we know that

$$\|x^* - x_n\| \leq t^* - t_n \quad \text{and} \quad \|x^* - x_n\| \leq s^* - s_n, \quad n \geq 0,$$

by Theorems 2.59 and 2.50, respectively.

Observe in Table 2.9 that the a priori error estimates obtained from the function $\phi(t)$ are much better than those obtained from the Kantorovich polynomial $p(t)$.

Finally, we observe that the domain of existence of solution given by Theorem 2.59, $B((x_0, y_0), 0.3751\ldots)$, is better than that given by Theorem 2.50, $B((x_0, y_0), 0.4734\ldots)$.

2.2.3.2 Global Convergence

The existence condition of the operator degree of logarithmic convexity L_F in a certain set, given in Theorem 2.39 of Sect. 2.2.2.2, seems very strong, since it implies the existence of the inverse operator $[F'(x)]^{-1}$ at each point in the set. However, by requiring conditions

Table 2.9 A priori error bounds from Example 2.63

n	$t^* - t_n$	$s^* - s_n$
0	$3.7510\ldots \times 10^{-1}$	$4.7340\ldots \times 10^{-1}$
1	$4.1773\ldots \times 10^{-2}$	$1.4006\ldots \times 10^{-1}$
2	$8.3811\ldots \times 10^{-4}$	$2.1020\ldots \times 10^{-2}$
3	$3.6072\ldots \times 10^{-7}$	$6.3554\ldots \times 10^{-4}$
4	$6.6900\ldots \times 10^{-14}$	$6.1717\ldots \times 10^{-7}$

2.2 Operator Equations in Banach Spaces

(C1)–(C2), the existence of L_F is guaranteed in a domain, as we can see in the following result.

Theorem 2.64 *Let $\widetilde{x} \in \Omega$, and suppose that conditions (C1)–(C2) hold. Then, there exists $L_F(x)$ for all $x \in B\left(\widetilde{x}, \frac{1}{M\beta}\right)$ and $\|L_F(x)\| \leq \dfrac{M\beta \,\|\widetilde{\Gamma} F(x)\|}{(1 - M\beta \|x - \widetilde{x}\|)^2}$.*

Proof Observe that

$$\widetilde{\Gamma} \int_{\widetilde{x}}^{x} F''(v)\, dv = \widetilde{\Gamma} F'(x) - I$$

and

$$\|\widetilde{\Gamma} F'(x) - I\| = \left\| \widetilde{\Gamma} \int_{\widetilde{x}}^{x} F''(v)\, dv \right\| \leq M\beta \|x - \widetilde{x}\| < 1.$$

Then, from the Banach lemma on invertible operators, there exists $[\widetilde{\Gamma} F'(x)]^{-1}$ for all $x \in B\left(\widetilde{x}, \frac{1}{M\beta}\right)$ and

$$\|[\widetilde{\Gamma} F'(x)]^{-1}\| \leq \frac{1}{1 - M\beta \|x - \widetilde{x}\|}.$$

Now, from

$$L_F(x) = [\widetilde{\Gamma} F'(x)]^{-1} \widetilde{\Gamma} F''(x) [\widetilde{\Gamma} F'(x)]^{-1} \widetilde{\Gamma} F(x), \tag{2.46}$$

the proof is complete. ∎

Moreover, we observe that $B(\widetilde{x}, R) \subset B\left(\widetilde{x}, \frac{1}{M\beta}\right)$ if $R \in [R_-, R_+]$, where $R_- = \frac{1 - \sqrt{1 - 6M\beta\eta}}{3M\beta}$ and $R_+ = \frac{1 + \sqrt{1 - 6M\beta\eta}}{3M\beta}$, since $R_+ < \frac{1}{M\beta}$. As a consequence, we establish the following result.

Theorem 2.65 *Let $\widetilde{x} \in \Omega$, and suppose conditions (C1)–(C2). Then, the operator $L_F(x)$ exists, for all $x \in B(\widetilde{x}, R)$, with $R \in [R_-, R_+]$, where $R_- = \frac{1 - \sqrt{1 - 6M\beta\eta}}{3M\beta}$ and $R_+ = \frac{1 + \sqrt{1 - 6M\beta\eta}}{3M\beta}$, and*

$$\|L_F(x)\| \leq \frac{(M\beta R)^2 + 2(M\beta R) + 2M\beta\eta}{2(1 - M\beta R)^2}.$$

Proof We have seen above that $\|\Gamma F'(\widetilde{x})\| \leq \frac{1}{1-M\beta R}$ and $\|\widetilde{\Gamma} F(x)\| \leq \frac{M}{2}\beta R^2 + R + \eta$. Now, from (2.46), it follows that

$$\|L_F(x)\| \leq \|\Gamma F'(\widetilde{x})\|\|\widetilde{\Gamma}\|\|F''(x)\|\|\Gamma F'(\widetilde{x})\|\|\widetilde{\Gamma} F(x)\| \leq \frac{M\beta\left(\frac{M}{2}\beta R^2 + R + \eta\right)}{2(1-M\beta R)^2},$$

for all $x \in B(\widetilde{x}, R)$, and the proof is complete. ∎

In addition, as $N_F'(x) = L_F(x)$, for $x \in B(\widetilde{x}, R)$ and $R \in [R_-, R_+]$, we obtain the next result.

Theorem 2.66 *Let $\widetilde{x} \in \Omega$, and suppose conditions (C1)–(C2). If $M\beta\eta \leq \frac{-5+4\sqrt{2}}{4} = 0.1642\ldots$, then there exists $R > 0$ such that $R \in [R_-, \sigma_-]$, where $\widetilde{R}_- = \frac{1-\sqrt{1-6Mba}}{3Mb}$, $\sigma_- = \frac{3-\sqrt{7+2M\beta\eta}}{M\beta}$, $\|[F'(\widetilde{y})]^{-1}\| \leq b$, and $\|[F'(\widetilde{y})]^{-1}F(\widetilde{y})\| \leq a$, the operator $N_F : \overline{B(\widetilde{x}, R)} \to \overline{B(\widetilde{x}, R)}$ has a unique fixed point x^*, and the Newton method starting at $x_0 \in \overline{B(\widetilde{x}, R)}$ converges quadratically to x^*.*

Proof From Theorem 2.35, we know that $N_F(x) \in \overline{B(\widetilde{x}, R)}$, for all $x \in \overline{B(\widetilde{x}, R)}$. Then, from Theorem 2.64, we have

$$\|N_F'(x)\| \leq \frac{(M\beta R)^2 + 2(M\beta R) + 2M\beta\eta}{2(1-M\beta R)^2},$$

since $N_F'(x) = L_F(x)$. Thus, $N_F(x)$ is a contraction if

$$\frac{(M\beta R)^2 + 2(M\beta R) + 2M\beta\eta}{2(1-M\beta R)^2} < 1.$$

As $M\beta\eta \leq \frac{1}{6}$, the last inequality is true if $R \leq \sigma_-$.

After that, we observe that $R_- \leq \sigma_-$ if $M\beta\eta \leq \frac{1}{6}$, so that $[R_-, \sigma_-] \neq \emptyset$. Also, if $M\beta\eta \leq 0.1642\ldots$, we have $16(M\beta\eta)^2 + 40(M\beta\eta) - 7 \leq 0$, and, as a consequence, $R_+ > \sigma_-$.

Finally, as a consequence of the last, the operator N_F is a contraction. Therefore, from the application of Fixed Point Theorem 2.24 to the operator N_F in $\overline{B(\widetilde{x}, R)}$, the proof is complete. ∎

Now, the ball of location of the fixed point is $\overline{B(\widetilde{x}, R_-)}$, and the ball of uniqueness is $\overline{B(\widetilde{x}, \sigma_-)}$, which in turn is the most favorable ball of global convergence.

We illustrate the last with the Fredholm integral equation given in Example 2.40.

Example 2.67 Consider the integral equation given in (2.22),

$$x(s) = \sin(\pi s) + \frac{1}{4}\int_0^1 \cos(\pi s)\sin(\pi t)x(t)^2\,dt, \quad s \in [0,1].$$

As in Example 2.40, if we choose $\tilde{x}(s) = \sin \pi s$, then $\beta = \frac{\pi}{\pi-1}$, $\eta = \frac{1}{2(\pi-1)}$, and $M\beta\eta = \frac{1}{2(\pi-1)^2} = 0.1090\ldots$ Therefore, we can apply Theorems 2.38 and 2.66. So, for Theorem 2.38, we have that the operator N_F has a fixed point in $\overline{B(\sin(\pi s), R)}$ with $R \in [0.2940\ldots, 0.5253\ldots]$, and, for Theorem 2.66, the operator N_F has a fixed point in $\overline{B(\sin(\pi s), R)}$ with $R \in [0.2940\ldots, 0.6710\ldots]$, so that the domain of restricted global convergence obtained from Theorem 2.66 is greater than that obtained from Theorem 2.38. More precisely, the balls of location of fixed points are the same, but the ball of separation of fixed points (the ball of uniqueness) obtained from Theorem 2.66 is greater than that obtained from Theorem 2.38, so that the improvement obtained from Theorem 2.66 with respect to Theorem 2.38 is considerable. In addition, the hypothesis of Theorem 2.66 is slightly less restrictive than those of Theorem 2.38.

Next, we can proceed as before. Although the conditions on the quantity $M\beta\eta$ seem very restrictive, it is sufficient that the Kantorovich condition $M\beta\eta \leq \frac{1}{2}$ for the auxiliary point is satisfied, since the value of η tends to zero in the Newton sequence. So, once the Kantorovich condition has been verified at the auxiliary point, it is enough to take one or two iterations of the Newton method to fix an auxiliary point that satisfies the required condition [46]. So, we find, from an auxiliary point $\tilde{x}$ such that $M\beta\eta \leq \frac{1}{2}$, a new point $\tilde{y}$ and obtain global convergence balls of the form $\overline{B(\tilde{y}, R)}$ that locate a solution of the equation $F(x) = 0$ and separate it from other possible solutions, so that the condition $M\beta\eta \leq 0.1642\ldots$ is relaxed to $M\beta\eta \leq \frac{1}{2}$, and the previous aim is expanded. For this, we consider balls of the form $\overline{B(\tilde{y}, R)}$ and apply Fixed Point Theorem 2.24. So, we establish the following theorem.

Theorem 2.68 *Let $\tilde{x} \in \Omega$, and suppose conditions (C1)–(C2) and $h = M\beta\eta \leq \frac{1}{2}$. Then, there exist $\tilde{y} \in \overline{B(\tilde{x}, \rho^*)}$, where $\rho^* = \frac{1-\sqrt{1-2h}}{h}\eta$, and $R > 0$ such that $R \in [\tilde{R}_-, \tilde{\sigma}_-]$, where $\tilde{\sigma}_- = \frac{3-\sqrt{7+2Mba}}{Mb}$, $\|[F'(\tilde{y})]^{-1}\| \leq b$, $\|[F'(\tilde{y})]^{-1}F(\tilde{y})\| \leq a$, the operator $N_F: \overline{B(\tilde{y}, R)} \to \overline{B(\tilde{y}, R)}$ has a unique fixed point x^*, and the Newton method starting at any point in $\overline{B(\tilde{y}, R)}$ converges quadratically to x^*.*

Proof If $M\beta\eta \leq 0.1642\ldots$, then we choose $\tilde{y} = \tilde{x}$, so that $Mba \leq 0.1642\ldots$ In another case, we observe from (2.23) that if we take $x_0 = \tilde{x}$ for the Newton method, then we can consider

$$N_0 = 1 + \left\lceil \frac{\log\left(\frac{\log(0.3284\ldots)}{\log 2h}\right)}{\log 2} \right\rceil,$$

so that the N_0-th iterate of the method is such that $h_{N_0} \leq 0.1642\ldots$ Therefore, if we choose $\widetilde{y} = x_{N_0}$, it is clear that $Mba \leq 0.1642\ldots$

Then, from Theorem 2.41, we know that $N_F(x) \in \overline{B(\widetilde{y}, R)}$, for all $x \in \overline{B(\widetilde{y}, R)}$. Hence, from Theorem 2.65, we have

$$\|N'_F(x)\| \leq \frac{(MbR)^2 + 2(MbR) + 2Mba}{2(1 - MbR)^2},$$

since $N'_F(x) = L_F(x)$. Thus, $N_F(x)$ is a contraction if

$$\frac{(MbR)^2 + 2(MbR) + 2Mba}{2(1 - MbR)^2} < 1.$$

As $Mba \leq \frac{1}{6}$, the last inequality is true if $R \leq \widetilde{\sigma}_-$.

After that, we observe that $\widetilde{R}_- \leq \widetilde{\sigma}_-$ if $Mba \leq \frac{1}{6}$, so that $[\widetilde{R}_-, \widetilde{\sigma}_-] \neq \emptyset$. Also, if $Mba \leq \frac{-5+4\sqrt{2}}{4} = 0.1642\ldots$, we have $16(Mba)^2 + 40(Mba) - 7 \leq 0$, and, as a consequence, $\widetilde{R}_+ > \widetilde{\sigma}_-$.

Finally, from the above, the operator N_F is a contraction. Therefore, from the application of Fixed Point Theorem 2.24 to the operator N_F in $\overline{B(\widetilde{y}, R)}$, the proof is complete. ∎

Now, the ball of location of the fixed point is $\overline{B(\widetilde{y}, \widetilde{R}_-)}$, and the ball of uniqueness is $\overline{B(\widetilde{y}, \widetilde{\sigma}_-)}$, which in turn is the most favorable ball of global convergence.

Example 2.69 We illustrate the last with the Davis type integral equation given in (2.26) of Example 2.46 and used in [46]

$$x(s) = s + \frac{7}{5}\int_0^1 G(s,t)x(t)^2 dt, \quad s \in [0, 1],$$

where the kernel $G(s, t)$ is the Green function

$$G(s, t) = \begin{cases} (1-s)t, & t \leq s, \\ s(1-t), & s \leq t. \end{cases}$$

In Example 2.46, after a process of discretization of the integral equation, we see that $h = M\beta\eta = 0.3629\ldots > 0.1513\ldots$ if $\widetilde{\mathbf{x}} = (1, 1, \ldots, 1)^t$, so that we cannot apply Theorem 2.38, but we can apply Theorems 2.44 and 2.68.

2.2 Operator Equations in Banach Spaces

For Theorem 2.68, we have $N_0 = 2$ and $h_{N_0} = 0.000192\ldots < 0.1642\ldots$, so that $\widetilde{\mathbf{y}} = \mathbf{x}_2$, $R \in [0.0006\ldots, 1.1515\ldots]$, the best ball of location of solution is $\overline{B(\mathbf{x}_2, 0.0006\ldots)}$, and the biggest ball of convergence is $\overline{B(\mathbf{x}_2, 1.1515\ldots)}$, which is bigger than that given by Theorem 2.44.

From the last result, it is easy to conclude that the ball of location of solution is the same and it is very precise, and the balls of separation of solutions given by the two theorems are good enough. Observe that Theorem 2.68 offers a better result for the ball of convergence.

Remark 2.70 An important advantage of the restricted global convergence results is that, as a consequence, we can obtain semilocal and local convergence results for the Newton method, taking $\widetilde{x} = x_0$ or $\widetilde{x} = x^*$, respectively.

2.2.3.3 Local Convergence

In this section, we study the local convergence of the Newton method under conditions (G1)–(G2) of Dennis and Schnabel given in Sect. 2.2.2.5 and using the operator degree of logarithmic convexity. For this, we introduce the following technical result that is used later and whose proof is analogous to that of Theorem 2.10.

Theorem 2.71 *Let $F : \Omega \subseteq X \to Y$ be a twice continuously Fréchet differentiable operator on a non-empty open convex domain Ω of a Banach space X with values in a Banach space Y. Let x^* be a solution of $F(x) = 0$ such that (G1) is satisfied. If (G2) holds and $B\left(x^*, \frac{1}{M\gamma}\right) \subset \Omega$, then there exits $L_F(x)$ for all $x \in B\left(x^*, \frac{1}{M\gamma}\right)$ and*

$$\|L_F(x)\| \leq \frac{M\gamma \, \|[F'(x^*)]^{-1} F(x^*)\|}{(1 - M\gamma \|x - x^*\|)^2}.$$

After that, we establish the local convergence of the Newton method.

Theorem 2.72 *Let $F : \Omega \subseteq X \to Y$ be a twice continuously Fréchet differentiable operator on a non-empty open convex domain Ω of a Banach space X with values in a Banach space Y. Let x^* be a solution of $F(x) = 0$ such that (G1) is satisfied. If (G2) holds and $B\left(x^*, \frac{1}{M\gamma}\right) \subset \Omega$, then the Newton sequence (2.4) is well defined and converges to x^* from any $x_0 \in B\left(x^*, \frac{0.5646\ldots}{M\gamma}\right)$.*

Proof From the previous theorem, it follows that there exists $L_F(x)$ for all $x \in B\left(x^*, \frac{1}{M\gamma}\right)$, so that $L_F(x_0)$ exists; therefore $[F'(x_0)]^{-1}$ exists, and $x_1 = N_F(x_0)$ is then well defined. Moreover,

$$x_1 - x^* = N_F(x_0) - N_F(x^*) = \int_{x^*}^{x_0} L_F(v)\, dv = \int_0^1 L_F(x^* + t(x_0 - x^*))\, dt\, (x_0 - x^*)$$

and

$$\|x_1 - x^*\| \leq \int_0^1 \frac{M\gamma \|[F'(x^*)]^{-1} F(x^* + t(x_0 - x^*))\|}{(1 - M\gamma \|x_0 - x^*\|t)^2} dt \|x_0 - x^*\|. \quad (2.47)$$

Now, from Taylor's formula, we have

$$F(x^* + t(x_0 - x^*)) = F(x^*) + F'(x^*) t(x_0 - x^*) + \int_{x^*}^{x^* + t(x_0 - x^*)} (F'(v) - F'(x^*)) \, dv$$

$$= F'(x^*) t(x_0 - x^*) + \int_0^1 (F'(s(x^* + t(x_0 - x^*))) - F'(x^*)) \, ds \, t(x_0 - x^*)$$

and

$$[F'(x^*)]^{-1} F(x^* + t(x_0 - x^*)) = t(x_0 - x^*) + [F'(x^*)]^{-1} \int_0^1 (F'(s(x^* + t(x_0 - x^*))) - F'(x^*)) \, ds \, t(x_0 - x^*).$$

From (G1)–(G2), we obtain

$$\|[F'(x^*)]^{-1} F(x^* + t(x_0 - x^*))\| \leq t \|x_0 - x^*\| + M\gamma \int_0^1 t^2 s \, ds \|x_0 - x^*\|^2$$

$$= t \|x_0 - x^*\| + \frac{1}{2} M\gamma t^2 \|x_0 - x^*\|^2.$$

Next, from (2.47), it follows

$$\|x_1 - x^*\| \leq \int_0^1 \frac{M\gamma \|x_0 - x^*\| + \frac{1}{2} M^2 \gamma^2 \|x_0 - x^*\|^2 t^2}{(1 - M\gamma \|x_0 - x^*\|t)^2} dt \|x_0 - x^*\|$$

$$= \frac{1}{M\gamma} \int_0^{M\gamma \|x_0 - x^*\|} \frac{u + \frac{u^2}{2}}{(1 - u)^2} du$$

$$= \frac{1}{M\gamma} \left(\frac{M\gamma \|x_0 - x^*\| (M\gamma \|x_0 - x^*\| - 4)}{2(1 - M\gamma \|x_0 - x^*\|)} + \ln \left(1 - M\gamma \|x_0 - x^*\|\right)^2 \right),$$

so that $\|x_1 - x^*\| < \|x_0 - x^*\|$ if

$$\frac{M\gamma \|x_0 - x^*\| (M\gamma \|x_0 - x^*\| - 4)}{2(1 - M\gamma \|x_0 - x^*\|)} + \ln \left(1 - M\gamma \|x_0 - x^*\|\right)^2 < M\gamma \|x_0 - x^*\|,$$

2.2 Operator Equations in Banach Spaces

which is true provided that $\|x_0 - x^*\| < \dfrac{0.5646\ldots}{M\gamma}$. As a consequence, $x_1 \in B\left(x^*, \dfrac{0.5646\ldots}{M\gamma}\right)$.

After that, by mathematical induction on n, it is easy to prove that

$$\|x_n - x^*\| < \|x_{n-1} - x^*\| < \cdots < \|x_0 - x^*\| < \dfrac{0.5646\ldots}{M\gamma}.$$

Thus, $x_n \in B\left(x^*, \dfrac{0.5646\ldots}{M\gamma}\right)$, for all $n \geq 0$, and the Newton sequence (2.4) converges to x^*. ∎

Next, we return to Example 2.48 to illustrate the last result.

Example 2.73 For the system (2.28), we have $\gamma = 1$ and $\|F''(\mathbf{x})\| \leq \max\{2, e^{\|x\|}\} = M$. If we choose $\Omega = B(\mathbf{x}^*, r)$, then $B\left(\mathbf{x}^*, \dfrac{1}{M\gamma}\right) \subset \Omega$, provided that $r > \dfrac{1}{2}$, so that

$$B\left(x^*, \dfrac{0.5646\ldots}{M\gamma}\right) = \begin{cases} B(x^*, 0.2823\ldots), & r \in \left(\dfrac{1}{2}, \ln 2\right), \\ B\left(x^*, \dfrac{0.5646\ldots}{e^r}\right), & r \geq \ln 2. \end{cases}$$

If we compare this result with that given by Dennis and Schnabel in [28] (see Sect. 2.2.2.5) and Theorem 2.47, we can emphasize three things. The first is that Theorem 2.72 cannot be applied if $r \leq \dfrac{1}{2}$. The second is that Theorem 2.72 improves the results of Dennis and Schnabel and Theorem 2.47 if $r \in \left(\dfrac{1}{2}, \ln 2\right)$, so that the domains of starting points are improved in both cases. The third is that Theorem 2.72 improves the result given by Dennis and Schnabel if $r \geq \ln 2$, since $R = \dfrac{1}{2e^r}$ is obtained in this case, so that the domain of starting points is improved. Moreover, if $r < 0.9828\ldots$, Theorem 2.72 also improves Theorem 2.47, but Theorem 2.47 improves the results of Dennis and Schnabel and Theorem 2.72 if $r \geq 0.9828\ldots$

On the other hand, Rall in [78] and Rheinboldt in [82] give results similar to that given by Dennis and Schnabel, but with different radii for the balls of convergence. In [13], Argyros generalizes the previous conditions by modifying condition (G2) by means of a Lipschitz condition for the second derivative of F. In addition, results corresponding to that of Dennis and Schnabel with the bigger radii $r < \dfrac{\sqrt{2}-1}{\sqrt{2}\gamma M}$ and $r < \dfrac{2}{3\gamma M}$ are proved under the same conditions as, respectively, Rall and Rheinboldt. Comments analogous to those made about the result of Dennis and Schnabel are also valid for the results of Rall and Rheinboldt, since the radius of the ball of convergence depends on a bound for F'' or on the Lipschitz constant for F', which in turn depend on the value of r, so that we can obtain similar conclusions. For the Argyros result, the same comments as are made above still apply.

Accelerations of the Newton Method

3

Once we have seen the influence of convexity on the speed of convergence of the Newton method in the previous chapter, we now give three acceleration procedures for the Newton method, which are based on the influence of convexity and lead us to the three well-known third-order iterative methods of Chebyshev, Super-Halley, and Halley [30, 31]. Next, we study these three methods in the scalar case and in Banach spaces. In the scalar case, we see the influence that convexity has on each of the three methods, and from this, we guarantee that we can always solve a scalar equation by any of the three methods under certain restrictions on the degree of logarithmic convexity of the function involved in the equation to be solved. The study of these three methods in Banach spaces is carried out from the same point of view: the method of majoring sequences developed by Ortega for the Newton method [74], which simplifies the seminal majorant principle developed by Kantorovich to analyze the semilocal convergence of the Newton method in Banach spaces [67]. We illustrate all theoretical results with examples, and, as in Chap. 2 for the Newton method, we use the theoretical significance of the methods to draw conclusions about the existence of solution of the equation to be solved.

3.1 Acceleration Procedures of the Newton Method

After proving in Sect. 2.1.2 that the convexity of the function involved influences the speed of convergence of the Newton method, we obtain different sequences that converge faster than the Newton sequence. We call these sequences accelerations of the Newton method, and they are one-point iterative methods. That is, we obtain $y_{n+1} = N_g(x_n), n \geq 0$, where $N_g(u) = u - \frac{g(u)}{g'(u)}$ and x_0 is given, so that y_n is closer to the solution x^* of the equation $f(x) = 0$ to solve than $x_n = N_f(x_{n-1}) = x_{n-1} - \frac{f(x_{n-1})}{f'(x_{n-1})}$. It is a consequence of the fact that we can define new iterative methods of the form $z_{n+1} = N_h(z_n)$, $n \geq 0$, with $z_0 \in [a, b]$, from the function N_h.

Throughout this section, we consider that the function f is increasing and convex in the interval $[a, b]$. And, to simplify the study, we consider the Newton sequence

$$x_{n+1} = N_f(x_n) = x_n - \frac{f(x_n)}{f'(x_n)}, \quad n \geq 0, \qquad (3.1)$$

with $x_0 \in [a, b]$ and $f(x_0) > 0$. In another case, similar results can be obtained, as we see later.

Next, we present three acceleration procedures of the Newton method, which are based on the influence of convexity of the real function involved f. The first is a procedure called local approximation, which consists of approximating the curve $y = f(x)$ by the straight line $y = f'(x_n)(x_n - x^*)$ in a neighborhood of each point $(x_n, f(x_n))$. Note that straight lines are the curves with the less degree of logarithmic convexity. The second is a procedure called global approximation, where the curve $y = f(x)$ is approximated by means of the tangent line at $(x^*, 0)$. Finally, the third procedure consists of reducing directly the degree of logarithmic convexity of f: From the function f, a function with less degree of logarithmic convexity than f is given.

From the three acceleration procedures, we define later three independent one-point iterative methods and see which methods converge more quickly to a solution of a nonlinear scalar equation.

3.1.1 Local Convex Acceleration of the Newton Method

The first procedure of acceleration, which we call local approximation, consists of approximating the curve $y = f(x)$ by straight lines. For each point x_n of (3.1), the curve $y = f(x)$ is approximated by the line $y = f'(x_n)(x - x^*)$, where x^* is a solution of $f(x) = 0$, in a neighborhood of x_n, as we can see in Fig. 3.1.

Fig. 3.1 Local approximation

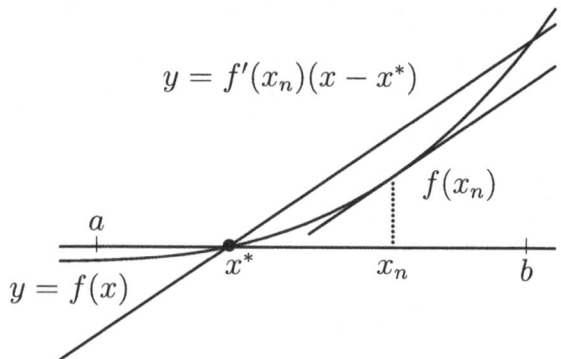

3.1 Acceleration Procedures of the Newton Method

We consider $f \in C^2([a, b])$ and Taylor's formula, approximate

$$f(x) \sim f(x_n) + f'(x_n)(x - x_n) + \frac{f''(x_n)}{2!}(x - x_n)^2, \qquad (3.2)$$

and choose $g(x) = f'(x_n)(x - x^*)$. Now, we have to approximate $g(x_n)$ and $g'(x_n)$ to obtain $y_{n+1} = x_n - \frac{g(x_n)}{g'(x_n)}$.

Taking into account (3.2) for $x = x^*$,

$$g(x_n) \sim f(x_n) + \frac{f''(x_n)}{2!}(x^* - x_n)^2,$$

and by (3.1), we have

$$(x^* - x_n)^2 \sim (x_{n+1} - x_n)^2 = \left(\frac{f(x_n)}{f'(x_n)}\right)^2,$$

and

$$g(x_n) \sim f(x_n)\left(1 + \frac{1}{2}L_f(x_n)\right).$$

On the other hand, it is obvious that $g'(x_n) = f'(x_n)$. Therefore,

$$x_n - \frac{g(x_n)}{g'(x_n)} \sim x_n - \frac{f(x_n)}{f'(x_n)}\left(1 + \frac{1}{2}L_f(x_n)\right),$$

and we then define the sequence $\{y_n\}$ from

$$y_{n+1} = x_n - \frac{f(x_n)}{f'(x_n)}\left(1 + \frac{1}{2}L_f(x_n)\right). \qquad (3.3)$$

To prove that (3.3) is an acceleration of (3.1), it suffices to see that

$$\lim_n \frac{|y_n - x^*|}{|x_n - x^*|} = 0.$$

For this, it is enough to consider

$$\lim_{x \to x^*} \frac{1}{x - x^*}\left(x - \frac{f(x)}{f'(x)}\left(1 + \frac{1}{2}L_f(x)\right)\right)$$

and apply L'Hôpital's rule.

Example 3.1 Given the function $f(x) = \dfrac{e^x - 5x}{x}$, we show in Table 3.1 that the sequence $\{y_n\}$ defined by (3.3) converges to $x^* = 2.542641357773526\ldots$ of $f(x) = 0$ in $[1, 4]$ faster than the Newton sequence $\{x_n\}$ with $x_0 = y_0 = 3.5$.

3.1.2 Global Convex Acceleration of the Newton Method

The global convex acceleration of the Newton method is introduced by Hernández in [56]. The curve $y = f(x)$ is approximated by the tangent line at $(x^*, 0)$, as we can see in Fig. 3.2. For this, it suffices to consider $f \in \mathcal{C}^2([a, b])$ and approximate $f(x)$ by the following Taylor's formula of f in a neighborhood of x^*:

$$f(x) \sim f(x^*) + f'(x^*)(x - x^*) + \frac{f''(x^*)}{2!}(x - x^*)^2.$$

Thus, $f'(x^*)(x - x^*) \sim f(x) - \frac{f''(x^*)}{2!}(x - x^*)^2$, and we then consider the function

$$g(x) = f(x) - \frac{f''(x^*)}{2!}(x - x^*)^2. \tag{3.4}$$

Table 3.1 Sequences (3.1) and (3.3) from Example 3.1

n	x_n	y_n
0	3.500000000000000…	3.500000000000000…
1	2.839835893846803…	2.659283282924826…
2	2.577023717097117…	2.543020336792808…
3	2.543144242829421…	2.542641357787998…
4	2.542641466706540…	2.542641357773526…

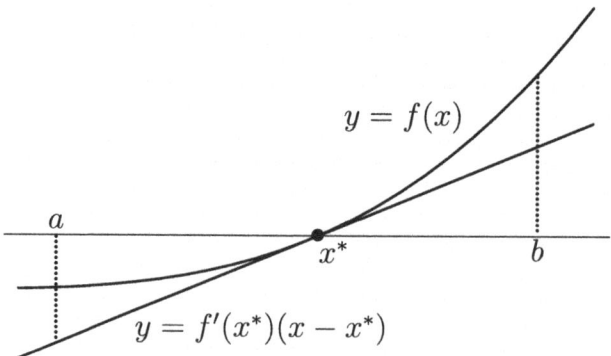

Fig. 3.2 Global approximation

3.1 Acceleration Procedures of the Newton Method

Theoretically, $L_f(x) > L_g(x)$ and $f(x^*) = g(x^*)$. Therefore, from Theorem 2.6, the Newton method defined from g converges to x^* faster than the Newton method defined from f. But, function g falls into a problem, since x^* is an unknown value.

As our purpose is to accelerate the Newton sequence (3.1), we must obtain from x_0 a new value $y_{n+1} = x_n - \frac{g(x_n)}{g'(x_n)}$, $n \geq 0$, closer to x^* than x_{n+1}. Hence, we have to evaluate $g(x)$ and $g'(x)$ at x_n, so that we approximate

- $f''(x^*) \sim f''(x_n)$
- $(x_n - x^*)^k \sim (x_n - x_{n+1})^k$ for $k = 1, 2$

since $\lim_n \frac{(x_n - x^*)^k}{(x_n - x_{n+1})^k} = 1$, as a consequence of $\lim_{x \to x^*} \frac{(x - x^*)^k}{(x - N_f(x))^k} = \lim_{x \to x^*} \left(\frac{1}{1 - N'_f(x)}\right)^k = \left(\lim_{x \to x^*} \frac{1}{1 - L_f(x)}\right)^k = 1$. Therefore, from (3.4) and the abovementioned equations, we consider

$$g(x_n) \sim f(x_n) - \frac{f''(x_n)}{2!}\left(\frac{f(x_n)}{f'(x_n)}\right)^2 \quad \text{and} \quad g'(x_n) \sim f'(x_n) - f''(x_n)\frac{f(x_n)}{f'(x_n)},$$

so that

$$x_n - \frac{g(x_n)}{g'(x_n)} \sim x_n - \frac{f(x_n)}{2f'(x_n)}\left(1 + \frac{1}{1 - L_f(x_n)}\right),$$

and we then define the sequence $\{y_n\}$ as

$$y_{n+1} = x_n - \frac{f(x_n)}{2f'(x_n)}\left(1 + \frac{1}{1 - L_f(x_n)}\right). \tag{3.5}$$

Using the same argument as for the sequence (3.3), we can prove that (3.5) is an acceleration of the Newton sequence (3.1), since that

$$\lim_n \frac{|y_n - x^*|}{|x_n - x^*|} = 0.$$

Example 3.2 We consider again $f(x) = \frac{e^x - 5x}{x}$ and see in Table 3.2 that the sequence $\{y_n\}$ given by (3.5) converges to the solution $x^* = 2.542641357773526\ldots$ of $f(x) = 0$ in $[1, 4]$ faster than the Newton sequence $\{x_n\}$ with $x_0 = y_0 = 3.5$.

Table 3.2 Sequences (3.1) and (3.5) from Example 3.2

n	x_n	y_n
0	3.500000000000000...	3.500000000000000...
1	2.839835893846803...	2.441271065123373...
2	2.577023717097117...	2.542750966419476...
3	2.543144242829421...	2.542641357773588...
4	2.542641466706540...	2.542641357773526...

3.1.3 Direct Reduction of the Degree of Logarithmic Convexity of a Function

The idea of this section is to construct an acceleration of the Newton method from Theorem 2.6 and a function g with a lower degree of logarithmic convexity than the function f that defines the Newton sequence (3.1). For this, we consider an increasing and convex function f in $[a, b]$ such that $L_f(x) > L_g(x)$, for $x \in (x^*, x_0]$. In addition, we choose $x_0 \in [a, b]$ with $f(x_0) > 0$ and $\{x_n\} \subset (x^*, b]$.

If we consider the function

$$g(x) = \frac{f(x)}{\sqrt{f'(x)}},$$

which is introduced by Alefeld [1] to define the Halley method as a variant of the Newton method, it is clear that $g(x) > 0$ in $(x^*, b]$, since $f(x) > 0$ in $(x^*, b]$. Thus, we have

$$g'(x) = \sqrt{f'(x)}\left(1 - \frac{1}{2}L_f(x)\right) \quad \text{and} \quad g''(x) = \frac{f''(x)L_f(x)}{2\sqrt{f'(x)}}\left(\frac{3}{2} - L_{f'}(x)\right).$$

So, if $L_f(x) < 2$ and $L_{f'}(x) \leq \frac{3}{2}$ in $(x^*, x_0]$, then g is increasing and convex in $(x^*, x_0]$. As

$$L_g(x) = \frac{L_f(x)^2}{(L_f(x) - 2)^2}(3 - 2L_{f'}(x)),$$

then $L_f(x) > L_g(x) > 0$ in $(x^*, x_0]$ if and only if $(L_f(x)-2)^2 - L_f(x)(3-2L_{f'}(x)) > 0$. For $L_f(x) > L_g(x)$ in $(x^*, x_0]$, it is then enough to choose x_0 so that

$$L_f(x) < \frac{1}{2}\left(7 - 2m - \sqrt{(7-2m)^2 - 16}\right)$$

in $(x^*, x_0]$, where $m = \min_{x \in (x^*, b]}\{L_{f'}(x)\}$. Under these conditions, and in view of Theorem 2.6, we obtain the acceleration of the Newton method defined by

3.1 Acceleration Procedures of the Newton Method

Table 3.3 Sequences (3.1) and (3.6) from Example 3.3

n	x_n	y_n
0	2.0000000000000000...	2.0000000000000000...
1	0.8807970779778824...	−0.2070451959228786...
2	−0.0842749600983386...	−0.5683407447276397...
3	−0.5193066837383489...	−0.5671432903624338...
4	−0.5667232231976213...	−0.5671432904097839...

$$y_{n+1} = x_n - \frac{g(x_n)}{g'(x_n)} = x_n - \frac{f(x_n)}{f'(x_n)}\left(\frac{2}{2 - L_f(x_n)}\right), \quad n \geq 0. \tag{3.6}$$

Example 3.3 Consider the function $f(x) = e^x + x$. We see in Table 3.3 that the sequence $\{y_n\}$, defined by (3.6), converges to the solution $x^* = -0.5671432904097839\ldots$ of the equation $f(x) = 0$ in $[-2, 2]$ faster than the Newton sequence with $x_0 = y_0 = 2$. Besides, as

$$L_f(x) = \frac{e^x(e^x + x)}{(e^x + 1)^2}, \qquad L_{f'}(x) = \frac{1 + e^x}{e^x}, \qquad m = L_{f'}(2) = 1.13534\ldots$$

and

$$\frac{1}{2}\left(7 - 2m - \sqrt{(7 - 2m)^2 - 16}\right) = 1.10306\ldots,$$

it follows $L_f(x) > L_g(x)$ in $(x^*, x_0]$ and $L_f(x) < 1.10306\ldots$ in $\mathbb{R}$. Then, from Theorem 2.6, the sequence $\{y_n\}$ is an acceleration of the Newton sequence $\{x_n\}$.

3.1.4 Iterative Methods Obtained by Means of Convex Accelerations of the Newton Method

We have obtained three accelerations of the Newton method as a consequence of the study of the convexity of the function f in the previous section. The three have the form $y_{n+1} = \Phi(x_n)$, $n \geq 0$, with x_0 given, so that independent one-point iterative methods can then be constructed.

In his classic paper of 1870, Schröder [86] proves that an iterative method of the form $x_{n+1} = \Phi(x_n)$, $n \geq 0$, converging to s^*, has order of convergence p if

$$\Phi(s^*) = s^*, \quad \Phi^{(j)}(s^*) = 0, \quad j = 1, 2, \ldots, p-1, \quad \text{and} \quad \Phi^{(p)}(s^*) \neq 0,$$

with p continuous derivatives of Φ (see [88] for more details).

Then, we define below three one-point iterative methods with cubic convergence, do a study of the convergence according to the degree of logarithmic convexity of f, and conclude which is faster.

The local approximation procedure leads us to the iteration

$$x_{n+1} = T(x_n) = x_n - \left(1 + \frac{1}{2}L_f(x_n)\right)\frac{f(x_n)}{f'(x_n)}, \quad n \geq 0, \quad \text{with } x_0 \text{ given}, \qquad (3.7)$$

which is usually known as the Chebyshev method or the method of tangent parabolas [9, 19].

By the global approximation procedure, we obtain the Super-Halley method [22,54,56]:

$$x_{n+1} = S(x_n) = x_n - \frac{1}{2}\left(1 + \frac{1}{1 - L_f(x_n)}\right)\frac{f(x_n)}{f'(x_n)}, \quad n \geq 0, \quad \text{with } x_0 \text{ given}. \qquad (3.8)$$

Third, the acceleration (3.6) provides the Halley method or the method of tangent hyperbolas [1, 2, 18, 21, 85, 92]:

$$x_{n+1} = H(x_n) = x_n - \frac{2}{2 - L_f(x_n)}\frac{f(x_n)}{f'(x_n)}, \quad n \geq 0, \quad \text{with } x_0 \text{ given}. \qquad (3.9)$$

On the other hand, and according to Gander [48], a method given by the expression

$$x_{n+1} = x_n - \Upsilon(L_f(x_n))\frac{f(x_n)}{f'(x_n)}, \quad n \geq 0, \quad \text{with } x_0 \text{ given},$$

where $\Upsilon(0) = 1$, $\Upsilon'(0) = \frac{1}{2}$, and $|\Upsilon''(0)| < \infty$, is of third order, so that the methods (3.7), (3.8), and (3.9) are cubically convergent.

Next, we give a global convergence result depending on L_f and $L_{f'}$ for the previous three iterative methods. For this, we consider that the function f is increasing, convex, differentiable enough in $[a, b]$ and such that $f(a)f(b) < 0$. Note that, under these conditions, there is only one solution x^* of $f(x) = 0$ in $[a, b]$.

Theorem 3.4 *Let $x_0 \in [a, b]$ with $f(x_0) > 0$ and $L_{f'}(x) \leq 0$ in $[a, b]$.*

(a) *The Chebyshev sequence (3.7) is decreasing and converges to x^*.*
(b) *If $L_f(x) < 1$ in $[a, b]$, then the Super-Halley sequence (3.8) is decreasing and converges to x^*.*
(c) *If $L_f(x) < 2$ in $[a, b]$, then the Halley sequence (3.9) is decreasing and converges to x^*.*

Proof We prove item (a). First, we see that $x_n \geq x^*$, for all $n \in \mathbb{N}$. From $f(x_0) > 0$, it follows $x_0 \geq x^*$. By the Mean Value Theorem, we have

$$x_1 - x^* = T(x_0) - T(x^*) = T'(\theta_0)(x_0 - x^*), \quad \text{with} \quad \theta_0 \in (x^*, x_0),$$

and taking into account

$$T'(x) = \frac{1}{2} L_f(x)^2 (3 - L_{f'}(x)), \tag{3.10}$$

we deduce that $T'(x) \geq 0$ in $[x^*, b]$, since $L_{f'}(x) \leq 0$ and $a \leq x^* \leq x_1$. In addition,

$$x_1 - x_0 = -\left(1 + \frac{1}{2} L_f(x_0)\right) \frac{f(x_0)}{f'(x_0)} \leq 0,$$

so that $x_1 \leq x_0 \leq b$. Therefore, $x_1 \in [a, b]$.

Now, by mathematical induction on n, it is easy to prove that $a \leq x^* \leq x_n$ and $x_n \leq x_{n-1} \leq b$, for all $n \in \mathbb{N}$, following the same idea as above for $n = 1$. Thus, the Chebyshev sequence (3.7) is a decreasing sequence in $[a, b]$.

As a consequence of the above, the sequence (3.7) is monotonous and bounded and then convergent, so that there exists $u \in [a, b]$ such that $u = \lim_n x_n$ and $u \geq x^*$. Now, by letting $n \to \infty$ in (3.7), we obtain

$$u = u - \left(1 + \frac{1}{2} L_f(u)\right) \frac{f(u)}{f'(u)}$$

and, as a result, $f(u) = 0$. But, under the hypotheses mentioned above, x^* is the unique solution of the equation $f(x) = 0$ in $[a, b]$, so that $u = x^*$ and item (a) is then proved.

The proofs of items (b) and (c) follow similarly to that of item (a) by taking into account

$$S'(x) = \frac{L_f(x)^2}{2(1 - L_f(x))^2} (L_f(x) - L_{f'}(x)) \quad \text{and}$$

$$H'(x) = \left(\frac{L_f(x)}{2 - L_f(x)}\right)^2 (3 - 2L_{f'}(x)), \tag{3.11}$$

respectively. ∎

Notice that the sequences (3.7), (3.8), and (3.9) are increasing and converge to x^* under the conditions of Theorem 3.4 if $f(x_0) < 0$.

Furthermore, we deduce the same result for the Chebyshev method if $L_{f'}(x) \leq 3$, for $x \in [a, b]$, and for the Halley method if $L_f(x) < 2$ and $L_{f'}(x) \leq \frac{3}{2}$ in $[a, b]$, as we can see later. Since the next aim is to compare the speed of convergence of the three methods, we require the same convergence conditions to obtain the simultaneous convergence of the three methods.

Theorem 3.5 *Suppose the conditions of Theorem 3.4 and $L_f(x) < 1$ in $[a, b]$. If the three methods start at the same initial approximation, the Super-Halley method is faster than the Halley method, and the Halley method is faster than the Chebyshev method.*

Proof Let $\{x_n\}$, $\{y_n\}$, and $\{z_n\}$ be defined, respectively, in (3.7), (3.8), and (3.9). Since $x_0 = y_0 = z_0$ and the three sequences are decreasing, we expect that $y_n \leq z_n \leq x_n$, for all $n \in \mathbb{N}$. This can be proved by mathematical induction on n. If $n = 1$, then

$$y_1 - z_1 = S(x_0) - H(x_0) = \left(\frac{2}{2 - L_f(x_0)} - \frac{2 - L_f(x_0)}{2(1 - L_f(x_0))}\right) \frac{f(x_0)}{f'(x_0)} \leq 0,$$

$$z_1 - x_1 = H(x_0) - T(x_0) = \left(\left(1 + \frac{1}{2}L_f(x_0)\right) - \frac{2}{2 - L_f(x_0)}\right) \frac{f(x_0)}{f'(x_0)} \leq 0.$$

After that, we assume that $y_{j-1} \leq z_{j-1} \leq x_{j-1}$ for $j = 1, 2, \ldots, n-1$. Taking now into account that T and H are nondecreasing functions in $[a, b]$, it is easy to see that

$$y_n - z_n = S(y_{n-1}) - H(z_{n-1}) \leq S(y_{n-1}) - H(y_{n-1}) \leq 0,$$

$$z_n - x_n = H(z_{n-1}) - T(x_{n-1}) \leq H(z_{n-1}) - T(z_{n-1}) \leq 0.$$

Then, the mathematical induction on n is complete. ∎

Example 3.6 Consider $f(x) = x - \cos x$, which is increasing and convex in $\left[0, \frac{\pi}{2}\right]$. Thus,

$$L_f(x) = \frac{(x - \cos x)\cos x}{(1 + \sin x)^2} \quad \text{and} \quad L_{f'}(x) = \frac{\sin x}{\sin x - 1},$$

so that $L_f(x) < 1$ and $L_{f'}(x) \leq 0$ in $\left[0, \frac{\pi}{2}\right]$. Then, by Theorem 3.4, the sequences (3.7), (3.8), and (3.9) converge to the solution $x^* = 0.7390851332151606428\ldots$ of $f(x) = 0$ in $\left[0, \frac{\pi}{2}\right]$. In Table 3.4, we compare the speed of convergence of the three sequences.

Table 3.4 Sequences $\{x_n\}$, $\{y_n\}$, and $\{z_n\}$, defined, respectively, by (3.7), (3.8), and (3.9) from Example 3.6

n	y_n	z_n	x_n
0	1.000000000000000000...	1.000000000000000000...	1.000000000000000000...
1	0.7404989832636941698...	0.7408739950803435706...	0.7412215390677832763...
2	0.7390851334050131377...	0.7390851338775818840...	0.7390851348155419594...
3	0.7390851332151606428...	0.7390851332151606499...	0.7390851332151606451...
4		0.7390851332151606428...	0.7390851332151606435...
5			0.7390851332151606428...

3.2 The Chebyshev Method

One of the best known third-order iterative methods to find a solution of a scalar nonlinear equation $f(t) = 0$ is the Chebyshev method [9, 19], which is given by

$$t_{n+1} = T(t_n) = t_n - \left(1 + \frac{1}{2}L_f(t_n)\right)\frac{f(t_n)}{f'(t_n)}, \quad n \geq 0, \quad \text{with } t_0 \text{ given.} \tag{3.12}$$

Using the degrees of logarithmic convexity of the functions f and f', we study the convergence of the Chebyshev method [58]. Next, practical situations are analyzed, and we prove, depending of the convexity of f and f', that this method can always be applied to solve scalar equations. After that, we give a new type of conditions for the convergence of this method in Banach spaces and apply this study to solve a boundary value problem.

3.2.1 The Chebyshev Method and Convexity

Starting from convexity, we carry out an exhaustive study of the convergence conditions of the Chebyshev method. This study is novel since we do it in terms of the convexity of the functions f and f'. For this, first we obtain global convergence conditions for the iterative method, and then we analyze its practical application and obtain a practical procedure that allows us to always solve a scalar equation by applying this method.

In this section, we consider $f \in \mathcal{C}^m([a,b])$, with $m \geq 3$, that satisfies the Fourier conditions given in Sect. 2.1.1, so that there is a unique solution t^* of the equation $f(t) = 0$ in $[a, b]$. Without loss of generality, we suppose that f is a convex and strictly increasing function in $[a, b]$ with $f(a)f(b) < 0$. In other case, it is sufficient to change $f(t)$ by $f(-t)$, $-f(t)$, or $-f(-t)$. We denote

$$M(g) = \max\{g(t) : t \in [a,b]\} \quad \text{and} \quad m(g) = \min\{g(t) : t \in [a,b]\}.$$

We begin by giving a more general result than Theorem 3.4 (a).

Theorem 3.7 *Suppose that $t_0 \in [a, b]$ and $L_{f'}(t) \leq 3$ in $[a, b]$. If $f(t_0) > 0$, then the Chebyshev sequence (3.12) is decreasing and converges to t^*. If $f(t_0) < 0$ and $L_f(t) \geq -2$ in $[a, b]$, then (3.12) is increasing and converges to t^*.*

Proof First, we consider $t_0 \in [a, b]$ with $f(t_0) > 0$ and prove that $t_n \geq t^*$, for all $n \in \mathbb{N}$. As $f(t_0) > 0$, it follows that $t_0 - t^* > 0$. Moreover, from (3.10), it follows that $T'(t) \geq 0$ in $[a, b]$, since $L_{f'}(t) \leq 3$. In addition,

$$t_1 - t^* = T(t_0) - T(t^*) = T'(\theta_0)(t_0 - t^*)) \geq 0, \quad \text{with} \quad \theta_0 \in (t^*, t_0),$$

and $a \leq t^* \leq t_1$. Besides,

$$t_1 - t_0 = -\left(1 + \frac{1}{2}L_f(t_0)\right)\frac{f(t_0)}{f'(t_0)} \leq 0,$$

so that $t_1 \leq t_0 \leq b$. Therefore, $t_1 \in [a, b]$.

After that, by mathematical induction on n, we obtain $a \leq t^* \leq t_n$ and $t_n \leq t_{n-1} \leq b$, for all $n \geq 0$, so that the sequence $\{t_n\}$ lies in $[a, b]$, is decreasing, and is lower bounded by t^*. As a consequence, the sequence $\{t_n\}$ is convergent, and, by letting $n \to \infty$ in (3.12), we obtain, from the uniqueness of t^*, that $\{t_n\}$ converges to t^*.

The case $t_0 \in [a, b]$ with $f(t_0) < 0$ follows analogously. ∎

Notice that the condition $L_f(t) \geq -2$ in $[a, b]$ required in Theorem 3.7 is not very restrictive since $L_f(t^*) = 0$ and $L_f(t) > 0$ in $[t^*, b]$.

Now, if we look for the convergence of the Chebyshev method when $L_{f'}(t) > 3$, we lose the condition that the function T is increasing in $[a, b]$ and, therefore, the monotonous character of the Chebyshev sequence. Thus, we define what an oscillating sequence is.

Definition 3.8 Let $\{s_n\}$ be a sequence that converges to s^*. We say that $\{s_n\}$ oscillates about s^* if $s_{2(n-1)} \geq s_{2n} \geq s^* \geq s_{2n+1} \geq s_{2n-1}$ or $s_{2n-1} \geq s_{2n+1} \geq s^* \geq s_{2n} \geq s_{2(n-1)}$, for $n \in \mathbb{N}$.

Next, we analyze what happens when the sequence is oscillating.

Theorem 3.9 *Let the interval $[a, b]$ be such that $a + \frac{3f(b)}{2f'(b)} \leq b$ and $t_0 \in [a, b]$ such that $f(t_0) > 0$ and $t_0 \geq a + \frac{3f(b)}{2f'(b)}$.*

(a) *If $L_{f'}(t) \in (3, 5]$ and $|L_f(t)| < 1$ in $[a, b]$, then the Chebyshev sequence (3.12) converges to t^* and oscillates about t^*.*
(b) *If $L_{f'}(t) > 5$ and $|L_f(t)| < \sqrt{\frac{2}{M(L_{f'})-3}}$ in $[a, b]$, then the Chebyshev sequence (3.12) converges to t^* and oscillates about t^*.*

Proof We consider item (a). From the hypotheses, we have $t_1 = T(t_0) \leq t_0 \leq b$. Besides,

$$t_1 = T(t_0) > t_0 - \frac{3f(t_0)}{2f'(t_0)} \geq t_0 - \frac{3f(b)}{2f'(b)} \geq a,$$

since $L_f(t_0) \in (0, 1)$ and $\frac{f}{f'}$ is an increasing function in $[a, b]$. Therefore, $t_1 \in [a, b]$.

Moreover, from (3.10) and taking into account that $L_{f'}(t) \in (3, 5]$, it follows that $T'(t) < 0$ and $|T'(t)| < 1$ in $[a, b]$.

3.2 The Chebyshev Method

Furthermore, we observe that $t_0 \geq t^*$, since $f(t_0) > 0$, and

$$t_1 - t^* = T(t_0) - T(t^*) = T'(\theta_0)(t_0 - t^*) \leq 0, \quad \text{with} \quad \theta_0 \in (t^*, t_0),$$

so that $t_1 \leq t^*$ and $|t_1 - t^*| < M(|T'(t)|)|t_0 - t^*| < |t_0 - t^*|$.
In addition,

$$t_2 - t^* = T(t_1) - T(t^*) = T'(\theta_1)(t_1 - t^*) \leq 0, \quad \text{with} \quad \theta_1 \in (t_1, t^*),$$

so that $t_2 \geq t^*$ and $|t_2 - t^*| < M(|T'(t)|)|t_1 - t^*| < M(|T'(t)|)^2|t_0 - t^*| < |t_0 - t^*|$.

After that, by mathematical induction on n, it is easy to prove that $b \geq t_{2n} \geq t^* \geq a$, $a \leq t_{2n+1} \leq t^* \leq b$, and $|t_{n+1} - t^*| < M(|T'(t)|)^{n+1}|t_0 - t^*| < |t_0 - t^*|$, for all $n \geq 0$. As a result, we obtain that $\{|t_n - t^*|\}$ is a decreasing sequence that converges to zero and $\{t_n\} \subset [a, b]$, since $M(|T'(t)|) < 1$. Then, $\{t_n\}$ converges to t^* and oscillates about t^*.

For item (b), we first prove that $t_1 \in [a, b]$ as in Theorem 3.7. Next, from $L_f(t)^2 < \frac{2}{M(L_{f'})-3}$, it is easy to see that $|T'(t)| < 1$ in $[a, b]$. The rest of the proof is analogous to that done in Theorem 3.7. ∎

Remark 3.10 If $t_0 \in [a, b]$ with $f(t_0) < 0$, it is enough to require that $t_0 \geq b + \frac{f(a)}{f'(a)}$ and $a \leq b + \frac{f(a)}{f'(a)}$ be satisfied to obtain the same result as in Theorem 3.9.

Next, we see what happens in practice. That is, given the scalar equation $f(t) = 0$, can we always approximate a root by the Chebyshev method? Let us see if it is. According to the values of $M(L_{f'})$ and $m(L_{f'})$, the following situations, which are drawn in Fig. 3.3, can occur:

(I) $M(L_{f'}) \leq 3$.
(II) $M(L_{f'}) \leq 5$ and $m(L_{f'}) > 3$.
(III) $3 < M(L_{f'}) \leq 5$ and $m(L_{f'}) \leq 3$.
(IV) $M(L_{f'}) > 5$ and $m(L_{f'}) \leq 5$.
(V) $m(L_{f'}) > 5$.

Observe that, in cases (I), (II), and (V), the convergence of the Chebyshev sequence (3.12) follows from Theorems 3.7 and 3.9 under certain conditions on L_f in $[a, b]$. In cases (III) and (IV), we know nothing about the growth of the function T, so that we can only guarantee the convergence of (3.12) to the root t^*. Both cases are contemplated in Theorem 3.12, which is given below.

In any case, as in the case in which $\{t_n\}$ is a sequence that oscillates about t^*, we have to guarantee that the sequence $\{t_n\}$ lies in the interval $[a, b]$. Here, we can make two considerations. The first is the same as that made in the oscillating case. The second

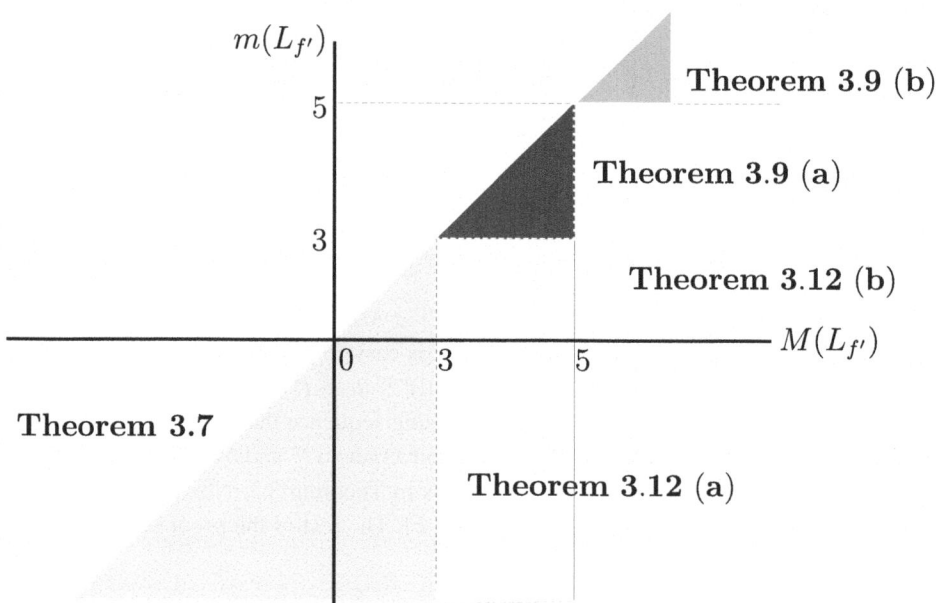

Fig. 3.3 Application of the Chebyshev method

consists of finding an interval in which any point on it can be the starting point for the sequence $\{t_n\}$. The following lemma refers to the last consideration.

Lemma 3.11 *Suppose $|T'(t)| < 1$ and $|L_f(t)| < 1$ in $[a, b]$.*

(a) *Let $t_0, \gamma \in [a, b]$ be such that $f(\gamma) < 0$ and $f(t_0) > 0$. If $\frac{3f(b)}{2f'(b)} \leq \gamma - a$, then $t_n \in [a, b]$, for all $n \in \mathbb{N}$.*
(b) *Let $t_0, \beta \in [a, b]$ be such that $f(\beta) > 0$ and $f(t_0) < 0$. If $\frac{f(a)}{f'(a)} \geq \beta - b$, then $t_n \in [a, b]$, for all $n \in \mathbb{N}$.*

Proof We consider item (a). First, it is clear that $t_1 < t_0$ and

$$t_1 = T(t_0) \geq t_0 - \frac{3f(t_0)}{2f'(t_0)} \geq t_0 - \frac{3f(b)}{2f'(b)} \geq t_0 - \gamma + a \geq a,$$

since $L_f(t) < 1$ and $\frac{f}{f'}$ is an increasing function in $[a, b]$. Thus, $t_1 \in [a, t_0) \subset [a, b]$. Next, we distinguish two cases: $t_1 < t^*$ and $t_1 > t^*$.

If $t_1 < t^*$, then $L_f(t_1) \in (-1, 0)$ and $t_1 < t_2$, since $f(t_1) < 0$. Besides, as $|T'(t)| < 1$, we have $|t_2 - t^*| < |t_1 - t^*| < |t_0 - t^*|$. Hence, $t_2 \in (t_1, t_0) \subset [a, b]$. After that, by mathematical induction on n, it follows that $t_n \in (t_1, t_0) \subset [a, b]$, for all $n \in \mathbb{N}$.

If $t_1 > t^*$, then $t^* < t_1 < t_0$ and

3.2 The Chebyshev Method

$$t_2 = T(t_1) \geq t_1 - \frac{3f(t_1)}{2f'(t_1)} \geq t_1 - \frac{3f(b)}{2f'(b)} \geq t_1 - \gamma + a \geq a,$$

so that $t_2 \in [a, t_1) \subset [a, t_0) \subset [a, b]$. Now, we again distinguish two cases.

On the one hand, if there exists any $t_k \in (t^*, b]$ with $k = 1, 2, \ldots, n$, we can prove analogously to the previous case that $t_{n+1} \in [a, t_n) \subset [a, t_0) \subset [a, b]$ and, as a result, $t_n \in [a, b]$, for all $n \in \mathbb{N}$.

On the other hand, if there exists any $t_n \in [a, t^*)$ with $t_k \in (t^*, b]$, for $k = 1, 2, \ldots, n-1$, we can prove as before that $t_{n+1} \in (t_n, t_{n-1}) \subset (t_n, t_0) \subset [a, b]$, so that $t_n \in [a, b]$, for all $n \in \mathbb{N}$.

Item (b) follows analogously. ∎

Theorem 3.12 *Let $t_0, \gamma, \beta \in [a, b]$ satisfy the conditions of Lemma 3.11. Suppose $M(|L_f|) < \sqrt{\frac{2}{\ell}}$ with $\ell = \max\{|3 - M(L_{f'})|, |3 - m(L_{f'})|\}$.*

(a) *If $M(L_{f'}) \in (3, 5]$, $m(L_{f'}) \leq 3$, and $|L_f(t)| < 1$ in $[a, b]$, then the Chebyshev sequence (3.12) converges to t^*.*
(b) *If $M(L_{f'}) > 5$, $m(L_{f'}) \leq 5$ in $[a, b]$, then the Chebyshev sequence (3.12) converges to t^*.*

Proof We consider item (a). First, we see that the sequence $\{t_n\}$ lies in $[a, b]$. For this, we prove that $|T'(t)| < 1$ in $[a, b]$ and apply Lemma 3.11. Indeed, $|T'(t)| < \frac{\ell}{2} L_f(t)^2 < 1$ in $[a, b]$ by hypotheses. Moreover, $|t_1 - t^*| < M(|T'(t)|)|t_0 - t^*| < |t_0 - t^*|$. If we suppose that

$$|t_k - t^*| < M(|T'(t)|)^k |t_0 - t^*|, \quad \text{for} \quad k = 1, 2, \ldots, n-1,$$

then

$$|t_n - t^*| < M(|T'(t)|)|t_{n-1} - t^*| < M(|T'(t)|)^n |t_0 - t^*| < |t_0 - t^*|.$$

In addition, $\lim_n |t_n - t^*| = 0$, since $M(|T'(t)|) < 1$, and $\{t_n\}$ converges to t^*.

To prove item (b), it is enough to take into account that $\ell \geq 2$ and, therefore, $|L_f(t)| < 1$ in $[a, b]$. The convergence of the sequence $\{t_n\}$ follows as in item (a). ∎

The novelty of the previous result is the characterization of the convergence under more complicated situations, such as the ignorance of the growth of T and obtaining conditions that lead us to $|T'(t)| < 1$ in an environment of the root t^*. From this result, we see that we can always solve the scalar equation $f(t) = 0$ by the Chebyshev method for all values of $M(L_{f'})$ and $m(L_{f'})$ under certain restrictions on L_f in $[a, b]$.

Next, we illustrate the previous analysis of the convergence of the Chebyshev method with an example.

Table 3.5 The Chebyshev method starting at $t_0 = 5$ from Example 3.13

n	t_n
0	5
1	3.026590097604989...
2	2.219791233475883...
3	2.095410600600519...
4	2.094551481886922...
5	2.094551481542327...

Table 3.6 The Chebyshev method starting at $t_0 = 1.5$ from Example 3.13

n	t_n
0	1.5
1	1.575521213004811...
2	1.828109473941022...
3	2.076617885516301...
4	2.094548236158903...
5	2.094551481542327...

Example 3.13 Consider the cubic polynomial $f(t) = t^3 - 2t - 5$ introduced by Ostrowski in [76].

If we choose $[a, b] = [1, 5]$, we observe that $L_{f'}(t) < \frac{1}{2}$. Taking then $t_0 = 5$ for Theorem 3.7, we obtain that the Chebyshev method converges to the solution $t^* = 2.094551481542327\ldots$ of $f(t) = 0$, as we can see in Table 3.5.

On the other hand, if $t_0 = 1.5$, it is then necessary that $L_f(t) \geq -2$ in $[a, b]$. So, in this situation, we consider $[a, b] = [1.5, 5]$ and obtain the iterations given in Table 3.6.

3.2.2 The Chebyshev Method in Banach Spaces

In this section, we develop the technique used in Sect. 2.2.2.1 to prove the semilocal convergence of the Newton method, which is based on the method of majorizing sequences, for the Chebyshev method when it is applied to solve an operator equation $F(x) = 0$ in Banach spaces. For this, we consider two Banach spaces X and Y and the nonlinear operator $F : \Omega \subset X \to Y$, which is three times continuously Fréchet differentiable and defined on a non-empty open convex domain Ω of the Banach space X. The Chebyshev method is defined in Banach spaces by

$$x_{n+1} = T(x_n) = x_n - \left(I + \frac{1}{2}L_F(x_n)\right)[F'(x_n)]^{-1}F(x_n), \quad n \geq 0, \text{ with } x_0 \text{ given in } \Omega, \tag{3.13}$$

where I denotes the identity operator defined on X.

3.2 The Chebyshev Method

Remember that the idea of the method of majorizing sequences is to prove the convergence of a sequence in a Banach space from the convergence of a scalar sequence. In Sect. 2.2.2.1, we see the importance of the scalar function involved in the definition of a majorizing sequence for the Newton method. For this, we introduced a new concept of majorizing function, which allows defining majorizing sequences from majorizing functions. If we now consider the Chebyshev method, we can define the majorant function as follows.

Definition 3.14 Let $f \in \mathcal{C}^3([t_0, +\infty))$ be a scalar function and the Chebyshev method (3.12) in $\mathbb{R}$. If (3.12) is a majorizing sequence of the Chebyshev sequence (3.13) defined in the Banach space X for an operator F, then the scalar function f involved in (3.12) is a majorant function of the operator F.

Once the majorant function is defined, we give first a result that provides conditions that a scalar function f must satisfy to be a majorant function of the operator F for the Chebyshev method. Thus, the main idea is to construct a scalar function $f \in \mathcal{C}^3([t_0, +\infty))$, with $t_0 \in \mathbb{R}$, that satisfies:

(L1) There exists $\Gamma_0 = [F'(x_0)]^{-1} \in \mathcal{L}(Y, X)$, for some $x_0 \in \Omega$, with $\|\Gamma_0\| \leq -\frac{1}{f'(t_0)}$ and $\|\Gamma_0 F(x_0)\| \leq -\frac{f(t_0)}{f'(t_0)}$. Besides, $\|F''(x_0)\| \leq f''(t_0)$.

(L2) $\|F'''(x)\| \leq f'''(t)$, for $\|x - x_0\| \leq t - t_0$, for $x \in \Omega$ and $t \in [t_0, +\infty)$.

As we can see in [42] for the Newton method, the scalar equation $f(t) = 0$ must have at least one solution t^* larger than t_0, so that the majorizing sequence constructed from f converges to t^* from t_0. So, we first study the function f and give some properties.

Theorem 3.15 *Let $f \in \mathcal{C}^3([t_0, +\infty))$ and $t_0 \in \mathbb{R}$ be such that $f(t_0) > 0$, $f'(t_0) < 0$, $f''(t_0) > 0$, and $f'''(t) > 0$.*

(a) *If there exists a root γ of $f'(t) = 0$ such that $\gamma > t_0$, then γ is the unique minimum of $f(t)$ in $[t_0, +\infty)$ and $f(t)$ is nonincreasing in $[t_0, \gamma)$.*

(b) *If $f(\gamma) \leq 0$, then $f(t) = 0$ has at least one root in $[t_0, +\infty)$. Besides, if t^* is the smallest root of $f(t) = 0$ in $[t_0, +\infty)$, then $t_0 < t^* \leq \gamma$.*

Proof As $f'''(t) > 0$ in $[t_0, +\infty)$, then $f''(t)$ is nondecreasing in $[t_0, +\infty)$. Moreover, as $f''(t_0) > 0$, then $f''(t) > 0$ in $[t_0, +\infty)$ and f is convex in $[t_0, +\infty)$. Furthermore, since $f'(\gamma) = 0$ with $\gamma > t_0$, then γ is the unique minimum of f in $[t_0, +\infty)$.

Now, as $f''(t) > 0$ in $[t_0, +\infty)$, then f' is nondecreasing in $[t_0, \gamma)$. In addition, $t_0 < \gamma$ and $f'(t) < 0$ in $[t_0, \gamma)$, since $f'(t_0) < 0$ and $f'(\gamma) = 0$. Thus, f is nonincreasing in $[t_0, \gamma)$.

After that, as $f(t_0) > 0$, f has at least one zero t^* in (t_0, γ) if $f(\gamma) < 0$. Besides, t^* is the unique zero of f in $[t_0, \gamma)$, since f is nonincreasing in $[t_0, \gamma)$. ∎

Further, we establish the convergence of the sequence (3.12) in the following theorem.

Theorem 3.16 *Consider the scalar sequence (3.12) with $f \in C^3([t_0, +\infty))$ and $t_0 \in \mathbb{R}$. Suppose that the conditions (L1)–(L2) are satisfied and there exists a root $\gamma \in (t_0, +\infty)$ of $f'(t) = 0$ such that $f(\gamma) \leq 0$. Then, the sequence (3.12) is nondecreasing and converges to t^*.*

Proof First, we prove that the scalar sequence (3.12) is nondecreasing and convergent to t^*. For this, we see that (3.12) is monotonous and bounded.

As $f(t) > 0$, $f'(t) < 0$, and $f''(t) > 0$ in (t_0, γ), we see by mathematical induction on n that $t_n \leq t^*$, for all $n \geq 0$. Since $f(t_0) > 0$, then $t_0 \leq t^*$ and

$$t_1 - t^* = T(t_0) - T(t^*) = T'(\xi_0)(t_0 - t^*), \quad \xi_0 \in (t_0, t^*).$$

Taking now into account that $T'(t) = \frac{1}{2} L_f(t)^2 \left(3 - L_{f'}(t)\right)$ in $[t_0, t^*)$ as a consequence of $f(t) > 0$ and $f''(t) > 0$ in $[t_0, t^*)$, we obtain $t_1 < t^*$. Then, we suppose that $t_n < t^*$, so that

$$t_{n+1} - t^* = T(t_n) - T(t^*) = T'(\xi_n)(t_n - t^*), \quad \xi_n \in (t_n, t^*),$$

and $t_{n+1} < t^*$. Therefore, $t_n < t^*$ is true for all positive integers n by mathematical induction.

Besides,

$$t_{n+1} - t_n = -\left(1 + \frac{1}{2} L_f(t_n)\right) \frac{f(t_n)}{f'(t_n)} \geq 0,$$

since $f(t) > 0$, $f'(t) < 0$, and $f''(t) > 0$ in $[t_0, t^*)$, so that the sequence (3.12) is nondecreasing.

Finally, we infer that there exists $s = \lim_n t_n \in [t_0, t^*]$ and, by the continuity of T, s is a root of $f(t) = 0$. In addition, $s = t^*$, since t^* is the unique zero of f in $[t_0, t^*]$. ∎

Now, we prove that (3.12) is a majorizing sequence of the Chebyshev method (3.13), defined in the Banach space X, and this sequence is well defined, provided that $B(x_0, t^* - t_0) \subseteq \Omega$.

Lemma 3.17 *Suppose that there exists $f \in C^3([t_0, +\infty))$ such that conditions (L1)-(L2) are satisfied with $t_0 \in \mathbb{R}$. Suppose that $x_n \in \Omega$, for all $n \geq 0$, and $f(\gamma) \leq 0$, where γ is a root of $f'(t) = 0$ in $(t_0, +\infty)$. Then, for all $n \in \mathbb{N}$, we have:*

(i_n) *There exists $\Gamma_n = [F'(x_n)]^{-1}$ and $\|\Gamma_n\| \leq -\dfrac{1}{f'(t_n)}$.*

(ii_n) *$\|F''(x_n)\| \leq f''(t_n)$.*

3.2 The Chebyshev Method

(iii_n) $\|F(x_n)\| \leq f(t_n)$.
(iv_n) $\|x_{n+1} - x_n\| \leq t_{n+1} - t_n$.

Proof We prove (i_n)-(ii_n)-(iii_n)-(iv_n) by mathematical induction on n. As the step $n = 1$ and the inductive step are analogous, we only develop the inductive step. So, we suppose that (i_j)-(ii_j)-(iii_j)-(iv_j) are true for $j = 1, 2, \ldots, n$ and prove that (i_{n+1})-(ii_{n+1})-(iii_{n+1})-(iv_{n+1}) are also true.

Taking into account Taylor's formula

$$F'(x_{n+1}) = F'(x_n) + F''(x_n)(x_{n+1} - x_n) + \int_{x_n}^{x_{n+1}} F'''(x)(x_{n+1} - x)\, dx$$

and denoting $x = x_n + \tau(x_{n+1} - x_n)$ and $t = t_n + \tau(t_{n+1} - t_n)$ with $\tau \in [0, 1]$, so that

$$\begin{aligned}\|x - x_0\| &\leq \tau \|x_{n+1} - x_n\| + \|x_n - x_{n-1}\| + \cdots + \|x_1 - x_0\| \\ &\leq \tau(t_{n+1} - t_n) + t_n - t_{n-1} + \cdots + t_1 - t_0 \\ &= t - t_0,\end{aligned}$$

we obtain

$$\begin{aligned}\|I - \Gamma_n F'(x_{n+1})\| &\leq \|\Gamma_n\| \Big(\|F''(x_n)\| \|x_{n+1} - x_n\| \\ &\quad + \int_0^1 \|F'''(x_n + \tau(x_{n+1} - x_n))\| \|x_{n+1} - x_n\|^2 (1 - \tau)\, d\tau \Big) \\ &\leq -\frac{1}{f'(t_n)} \Big(f''(t_n)(t_{n+1} - t_n) \\ &\quad + \int_0^1 f'''(t_n + \tau(t_{n+1} - t_n))(t_{n+1} - t_n)^2 (1 - \tau)\, d\tau \Big) \\ &= -\frac{f'(t_{n+1}) - f'(t_n)}{f'(t_n)} \\ &= 1 - \frac{f'(t_{n+1})}{f'(t_n)} \\ &< 1,\end{aligned}$$

since $t_n < t_{n+1} \leq t^*$, $f''(t) > 0$ in $[t_0, +\infty)$ and $\frac{f'(t_{n+1})}{f'(t_n)} < 1$. Then, by the Banach lemma on invertible operators, the operator Γ_{n+1} exists and is such that $\|\Gamma_{n+1}\| \leq -\frac{1}{f'(t_{n+1})}$.

Taking now into account that

$$F''(x_{n+1}) = F''(x_n) + \int_0^1 F'''(x_n + \tau(x_{n+1} - x_n))(x_{n+1} - x_n)\,d\tau,$$

it follows

$$\|F''(x_{n+1})\| \leq f''(t_n) + \int_0^1 f'''(t_n + \tau(t_{n+1} - t_n))(t_{n+1} - t_n)\,d\tau = f''(t_{n+1}),$$

since $\|x - x_0\| \leq t - t_0$ with $x = x_n + \tau(x_{n+1} - x_n)$, $t = t_n + \tau(t_{n+1} - t_n)$, and $\tau \in [0, 1]$. Taking next into account Taylor's formula

$$F(x_{n+1}) = F(x_n) + F'(x_n)(x_{n+1} - x_n) + \frac{1}{2}F''(x_n)(x_{n+1} - x_n)^2$$
$$+ \frac{1}{2}\int_{x_n}^{x_{n+1}} F'''(x)(x_{n+1} - x)^2\,dx$$

and (3.13), we have

$$F(x_{n+1}) = \frac{1}{2}F''(x_n)\Gamma_n F(x_n)L_F(x_n)\Gamma_n F(x_n) + \frac{1}{8}F''(x_n)\left(L_F(x_n)\Gamma_n F(x_n)\right)^2$$
$$+ \frac{1}{2}\int_0^1 F'''(x_n + \tau(x_{n+1} - x_n))(x_{n+1} - x_n)^3(1 - \tau)^2\,d\tau.$$

Since $\|L_F(x_n)\| \leq \|\Gamma_n\|\|F''(x_n)\|\|\Gamma_n\|\|F(x_n)\| \leq L_f(t_n)$ and $\|x - x_0\| \leq t - t_0$, where $x = x_n + \tau(x_{n+1} - x_n)$, $t = t_n + \tau(t_{n+1} - t_n)$, and $\tau \in [0, 1]$, then

$$\|F(x_{n+1})\| = \frac{1}{2}f(t_n)L_f(t_n)^2 + \frac{1}{8}f(t_n)L_f(t_n)^3$$
$$+ \frac{1}{2}\int_0^1 f'''(t_n + \tau(t_{n+1} - t_n))(t_{n+1} - t_n)^3(1 - \tau)^2\,d\tau$$
$$= f(t_{n+1}).$$

Moreover,

$$\|x_{n+2} - x_{n+1}\| \leq \left(1 + \frac{1}{2}\|L_F(x_{n+1})\|\right)\|\Gamma_{n+1}\|\|F(x_{n+1})\|$$
$$\leq -\left(1 + \frac{1}{2}L_f(t_{n+1})\right)\frac{f(t_{n+1})}{f'(t_{n+1})}$$
$$= t_{n+2} - t_{n+1}.$$

3.2 The Chebyshev Method

Thus, by mathematical induction on n, we conclude that the items (i_n)-(ii_n)-(iii_n)-(iv_n) are true for all positive integers n. ∎

After that, we prove the semilocal convergence of the Chebyshev method (3.13) in X from the fact that the sequence (3.12) is a majorizing sequence of (3.13).

Theorem 3.18 (General Semilocal Convergence) *Let $F : \Omega \subseteq X \longrightarrow Y$ be a three times continuously Fréchet differentiable operator defined on a non-empty open convex domain Ω of a Banach space X with values in a Banach space Y. Suppose that there exists $f \in \mathcal{C}^3([t_0, +\infty))$ such that (L1)-(L2) are satisfied. Suppose also that there exists $\gamma \in (t_0, +\infty)$, such that $f'(\gamma) = 0$ and $f(\gamma) \leq 0$, and $\overline{B(x_0, t^* - t_0)} \subset \Omega$. Then, the Chebyshev sequence (3.13) starting at x_0 converges to a root x^* of $F(x) = 0$. Moreover, $x_n, x^* \in \overline{B(x_0, t^* - t_0)}$ and*

$$\|x^* - x_n\| \leq t^* - t_n, \quad \text{for all} \quad n \geq 0,$$

where $\{t_n\}$ is the Chebyshev sequence (3.12).

Proof It is easy to see that x_1 is well defined and $\|x_1 - x_0\| = t_1 - t_0 < t^* - t_0$ from (L1)-(L2). Thus, $x_1 \in B(x_0, t^* - t_0) \subset \Omega$. We now suppose that x_i is well defined and $x_i \in B(x_0, t^* - t_0)$ for $i = 1, 2, \ldots, n$.

From Lemma 3.17, we have that there exists Γ_n, $\|\Gamma_n\| \leq -\dfrac{1}{f'(t_n)}$, $\|F''(x_n)\| \leq f''(t_n)$, and $\|L_F(x_n)\| \leq L_f(t_n)$. Hence, x_{n+1} is well defined. Now, as $\|x_{i+1} - x_i\| \leq t_{i+1} - t_i$, for $i = 1, 2, \ldots, n$, we obtain

$$\|x_{n+1} - x_0\| \leq \sum_{j=0}^{n+1} \|x_{j+1} - x_j\| \leq \sum_{j=0}^{n+1}(t_{j+1} - t_j) = t_{n+1} - t_0 < t^* - t_0,$$

and then $x_{n+1} \in B(x_0, t^* - t_0)$. As a consequence, the Chebyshev sequence (3.13) is well defined and $x_n \in B(x_0, t^* - t_0)$ for all $n \in \mathbb{N}$.

In addition, from Lemma 3.17, it follows that

$$\|x_{n+1} - x_n\| \leq \left\|I + \frac{1}{2}L_F(x_n)\right\| \|\Gamma_n\| \|F(x_n)\| \leq -\left(1 + \frac{1}{2}L_f(t_n)\right)\frac{f(t_n)}{f'(t_n)} = t_{n+1} - t_n,$$

for all $n \geq 0$, since $\|F(x_n)\| \leq f(t_n)$ and $\|L_F(x_n)\| \leq L_f(t_n)$. Hence, $\{t_n\}$ is a majorizing sequence of $\{x_n\}$ and convergent. Besides, as $\lim_n t_n = t^*$, we have $\|x^* - x_n\| \leq t^* - t_n$, for all $n \geq 0$, if $x^* = \lim_n x_n$. Finally, from $\|F(x_n)\| \leq f(t_n)$, for all $n \geq 0$, it follows that $F(x^*) = 0$ by letting $n \to +\infty$ and the continuity of F. ∎

We have supposed above that the function $f \in \mathcal{C}^3([t_0, +\infty))$ exists and satisfies conditions (L1)-(L2) with $t_0 \in \mathbb{R}$. Below, we see the existence of such a function, so that we can prove the semilocal convergence of the Chebyshev method (3.13), in particular for operators with ω-bounded third derivative. For this, we first suppose the following conditions:

(N1) There exists $\Gamma_0 = [F'(x_0)]^{-1} \in \mathcal{L}(Y, X)$, for some $x_0 \in \Omega$, with $\|\Gamma_0\| \leq \beta$ and $\|\Gamma_0 F(x_0)\| \leq \eta$. Besides, $\|F''(x_0)\| \leq \delta$.

(N2) There exists a continuous and nondecreasing function $\omega : [0, +\infty) \longrightarrow \mathbb{R}$ such that $\|F'''(x)\| \leq \omega(\|x\|)$, for $\|x - x_0\| \leq t - t_0$, and $\omega(0) = 0$.

Now, from (N2), we have

$$\|F'''(x)\| \leq \omega(\|x\|) \leq \omega(t - t_0 + \|x_0\|),$$

since $\|x\| \leq \|x_0\| + \|x - x_0\|$. From this point on, we consider $\varpi(t; \|x_0\|, t_0) = \omega(t - t_0 + \|x_0\|)$, for $\|x - x_0\| \leq t - t_0$, where $\varpi : [t_0, +\infty) \longrightarrow \mathbb{R}$ is a continuous nondecreasing function such that $\varpi(t_0; \|x_0\|, t_0) \geq 0$.

After that, we find a scalar function f from conditions (N1)-(N2) that solves the next initial value problem:

$$\begin{cases} y'''(t) = \varpi(t; \|x_0\|, t_0), \\ y(t_0) = \dfrac{\eta}{\beta}, \quad y'(t_0) = -\dfrac{1}{\beta}, \quad y''(t_0) = \delta. \end{cases} \quad (3.14)$$

We give the solution function of (3.14) in the next theorem.

Theorem 3.19 *For any nonnegative scalar numbers $\beta \neq 0$, η, and δ, there exists only one solution function $f(t)$ of the initial value problem (3.14) in $[t_0, +\infty)$; that is,*

$$f(t) = \int_{t_0}^{t} \int_{t_0}^{v} \int_{t_0}^{u} \varpi(s; \|x_0\|, t_0) \, ds \, du \, dv + \frac{\delta}{2!}(t - t_0)^2 - \frac{t - t_0}{\beta} + \frac{\eta}{\beta}. \quad (3.15)$$

Note that (3.15) is a polynomial of degree three if ϖ is constant. In particular, if $\varpi(s; \|x_0\|, t_0) = \ell = $ constant, (3.15) is reduced to

$$f(t) = \frac{\ell}{3!}(t - t_0)^3 + \frac{\delta}{2!}(t - t_0)^2 - \frac{t - t_0}{\beta} + \frac{\eta}{\beta}.$$

Next, in order to apply Theorem 3.18, the scalar equation $f(t) = 0$ must have at least one solution larger than the initial value t_0. Thus, we can guarantee the convergence of the Chebyshev sequence (3.12) starting at t_0 to this solution. Then, we analyze first the function (3.15).

3.2 The Chebyshev Method

Theorem 3.20 *Let f be the scalar function (3.15).*

(a) If there exists only one positive root $\gamma > t_0$ of the equation $f'(t) = 0$, then γ is the unique minimum of $f(t)$ in $[t_0, +\infty)$ and $f(t)$ is nonincreasing in $[t_0, \gamma)$.

(b) If $f(\gamma) \leq 0$, then the equation $f(t) = 0$ has at least one root in $[t_0, +\infty)$. Besides, if t^* is the smallest root of $f(t) = 0$ in $[t_0, +\infty)$, we have $t_0 < t^* \leq \gamma$.

As the scalar function (3.15) satisfies conditions (L1)-(L2), the proof of Theorem 3.20 follows immediately from Theorem 3.15.

Observe that the scalar function (3.15) satisfies the conditions of Theorem 3.18, so that the convergence of the Chebyshev method is guaranteed in the Banach space X. In particular, we establish the next theorem, whose proof is immediate from Theorem 3.18. First, we suppose that:

(N3) $f(\gamma) \leq 0$, where f is the scalar function (3.15) and γ is the unique positive root of $f'(t) = 0$ such that $\gamma > t_0$, and $\overline{B(x_0, t^* - t_0)} \subset \Omega$, where t^* is the smallest positive root of $f(t) = 0$ in $[t_0, +\infty)$.

Theorem 3.21 *Let $F : \Omega \subseteq X \longrightarrow Y$ be a three times continuously Fréchet differentiable operator defined on a non-empty open convex domain Ω of a Banach space X with values in a Banach space Y and $f(t)$ the scalar function (3.15). Suppose that conditions (N1)-(N2)-(N3) are satisfied. Then, the Chebyshev sequence (3.13) starting at x_0 converges to a root x^* of $F(x) = 0$. Besides, $x_n, x^* \in \overline{B(x_0, t^* - t_0)}$ and*

$$\|x^* - x_n\| \leq t^* - t_n, \quad \text{for all} \quad n \geq 0,$$

where $\{t_n\}$ is the Chebyshev sequence (3.12).

Finally, we note that we can do a study analogous to the previous one if the function ω is nonincreasing. In [42], we can see this study for the Newton method.

Next, we illustrate the previous study with a boundary value problem.

Example 3.22 Consider the boundary value problem given by

$$y''(t) + y(t)^3 = 2, \quad y(0) = 0 = y(1). \tag{3.16}$$

We divide the interval $[0, 1]$ into $m + 1$ subintervals of length $h = \frac{1}{m+1}$ and introduce the points $t_j = jh$, with $j = 0, 1, \ldots, m + 1$, to solve the problem numerically. The corresponding values of the function are

$$y_0 = y(t_0), y_1 = y(t_1), \ldots, y_{m+1} = y(t_{m+1}).$$

A standard approximation of the first derivative at these points is

$$y''(t_j) \approx \frac{y_{j-1} - 2y_j + y_{j+1}}{h^2}, \qquad j = 1, 2, \ldots, m,$$

so that, by following a process of discretization, we obtain the following nonlinear system:

$$-2y_1 + h^2 y_1^3 + y_2 = 0,$$
$$y_{i-1} - 2y_i + h^2 y_i^3 + y_{i+1} = 0, \qquad i = 2, 3, \ldots m-1,$$
$$y_{m-1} - 2y_m + h^2 y_m^3 = 0.$$

Observe that $y_0 = y_{m+1} = 0$, since y_0 and y_{m+1} are determined by the boundary conditions. Thus, the unknowns are $y_1, y_2, \ldots, y_m$.

We then define the function $F : \mathbb{R}^m \to \mathbb{R}^m$ by

$$F(\mathbf{y}) = A\mathbf{y} + h^2 \hat{\mathbf{y}} - 2h^2 \mathbf{u},$$

where

$$A = \begin{pmatrix} -2 & 1 & 0 & \cdots & 0 \\ 1 & -2 & 1 & \cdots & 0 \\ 0 & 1 & -2 & \cdots & 0 \\ \vdots & \vdots & \vdots & \ddots & \vdots \\ 0 & 0 & 0 & \cdots & -2 \end{pmatrix}_{m \times m},$$

$\mathbf{y} = (y_1, y_2, \ldots, y_m)^T$, $\hat{\mathbf{y}} = (y_1^3, y_2^3, \ldots, y_m^3)^T$, and $\mathbf{u} = (1, 1, \ldots, 1)^T$.

In addition,

$$F'(\mathbf{y}) = A + 3h^2 \operatorname{diag}\left\{y_1^2, y_2^2, \ldots, y_m^2\right\},$$

$$F''(\mathbf{y})\mathbf{z}\mathbf{w} = 6h^2(y_1 z_1 w_1, y_2 z_2 w_2, \ldots, y_m z_m w_m),$$

$$F'''(\mathbf{y})\mathbf{z}\mathbf{w}\mathbf{p} = 6h^2(z_1 w_1 p_1, z_2 w_2 p_2, \ldots, z_m w_m p_m),$$

where $\mathbf{z} = (z_1, z_2, \ldots, z_m)$, $\mathbf{w} = (w_1, w_2, \ldots, w_m)$, and $\mathbf{p} = (p_1, p_2, \ldots, p_m)$. Taking the norms of $\mathbf{x} \in \mathbb{R}^m$ and $A \in \mathbb{R}^m \times \mathbb{R}^m$ as

$$\|x\| = \max_{1 \leq i \leq m} |x_i|, \qquad \|A\| = \max_{1 \leq i \leq m} \sum_{k=1}^{m} |a_{ik}|, \qquad (3.17)$$

respectively, we have

3.2 The Chebyshev Method

$$\|F'''(\mathbf{y})\| = \sup_{\|\mathbf{z}\|=1, \|\mathbf{w}\|=1, \|\mathbf{p}\|=1} \|F'''(\mathbf{y})\mathbf{zwp}\| \leq 6h^2,$$

so that $\omega(t) = 6h^2$ and $\varpi(t; \|x_0\|, t_0) = \omega(t - t_0 + \|x_0\|) = 6h^2$.

We choose the initial approximation $\frac{\sin \pi x}{3}$ which gives the starting vector

$$\mathbf{x}_0 = \begin{pmatrix} 0.1030\ldots \\ 0.1959\ldots \\ 0.2696\ldots \\ 0.3170\ldots \\ 0.3333\ldots \\ 0.3170\ldots \\ 0.2696\ldots \\ 0.1959\ldots \\ 0.1030\ldots \end{pmatrix}$$

for the Chebyshev method if $\mathfrak{m} = 9$. We then have

$$\beta = \|[F'(\mathbf{x}_0)]^{-1}\| = 12.8464\ldots, \quad \eta = \|[F'(\mathbf{x}_0)]^{-1} F(\mathbf{x}_0)\| = 0.5963\ldots,$$

$$\delta = \|F''(\mathbf{x}_0)\| = 0.02.$$

Besides, if $t_0 = 0$, the function (3.15) is reduced to the third-degree polynomial

$$f(t) = \frac{t^3}{100} + \frac{t^2}{100} - (0.0778\ldots)t + (0.0464\ldots),$$

which has two positive zeros $t^* = 0.7052\ldots$ and $t^{**} = 1.8507\ldots$ Moreover, the unique minimum of $f(t)$ in $[0, +\infty)$ is $\gamma = 1.3116\ldots$, and it is such that $f(\gamma) = -0.0159\ldots \leq 0$. Hence, the convergence of the Chebyshev method to a solution $\mathbf{y}^*$ is guaranteed from Theorem 3.21. Furthermore, the domain of existence of solution is

$$\{\mathbf{v} \in \mathbb{R}^9 : \|\mathbf{v} - \mathbf{x}_0\| \leq t^* = 0.7052\ldots\}.$$

In particular, the Chebyshev method converges to the solution $\mathbf{y}^* = (y_1^*, y_2^*, \ldots, y_9^*)^T$ given in Table 3.7 after three iterations and using the stopping criterion $\|\mathbf{y}_n - \mathbf{y}_{n-1}\| < 10^{-16}$.

In Table 3.8 we show the errors $\|\mathbf{y}^* - \mathbf{y}_n\|$ using the stopping criterion $\|\mathbf{y}^* - \mathbf{y}_n\| < 10^{-16}$ and the sequence $\{\|F(\mathbf{y}_n)\|\}$. From the last, we notice that the vector shown in Table 3.7 is a good approximation of the solution of the system $F(\mathbf{y}) = 0$ with $\mathfrak{m} = 9$, since $\|F(\mathbf{y}^*)\| \leq \text{constant} \times 10^{-16}$. See the sequence $\{\|F(\mathbf{y}_n)\|\}$ in Table 3.8.

Table 3.7 The numerical solution $\mathbf{y}^* = (y_1^*, y_2^*, \ldots, y_9^*)^T$ of the system $F(\mathbf{y}) = 0$ from Example 3.22

i	y_i^*	i	y_i^*	i	y_i^*
1	$-0.0903626\ldots$	4	$-0.2412515\ldots$	7	$-0.2110317\ldots$
2	$-0.1607179\ldots$	5	$-0.2513309\ldots$	8	$-0.1607179\ldots$
3	$-0.2110317\ldots$	6	$-0.2412515\ldots$	9	$-0.0903626\ldots$

Table 3.8 Absolute errors, a priori error estimates, and $\{\|F(\mathbf{y}_n)\|\}$ from Example 3.22

n	$\|\mathbf{y}^* - \mathbf{y}_n\|$	$\|t^* - t_n\|$	$\|F(\mathbf{y}_n)\|$
0	$5.8466\ldots \times 10^{-1}$	$7.0529\ldots \times 10^{-1}$	$3.2258\ldots \times 10^{-2}$
1	$1.7810\ldots \times 10^{-2}$	$6.3292\ldots \times 10^{-2}$	$2.1641\ldots \times 10^{-3}$
2	$4.6426\ldots \times 10^{-7}$	$2.0537\ldots \times 10^{-4}$	$6.0856\ldots \times 10^{-8}$
3	$8.1601\ldots \times 10^{-21}$	$8.8269\ldots \times 10^{-12}$	$1.0891\ldots \times 10^{-21}$

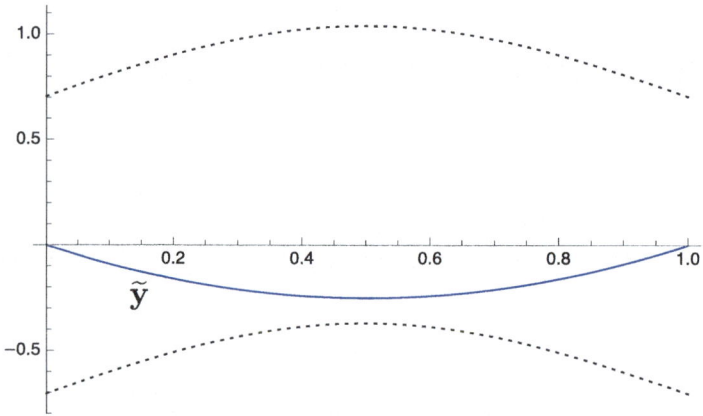

Fig. 3.4 The approximate solution $\widetilde{\mathbf{y}}$ of problem (3.16) (solid line) and the domain of existence of solution (dotted lines)

Finally, by interpolating the values of Table 3.7 and taking into account the boundary conditions, we obtain the solution drawn in Fig. 3.4 which is denoted by $\widetilde{\mathbf{y}}$. Observe that the approximate solution $\widetilde{\mathbf{y}}$ lies within the domain of existence of solution $B\left(\frac{\sin \pi x}{3}, 0.7052\ldots\right)$.

3.3 The Super-Halley Method

To solve the nonlinear scalar equation $f(t) = 0$, we consider the Super-Halley method

3.3 The Super-Halley Method

$$t_{n+1} = S(t_n) = t_n - \left(1 + \frac{L_f(t_n)}{2(1 - L_f(t_n))}\right) \frac{f(t_n)}{f'(t_n)}, \quad n \geq 0, \quad \text{with } t_0 \text{ given,} \quad (3.18)$$

which has cubic convergence [56]. Moreover, for quadratic equations, (3.18) has R-order of convergence at least four [22, 54]. The convergence of this method is analyzed by means of the convexity of the scalar functions f and f' [33]. We analyze that we can always consider it to solve any scalar equation. We also study this method in Banach spaces for operators with ω-bounded third derivative. Finally, we illustrate the previous study with an application to a boundary value problem.

3.3.1 The Super-Halley Method and Convexity

As for the Chebyshev method, we do an exhaustive study of the convergence conditions of the Super-Halley method in terms of the convexity of the functions f and f'. For this, we first obtain global convergence conditions for the iterative method, then analyze its practical application, and obtain a practical procedure that allows us to always solve a scalar equation by the method.

Then, we consider $f \in \mathcal{C}^m([a, b])$, with $m \geq 3$, which satisfies the Fourier conditions given in Sect. 2.1.1, so that there is a unique solution t^* of the equation $f(t) = 0$ in $[a, b]$. Without loss of generality, we suppose that f is a convex and strictly increasing function in $[a, b]$ with $f(a)f(b) < 0$. In the other case, it is sufficient to change $f(t)$ by $f(-t)$, $-f(t)$ or $-f(-t)$. We denote

$$M(g) = \max\{g(t) : t \in [a, b]\} \quad \text{and} \quad m(g) = \min\{g(t) : t \in [a, b]\}.$$

In Theorem 3.4 (b), we give a first global convergence result of the Super-Halley method that depends on the degree of logarithmic convexity of f' in $[a, b]$; in particular, $L_{f'}(t) \leq 0$ in $[a, b]$. Next, we extend this result, so that the following results reduce to those obtained for $L_{f'}(t) > 0$ in $[a, b]$. We first give the following lemma.

Lemma 3.23 *In the conditions mentioned above, if $|L_f(t)| \leq \frac{1}{k}$ with $k > 1.754877\ldots$ in $[a, b]$, then*

$$\left[\frac{1}{k}, 2(k-1)^2 - \frac{1}{k}\right) \neq \emptyset \quad \text{and} \quad \frac{L_f(t)^2}{2(1 - L_f(t))^2} \leq \frac{1}{2(k-1)^2}.$$

Proof First, to prove that the interval $\left[\frac{1}{k}, 2(k-1)^2 - \frac{1}{k}\right)$ is non-empty, we have to check that the inequality $\frac{1}{k} < 2(k-1)^2 - \frac{1}{k}$ is satisfied, that is, $k^3 - 2k^2 + k - 1 > 0$. The last cubic polynomial has a single real root, $k = 1.754877\ldots$ [70], so that $\frac{1}{k} < 2(k-1)^2 - \frac{1}{k}$ if $k > 1.754877\ldots$

Second, to prove $\frac{L_f(t)^2}{2(1-L_f(t))^2} \leq \frac{1}{2(k-1)^2}$, we consider $h(z) = \frac{z^2}{2(1-z)^2}$ with $z = L_f(t)$. Then, $h'(z) = \frac{z}{(1-z)^3}$. As $|L_f(t)| \leq \frac{1}{k}$ in $[a, b]$, we have that $h'(z) \geq 0$ if $z \in \left[0, \frac{1}{k}\right]$ and $h'(z) \leq 0$ if $z \in \left[-\frac{1}{k}, 0\right]$. As a consequence,

$$\max\left\{h(z) : |z| \leq \frac{1}{k}\right\} = \max\left\{\frac{\left(\frac{1}{k}\right)^2}{2\left(1-\frac{1}{k}\right)^2}, \frac{\left(-\frac{1}{k}\right)^2}{2\left(1+\frac{1}{k}\right)^2}\right\} = \frac{\left(\frac{1}{k}\right)^2}{2\left(1-\frac{1}{k}\right)^2} = \frac{1}{2(k-1)^2},$$

and the thesis follows. ∎

After that, we can establish the following result, whose proof is analogous to that of Theorem 3.9 taking now into account (3.11).

Theorem 3.24 *Let $k > 1.754877\ldots$ Suppose that the interval $[a, b]$ is such that $a + \frac{2k-1}{2(k-1)} \frac{f(b)}{f'(b)} \leq b$ and the starting point $t_0 \in [a, b]$ such that $f(t_0) > 0$ and $t_0 \geq a + \frac{2k-1}{2(k-1)} \frac{f(b)}{f'(b)}$. If $L_{f'}(t) \in \left[\frac{1}{k}, 2(k-1)^2 - \frac{1}{k}\right)$ and $|L_f(t)| \leq \frac{1}{k}$ in $[a, b]$, the Super-Halley sequence (3.18) converges to t^* and oscillates about t^*.*

Remark 3.25 Observe that the previous result also follows if $t_0 \in [a, b]$ with $f(t_0) < 0$ and $t_0 \geq b + \frac{f(a)}{f'(a)}$ and the interval $[a, b]$ is such that $a \leq b + \frac{f(a)}{f'(a)}$.

Next, we illustrate the last result with the following example.

Example 3.26 Consider the real equation $f(t) = t^4 - \mu = 0$ with $\mu \in (0, 1)$. Then,

$$L_f(t) = \frac{3(t^4 - \mu)}{4t^4} \quad \text{and} \quad L_{f'}(t) = \frac{2}{3}.$$

If $k > 1.78339\ldots$, we have $\frac{1}{k} \leq L_{f'}(t) = \frac{2}{3} < 2(k-1)^2 - \frac{1}{k}$. We then choose, for example, $k = 1.79$. As a consequence, $|L_f(t)| \leq 0.5607\ldots$ in $[0.731355, 1.18644]$ for $\mu > 0.499839\ldots$ Now, we choose $\mu = 0.5$ and $[a, b] = [0.731355, 1.18644]$, which checks

$$1.09 = a + \frac{2k-1}{2(k-1)} \frac{f(b)}{f'(b)} \leq b = 1.18644.$$

After that, we choose t_0 such that $t_0 \geq 1.09$; for example, $t_0 = 1.18$. And we then obtain the iterates of the Super-Halley method given in Table 3.9 to approximate the solution $t^* = 0.8408964152537145430$ with 20 significant digits.

3.3 The Super-Halley Method

Table 3.9 The Super-Halley method from Example 3.26

n	t_n
0	1.1800000000000000000
1	0.8236846056354870351
2	0.8409040229165229872
3	0.8408964152537139204
4	0.8408964152537145430

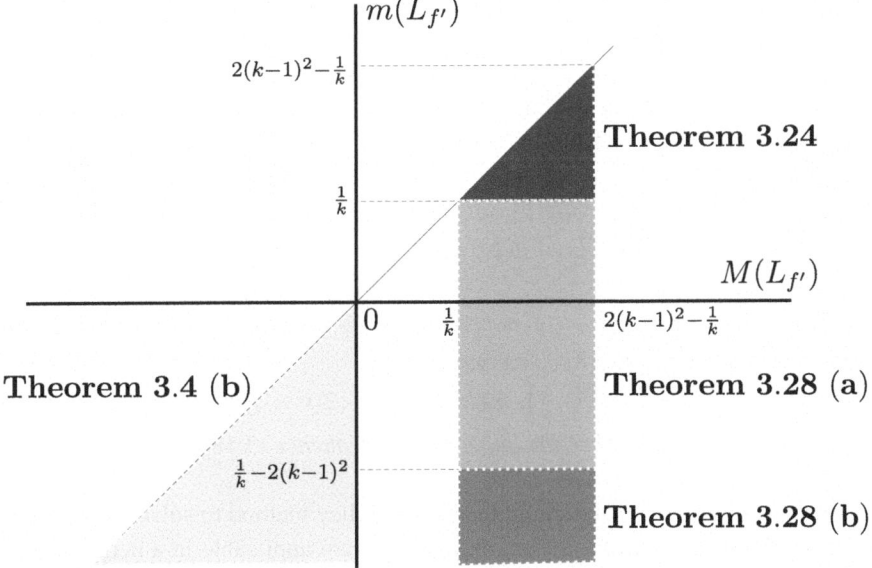

Fig. 3.5 Application of the Super-Halley method for $k > 1.754877\ldots$

In practice, we can always apply the Super-Halley method to solve $f(t) = 0$. Taking into account the values of $m(L_{f'})$ and $M(L_{f'})$, we have the following situations that are drawn in Fig. 3.5:

(I) $M(L_{f'}) \leq 0$.
(II) $M(L_{f'}) < 2(k-1)^2 - \frac{1}{k}$ and $m(L_{f'}) \geq \frac{1}{k}$.
(III) $\frac{1}{k} < M(L_{f'}) < 2(k-1)^2 - \frac{1}{k}$ and $m(L_{f'}) < \frac{1}{k}$:
  (III$_1$) $\frac{1}{k} \leq M(L_{f'}) < 2(k-1)^2 - \frac{1}{k}$ and $\frac{1}{k} - 2(k-1)^2 < m(L_{f'}) < \frac{1}{k}$.
  (III$_2$) $\frac{1}{k} \leq M(L_{f'}) < 2(k-1)^2 - \frac{1}{k}$ and $m(L_{f'}) \leq \frac{1}{k} - 2(k-1)^2$.

In cases (I) and (II), the convergence of the Super-Halley method (3.18) follows from Theorem 3.4 (b) and Theorem 3.24, respectively. Case (III), which is studied below, completes all the regions of the semiplane $M(L_{f'}) \geq m(L_{f'})$. For this, it suffices to take k varying on $\mathbb{R}_+$.

For case (III), we need first the following result to guarantee that the Super-Halley sequence (3.18) lies in $[a, b]$, whose proof is similar to that of Lemma 3.11.

Lemma 3.27 *Suppose $|S'(t)| < 1$ and $|L_f(t)| \leq \frac{1}{k}$, with $k > 1.754877\ldots$, in $[a, b]$.*

(a) *Let $t_0, \gamma \in [a, b]$ be such that $f(\gamma) < 0$ and $f(t_0) > 0$. If $\frac{2k-1}{2(k-1)} \frac{f(b)}{f'(b)} \leq \gamma - a$, then $t_n \in [a, b]$, for all $n \in \mathbb{N}$.*
(b) *Let $t_0, \beta \in [a, b]$ be such that $f(\beta) > 0$ and $f(t_0) < 0$. If $\frac{f(a)}{f'(a)} \geq \beta - b$, then $t_n \in [a, b]$, for all $n \in \mathbb{N}$.*

Then, we are ready to study case (III). For this, we establish the following result, whose proof is analogous to that of Theorem 3.12.

Theorem 3.28 *Let $t_0, \gamma, \beta \in [a, b]$ satisfy the conditions of Lemma 3.27. Suppose $|L_f(t)| \leq \frac{1}{k}$, with $k > 1.754877\ldots$, in $[a, b]$.*

(a) *If $M(L_{f'}) \in \left[\frac{1}{k}, 2(k-1)^2 - \frac{1}{k}\right)$ and $m(L_{f'}) \in \left(\frac{1}{k} - 2(k-1)^2, \frac{1}{k}\right)$ in $[a, b]$, then the Super-Halley sequence (3.18) converges to t^*.*
(b) *If $M(L_{f'}) \in \left[\frac{1}{k}, 2(k-1)^2 - \frac{1}{k}\right)$, $m(L_{f'}) \leq \frac{1}{k} - 2(k-1)^2$, and $M(|L_f|) \leq 2(k-1)^2 - |m(L_{f'})|$ in $[a, b]$, then the Super-Halley sequence (3.18) converges to t^*.*

As a summary, we can always apply the Super-Halley method to solve $f(t) = 0$ under certain restrictions on L_f in $[a, b]$, but these results are applicable in a neighborhood of t^*, since $L_f(t^*) = 0$.

Remark 3.29 Observe that, in practical situations, it suffices that $m(L_{f'})$ satisfies the hypotheses of item (a) of Theorem 3.28, and even that $m(L_{f'}) \in \left[\frac{1}{k} - 2(k-1)^2, 0\right]$, since we can always find a finite and sufficiently large value of k that allows taking the infinite region of the plane that is necessary.

Now, we illustrate the last result with the following example.

Example 3.30 Consider the nonlinear scalar equation $f(t) = -1 + \arcsin(1+t) = 0$. Then,

$$L_f(t) = \frac{(1+t)(-1 + \arcsin(1+t))}{\sqrt{1 - (1-t)^2}} \quad \text{and} \quad L_{f'}(t) = \frac{2t^2 + 4t + 3}{(1+t)^2}.$$

As the function $L_{f'}$ is nondecreasing in $[-0.9, -0.01]$, we have $M(L_{f'}) = L_{f'}(-0.9) = 102$ and $m(L_{f'}) = L_{f'}(-0.01) = 3.0203$. If $k > 8.14572\ldots$, from Theorem 3.28, we have

3.3 The Super-Halley Method

Table 3.10 The Super-Halley method from Example 3.30

n	t_n
0	-0.2100000000000000
1	-0.1581347287092053
2	-0.1585290154830309
3	-0.1585290151921035

$$\frac{1}{k} \leq 102 < 2(k-1)^2 - \frac{1}{k} \quad \text{and} \quad \frac{1}{k} - 2(k-1)^2 \leq 3.0203 < \frac{1}{k}.$$

Hence, we can choose $k = 8.15$, so that $|L_f(t)| \leq \frac{1}{k} = 0.122699\ldots$ in $[-0.214634, -0.123979]$. Moreover,

$$\frac{f(-0.214634)}{f'(-0.214634)} = -0.0598683\ldots \geq \beta + 0.123979$$

for some $\beta \in [-0.214634, -0.183847]$ with $f(\beta) > 0$. Therefore, we can choose $[a, b] = [-0.214634, -0.183847]$ and $x_0 = -0.21$ and use the Super-Halley method to approximate the solution $t^* = -0.1585290151921035$, with 17 significant digits. The approximations obtained are given in Table 3.10.

Remark 3.31 Despite the fact that the value of k can be as large as we want, it is interesting to choose the lowest value of k that verifies the conditions required in Theorem 3.28, since the smaller the value of k, the lager the interval over which L_f takes values. Therefore, the condition on L_f is milder in $[a, b]$.

3.3.2 The Super-Halley Method in Banach Spaces

In this section, we study the semilocal convergence of the Super-Halley method for solving operator equations in Banach spaces. For this, we use the method of majorizing sequences introduced in Sect. 3.2.2 and suppose the same initial convergence conditions (L1)–(L2) as for the Chebyshev method. Then, we consider a nonlinear operator $F : \Omega \subseteq X \to Y$ which is three times continuously Fréchet differentiable on a non-empty open convex domain Ω of a Banach space X with values in a Banach space Y. The Super-Halley method for approximating a solution x^* of the equation $F(x) = 0$ is given by

$$x_{n+1} = S(x_n) = x_n - \left(I + \frac{1}{2}L_F(x_n)[I - L_F(x_n)]^{-1}\right)[F'(x_n)]^{-1}F(x_n),$$

$$n \geq 0, \quad \text{with } x_0 \text{ given in } \Omega, \tag{3.19}$$

where I denotes the identity operator on X.

As in Sect. 3.2.2, the main idea is to construct a scalar function $f \in \mathcal{C}^3([t_0, +\infty))$, with $t_0 \in \mathbb{R}$, that satisfies conditions (L1)–(L2). In addition, Theorems 3.15 and 3.16 are given, where the function f is analyzed. Now, it is not difficult to prove that the scalar sequence (3.18) is a majorizing sequence of the Super-Halley method (3.19) defined in the Banach space X. Besides, the sequence (3.19) is well defined, provided that $B(x_0, t^* - t_0) \subseteq \Omega$. For this, it is enough to take into account a result totally analogous to Lemma 3.17. Items (i_n) and (ii_n) are followed exactly the same as in Lemma 3.17. For item (iii_n), we have to take into account the expression (3.19) of the Super-Halley method and the decomposition

$$F(x_{n+1}) = \frac{1}{8} F''(x_n) \left(L_F(x_n) \left(I - L_F(x_n)\right)^{-1} \Gamma_n F(x_n) \right)^2$$
$$+ \frac{1}{2} \int_0^1 F'''(x_n + \tau(x_{n+1} - x_n))(x_{n+1} - x_n)^3 (1 - \tau)^2 d\tau$$

instead of that given in the proof of Lemma 3.17 for the Chebyshev method. For item (iv_n), we only have to take into account the expression (3.19).

After that, the semilocal convergence of the Super-Halley method (3.19) in Banach spaces is followed exactly the same as in Theorem 3.18 for the Chebyshev method, as we can see in the following result.

Theorem 3.32 (General Semilocal Convergence) *Let $F : \Omega \subseteq X \longrightarrow Y$ be a three times continuously Fréchet differentiable operator defined on a non-empty open convex domain Ω of a Banach space X with values in a Banach space Y. Suppose that there exists $f \in \mathcal{C}^3([t_0, +\infty))$ such that (L1)–(L2) are satisfied. Suppose also that there exists $\gamma \in (t_0, +\infty)$ such that $f'(\gamma) = 0$ and $f(\gamma) \leq 0$, and $B(x_0, t^* - t_0) \subset \Omega$. Then, the Super-Halley sequence (3.19) starting at x_0 converges to a root x^* of $F(x) = 0$. Moreover, $x_n, x^* \in \overline{B(x_0, t^* - t_0)}$ and*

$$\|x^* - x_n\| \leq t^* - t_n, \quad \text{for all} \quad n \geq 0,$$

where $\{t_n\}$ is the Super-Halley sequence (3.18).

Next, we suppose conditions (N1)–(N2) and consider $\varpi(t; \|x_0\|, t_0) = \omega(t - t_0 + \|x_0\|)$, for $\|x - x_0\| \leq t - t_0$, where $\varpi : [t_0, +\infty) \longrightarrow \mathbb{R}$ is a continuous nondecreasing function such that $\varpi(t_0; \|x_0\|, t_0) \geq 0$, so that the function (3.15) is the solution of the initial value problem given in (3.14). Note that the majorizing function (3.15) thus obtained is the same as for the Chebyshev method, since this function is obtained from the convergence conditions required to the operator involved F, which are the same for both methods. Then, a result analogous to Theorem 3.20 is established, and, taking into account the additional condition (N3), the semilocal convergence of the Super-Halley method (3.19)

3.3 The Super-Halley Method

in the Banach space X is followed exactly the same as in Theorem 3.21 for the Chebyshev method, as we can see in the following result.

Theorem 3.33 *Let $F : \Omega \subseteq X \longrightarrow Y$ be a three times continuously Fréchet differentiable operator defined on a non-empty open convex domain Ω of a Banach space X with values in a Banach space Y and $f(t)$ the scalar function (3.15). Suppose that conditions (N1)-(N2)-(N3) are satisfied. Then, the Super-Halley sequence (3.19) starting at x_0 converges to a root x^* of $F(x) = 0$. Besides, $x_n, x^* \in \overline{B(x_0, t^* - t_0)}$ and*

$$\|x^* - x_n\| \leq t^* - t_n, \quad \text{for all} \quad n \geq 0,$$

where $\{t_n\}$ is the Super-Halley sequence (3.18).

To illustrate this study of the Super-Halley method in Banach spaces, we consider the following boundary value problem.

Example 3.34 Consider

$$y''(t) + y'(t) - y(t)^4 = 2, \quad y(0) = 0 = y(1). \quad (3.20)$$

We divide the interval $[0, 1]$ into $m + 1$ subintervals of length $h = \frac{1}{m+1}$ and introduce the points $t_j = jh$, with $j = 0, 1, \ldots, m + 1$, to solve the problem numerically. The corresponding values of the function are

$$y_0 = y(t_0), y_1 = y(t_1), \ldots, y_{m+1} = y(t_{m+1}).$$

Standard approximations for the first and second derivatives at these points are given, respectively, by

$$y'(t_j) \approx \frac{y_{j+1} - y_{j-1}}{2h}, \quad y''(t_j) \approx \frac{y_{j-1} - 2y_j + y_{j+1}}{h^2}, \quad j = 1, 2, \ldots, m,$$

so that, by following a process of discretization, we obtain the following nonlinear system:

$$-4y_1 + (2+h)y_2 - 2h^2 y_1^4 - 4h^2 = 0,$$
$$(2-h)y_{i-1} - 4y_i + (2+h)y_{i+1} - 2h^2 y_i^4 - 4h^2 = 0, \quad i = 2, 3, \ldots m-1,$$
$$(2-h)y_{m-1} - 4y_m - 2h^2 y_m^4 - 4h^2 = 0.$$

Observe that $y_0 = y_{m+1} = 0$, since y_0 and y_{m+1} are determined by the boundary conditions. Thus, the unknowns are $y_1, y_2, \ldots, y_m$.

We then define the function $F : \mathbb{R}^m \to \mathbb{R}^m$ by

$$F(\mathbf{y}) = 2A\,\mathbf{y} + hB\,\mathbf{y} - 2h^2\,\hat{\mathbf{y}} - 4h^2\,\mathbf{u},$$

where

$$A = \begin{pmatrix} -2 & 1 & 0 & \cdots & 0 \\ 1 & -2 & 1 & \cdots & 0 \\ 0 & 1 & -2 & \cdots & 0 \\ \vdots & \vdots & \vdots & \ddots & \vdots \\ 0 & 0 & 0 & \cdots & -2 \end{pmatrix}_{m \times m}, \quad B = \begin{pmatrix} 0 & 1 & 0 & \cdots & 0 \\ -1 & 0 & 1 & \cdots & 0 \\ 0 & -1 & 0 & \cdots & 0 \\ \vdots & \vdots & \vdots & \ddots & \vdots \\ 0 & 0 & 0 & \cdots & 0 \end{pmatrix}_{m \times m},$$

$\mathbf{y} = (y_1, y_2, \ldots, y_m)^T$, $\hat{\mathbf{y}} = (y_1^4, y_2^4, \ldots, y_m^4)^T$, and $\mathbf{u} = (1, 1, \ldots, 1)^T$.

In addition,

$$F'(\mathbf{y}) = 2A + hB - 8h^2 \operatorname{diag}\left\{y_1^3, y_2^3, \ldots, y_m^3\right\},$$

$$F''(\mathbf{y})\mathbf{zw} = -24h^2(y_1^2 z_1 w_1, y_2^2 z_2 w_2, \ldots, y_m^2 z_m w_m),$$

$$F'''(\mathbf{y})\mathbf{zwp} = -48h^2(y_1 z_1 w_1 p_1, y_2 z_2 w_2 p_2, \ldots, y_m z_m w_m p_m),$$

where $\mathbf{z} = (z_1, z_2, \ldots, z_m)$, $\mathbf{w} = (w_1, w_2, \ldots, w_m)$, and $\mathbf{p} = (p_1, p_2, \ldots, p_m)$. Taking the norms of $\mathbf{x} \in \mathbb{R}^m$ and $A \in \mathbb{R}^m \times \mathbb{R}^m$ as in (3.17), we have

$$\|F'''(\mathbf{y})\| = \sup_{\|\mathbf{z}\|=1, \|\mathbf{w}\|=1, \|\mathbf{p}\|=1} \|F'''(\mathbf{y})\mathbf{zwp}\| \leq 48h^2 \|\mathbf{y}\|,$$

so that $\omega(t) = 48h^2 t$ and $\varpi(t; \|x_0\|, t_0) = \omega(t - t_0 + \|x_0\|) = 48h^2(t - t_0 + \|x_0\|)$.

We choose the initial approximation $\frac{3}{10}$ which gives the starting vector

$$\mathbf{x}_0 = (0.3, 0.3, 0.3, 0.3, 0.3, 0.3, 0.3, 0.3, 0.3)^T,$$

for the Super-Halley method if $m = 9$. We then have

$$\beta = \|[F'(\mathbf{x}_0)]^{-1}\| = 6.0606\ldots, \quad \eta = \|[F'(\mathbf{x}_0)]^{-1} F(\mathbf{x}_0)\| = 0.5394\ldots,$$

$$\delta = \|F''(\mathbf{x}_0)\| = 0.0216.$$

Besides, if $t_0 = 0$, the function (3.15) is reduced to the fourth-degree polynomial

$$f(t) = \frac{t^4}{50} + (0.24)t^3 + (0.0108)t^2 - (0.1649\ldots)t + (0.0890\ldots), \tag{3.21}$$

which has two positive zeros $t^* = 0.6156\ldots$ and $t^{**} = 1.2831\ldots$ Moreover, the unique minimum of $f(t)$ in $[0, +\infty)$ is $\gamma = 0.9784\ldots$, and it is such that $f(\gamma) = -0.0212\ldots \leq$

3.4 The Halley Method

Table 3.11 The numerical solution $\mathbf{y}^* = (y_1^*, y_2^*, \ldots, y_9^*)^T$ of the system $F(\mathbf{y}) = 0$ from Example 3.34

i	y_i^*	i	y_i^*	i	y_i^*
1	$-0.1012745\ldots$	4	$-0.2435699\ldots$	7	$-0.1931094\ldots$
2	$-0.1738553\ldots$	5	$-0.2453911\ldots$	8	$-0.1425201\ldots$
3	$-0.2204672\ldots$	6	$-0.2279567\ldots$	9	$-0.0776972\ldots$

Table 3.12 Absolute errors, a priori error estimates, and $\{\|F(\mathbf{y}_n)\|\}$ from Example 3.34

n	$\|\mathbf{y}^* - \mathbf{y}_n\|$	$|t^* - t_n|$	$\|F(\mathbf{y}_n)\|$
0	$5.4539\ldots \times 10^{-1}$	$6.1564\ldots \times 10^{-1}$	$6.7016\ldots \times 10^{-1}$
1	$1.1600\ldots \times 10^{-2}$	$5.5664\ldots \times 10^{-2}$	$2.2189\ldots \times 10^{-3}$
2	$1.1818\ldots \times 10^{-7}$	$1.1792\ldots \times 10^{-4}$	$3.1731\ldots \times 10^{-8}$
3	0.0×10^{-23}	$1.1361\ldots \times 10^{-12}$	0.0×10^{-23}

0. Hence, the convergence of the Super-Halley method to a solution $\mathbf{y}^*$ is guaranteed from Theorem 3.33. Furthermore, the domain of existence of solution is

$$\{\mathbf{v} \in \mathbb{R}^9 : \|\mathbf{v} - \mathbf{x}_0\| \leq t^* = 0.6156\ldots\}.$$

In particular, the Super-Halley method converges to the solution $\mathbf{y}^* = (y_1^*, y_2^*, \ldots, y_9^*)^T$ given in Table 3.11 after three iterations and using the stopping criterion $\|\mathbf{y}_n - \mathbf{y}_{n-1}\| < 10^{-16}$.

In Table 3.12 we show the errors $\|\mathbf{y}^* - \mathbf{y}_n\|$ using the stopping criterion $\|\mathbf{y}^* - \mathbf{y}_n\| < 10^{-16}$ and the sequence $\{\|F(\mathbf{y}_n)\|\}$. From the last, we notice that the vector shown in Table 3.11 is a good approximation of the solution of the system $F(\mathbf{y}) = 0$ with $\mathfrak{m} = 9$, since $\|F(\mathbf{y}^*)\| \leq \text{constant} \times 10^{-16}$. See the sequence $\{\|F(\mathbf{y}_n)\|\}$ in Table 3.12.

Finally, by interpolating the values of Table 3.11 and taking into account the boundary conditions, we obtain the solution drawn in Fig. 3.6 which is denoted by $\widetilde{\mathbf{y}}$. Observe that the approximate solution $\widetilde{\mathbf{y}}$ lies within the domain of existence of solution $\overline{B(0.3, 0.6156\ldots)}$.

3.4 The Halley Method

In Sect. 3.1.3 we see that the Halley method is an acceleration of the Newton method, which is obtained from the direct reduction of the degree of logarithmic convexity of the function that defines the Newton method. In this section, we give conditions for the convergence of the Halley method for solving nonlinear scalar equations by using the degree of logarithmic convexity of f and f' [59]. Also, we do a practical study of the Halley method and prove that we can always apply the method to solve a nonlinear scalar equation according to the convexity of f and f'. Besides, we also study the Halley method

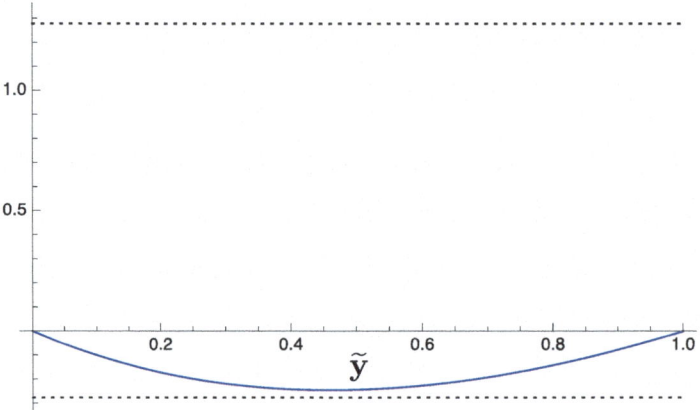

Fig. 3.6 The approximate solution $\widetilde{y}$ of problem (3.20) (solid line) and the domain of existence of solution (dotted lines)

in Banach spaces for operators with ω-bounded third derivative and using the technique of majorizing sequences introduced in Sect. 2.2.2.1. Finally, we apply the last study to approximate a solution of a boundary value problem.

3.4.1 The Halley Method and Convexity

As for the Chebyshev and the Super-Halley methods, we do an exhaustive study of the convergence conditions of the Halley method in terms of the convexity of the functions f and f'. For this, we first obtain global convergence conditions for the iterative method, then analyze its practical application, and obtain a practical procedure that allows us to always solve a scalar equation by the Halley method.

We start by considering a real function $f \in \mathcal{C}^m([a,b])$ with $m \geq 3$, that satisfies the Fourier conditions given in Sect. 2.1.1, so that there is a unique solution t^* of the equation $f(t) = 0$ in $[a,b]$. Without loss of generality, we can assume that f is a convex and strictly increasing function in $[a,b]$ with $f(a)f(b) < 0$. In other case, it is sufficient to change $f(t)$ by $f(-t)$, $-f(t)$, or $-f(-t)$. We denote

$$M(g) = \max\{g(t) : t \in [a,b]\} \quad \text{and} \quad m(g) = \min\{g(t) : t \in [a,b]\}.$$

Remember that a first global convergence result of the Halley method

$$t_{n+1} = H(t_n) = t_n - \left(\frac{2}{2 - L_f(t_n)}\right) \frac{f(t_n)}{f'(t_n)}, \quad n \geq 0, \quad \text{with } t_0 \text{ given} \qquad (3.22)$$

3.4 The Halley Method

is given in Theorem 3.4 (c). Notice that the condition $L_f(t) < 2$ in $[a, b]$ is not very restrictive since $L_f(t^*) = 0$ and $L_f(t) \geq 0$ in $[t^*, b]$. In addition, we extend this result in the following theorem, whose proof is analogous to that of Theorem 3.7 taking into account (3.11).

Theorem 3.35 *Let $t_0 \in [a, b]$. Suppose that $L_{f'}(t) \leq \frac{3}{2}$ in $[a, b]$. If $f(t_0) > 0$ and $L_f(t) < 2$ in $[a, b]$, then the Halley sequence (3.22) is decreasing and converges to t^*. If $f(t_0) < 0$, then the Halley sequence (3.22) is increasing and converges to t^*.*

Next, we see what happens if $L_{f'}(t) > \frac{3}{2}$ in $[a, b]$. As the Halley sequence (3.22) oscillates about t^*, we have to guarantee that $t_1 \in [a, b]$.

Theorem 3.36 *Let the interval $[a, b]$ be such that $a + 2\frac{f(b)}{f'(b)} \leq b$ and $t_0 \in [a, b]$ such that $f(t_0) > 0$ and $t_0 \geq a + 2\frac{f(b)}{f'(b)}$.*

(a) *If $L_{f'}(t) \in \left(\frac{3}{2}, 2\right]$ and $|L_f(t)| < 1$ in $[a, b]$, then the Halley sequence (3.22) converges to t^* and oscillates about t^*.*
(b) *If $L_{f'}(t) > 2$ and $|L_f(t)| < \frac{-1+\sqrt{2M(L_{f'})-3}}{M(L_{f'})-2}$ in $[a, b]$, then the Halley sequence (3.22) converges to t^* and oscillates about t^*.*

The proof of Theorem 3.36 is analogous to that of Theorem 3.9.

Remark 3.37 Observe that the last results also follow if $t_0 \in [a, b]$ with $f(t_0) < 0$ and $t_0 \geq b + \frac{f(a)}{f'(a)}$, and the interval $[a, b]$ is such that $a \leq b + \frac{f(a)}{f'(a)}$.

In practice, we can always apply the Halley method to solve $f(t) = 0$. According the values of $M(L_{f'})$ and $m(L_{f'})$, we have the following situations that are drawn in Fig. 3.7:

(I) $M(L_{f'}) \leq \frac{3}{2}$.
(II) $M(L_{f'}) \leq 2$ and $m(L_{f'}) > \frac{3}{2}$.
(III) $\frac{3}{2} < M(L_{f'}) \leq 2$ and $m(L_{f'}) \leq \frac{3}{2}$.
(IV) $M(L_{f'}) > 2$ and $m(L_{f'}) \leq 2$.
(V) $m(L_{f'}) > 2$.

In cases (I), (II), and (V), the convergence of the Halley sequence (3.22) follows from Theorems 3.35 and 3.39. Cases (III) and (IV) are studied in the next under some conditions on L_f. First, we give a lemma that is analogous to Lemma 3.11.

Lemma 3.38 *Suppose $|H'(t)| < 1$ and $|L_f(t)| < 1$ in $[a, b]$.*

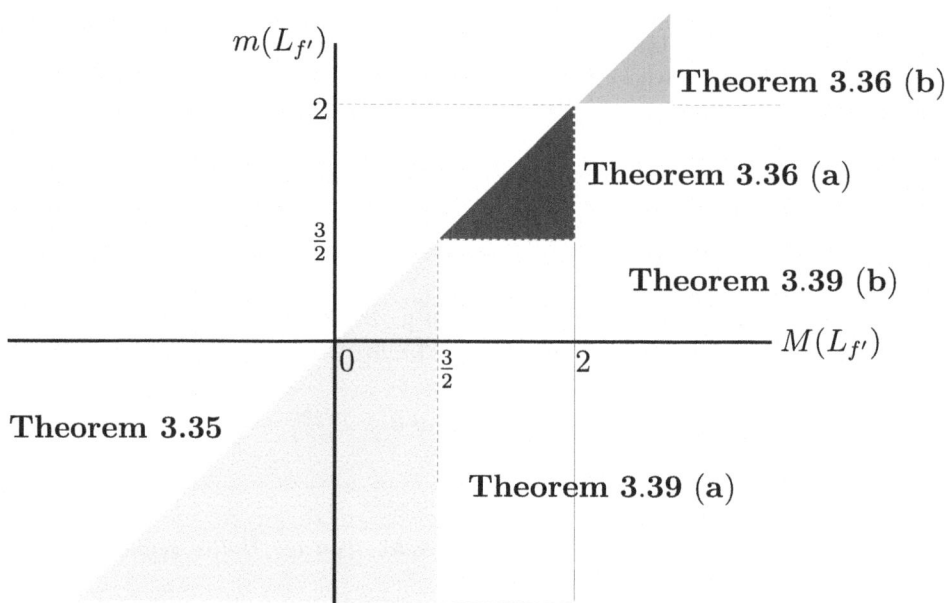

Fig. 3.7 Application of the Halley method

(a) Let $t_0, \gamma \in [a, b]$ such that $f(\gamma) < 0$ and $f(t_0) > 0$. If $2\frac{f(b)}{f'(b)} \leq \gamma - a$, then $t_n \in [a, b]$, for all $n \in \mathbb{N}$.
(b) Let $t_0, \beta \in [a, b]$ such that $f(\beta) > 0$ and $f(t_0) < 0$. If $\frac{f(a)}{f'(a)} \geq \beta - b$, then $t_n \in [a, b]$, for all $n \in \mathbb{N}$.

Theorem 3.39 *Let $t_0, \gamma, \beta \in [a, b]$ satisfy the conditions of Lemma 3.38. Suppose $M(|L_f|) < \frac{2}{1+\sqrt{\ell}}$ with $\ell = \max\{|3 - 2M(L_{f'})|, |3 - 2m(L_{f'})|\}$.*

(a) *If $M(L_{f'}) \in \left(\frac{3}{2}, 2\right]$, $m(L_{f'}) \leq \frac{3}{2}$, and $|L_f(t)| < 1$ in $[a, b]$, then the Halley sequence (3.22) converges to t^*.*
(b) *If $M(L_{f'}) > 2$ and $m(L_{f'}) \leq 2$ in $[a, b]$, then the Halley sequence (3.22) converges to t^*.*

The proof of Theorem 3.39 is analogous to that of Theorem 3.12.

Notice that the previous result is also true if $f(t_0) < 0$ and analogous conditions.

As a consequence of the abovementioned results, we can always apply the Halley method to solve $f(t) = 0$ with some restrictions on $L_f(t)$ in $[a, b]$, and the results given can always be applied in a neighborhood of the solution t^*, since $L_f(t^*) = 0$.

3.4 The Halley Method

Table 3.13 The Halley method from Example 3.40

n	t_n	n	t_n
0	4.000000...	4	2.846385...
1	3.741961...	5	2.563813...
2	3.465034...	6	2.467671...
3	3.165567...	7	2.465652...

Example 3.40 To illustrate the above, we consider the equation $f(t) = -8 + e^{t^2-4} = 0$ in $[1, 5]$. So,

$$L_f(t) = \frac{(1+2t^2)(-8+e^{t^2-4})}{2t^2 e^{t^2-4}} \quad \text{and} \quad L_{f'}(t) = \frac{2t^2(3+2t^2)}{(1+2t^2)^2}.$$

Observe that $L_{f'}(t) \leq \frac{3}{2}$ and $L_f(t) < 2$ in $[0, +\infty)$. Then, for example, we can choose any point of $[1, 5]$ as t_0 and such that $f(t_0) > 0$. Then, we choose $t_0 = 4$, and, by Theorem 3.35, we obtain that the Halley sequence (3.22) is decreasing and converges to the solution $t^* = 2.465652...$ of $f(t) = 0$, as we can see in Table 3.13.

3.4.2 The Halley Method in Banach Spaces

Now, we study the semilocal convergence of the Halley method for solving operator equations in Banach spaces. Again, we use the method of majorizing sequences introduced in Sect. 3.2.2 and suppose the same initial convergence conditions, (L1)–(L2), as for the Chebyshev and Super-Halley methods. So, we consider a nonlinear operator $F : \Omega \subseteq X \to Y$ which is three times continuously Fréchet differentiable on a non-empty open convex domain Ω of a Banach space X with values in a Banach space Y. The Halley method for approximating a solution x^* of the equation $F(x) = 0$ is given by

$$x_{n+1} = H(x_n) = x_n - \left(I - \frac{1}{2}L_F(x_n)\right)^{-1} [F'(x_n)]^{-1} F(x_n), \quad n \geq 0, \text{ with } x_0 \text{ given in } \Omega,$$
(3.23)

where I denotes the identity operator on X.

We have already seen in Sects. 3.2.2 and 3.3.2 that the main idea is to construct a scalar function $f \in \mathcal{C}^3([t_0, +\infty))$, with $t_0 \in \mathbb{R}$, that satisfies conditions (L1)–(L2), so that Theorems 3.15 and 3.16 are established. As a consequence, the scalar sequence (3.22) is a majorizing sequence of the Halley method (3.23) in the Banach space X, and the sequence (3.19) is well defined, provided that $B(x_0, t^* - t_0) \subseteq \Omega$. For this, it is enough to take into account a result totally analogous to Lemma 3.17. Items (i_n) and (ii_n) are followed exactly the same as in Lemma 3.17. For item (iii_n), we have to take into account the expression (3.23) of the Halley method and the decomposition

$$F(x_{n+1}) = \frac{1}{4}F''(x_n)\left(I - \frac{1}{2}L_F(x_n)\right)^{-1}\Gamma_n F(x_n)L_F(x_n)\left(I - \frac{1}{2}L_F(x_n)\right)^{-1}\Gamma_n F(x_n)$$
$$+ \frac{1}{2}\int_0^1 F'''(x_n + \tau(x_{n+1} - x_n))(x_{n+1} - x_n)^3(1-\tau)^2 d\tau$$

instead of that given in the proof of Lemma 3.17 for the Chebyshev method. For item (iv_n), we only have to take into account the expression (3.23).

Next, the semilocal convergence of the Halley method (3.23) in Banach spaces is followed exactly the same as in Theorem 3.18 for the Chebyshev method, as we can see in the following result.

Theorem 3.41 (General Semilocal Convergence) *Let* $F : \Omega \subseteq X \longrightarrow Y$ *be a three times continuously Fréchet differentiable operator defined on a non-empty open convex domain* Ω *of a Banach space* X *with values in a Banach space* Y. *Suppose that there exists* $f \in \mathcal{C}^3([t_0, +\infty))$ *such that (L1)–(L2) are satisfied. Suppose also that there exists* $\gamma \in (t_0, +\infty)$ *such that* $f'(\gamma) = 0$ *and* $f(\gamma) \leq 0$, *and* $B(x_0, t^* - t_0) \subset \Omega$. *Then, the Halley sequence (3.23) starting at* x_0 *converges to a root* x^* *of* $F(x) = 0$. *Moreover,* $x_n, x^* \in \overline{B(x_0, t^* - t_0)}$ *and*

$$\|x^* - x_n\| \leq t^* - t_n, \quad \text{for all} \quad n \geq 0,$$

where $\{t_n\}$ *is the Halley sequence (3.22).*

Then, as for the Chebyshev and Super-Halley methods, we suppose conditions (N1)–(N2) and consider $\varpi(t; \|x_0\|, t_0) = \omega(t - t_0 + \|x_0\|)$, for $\|x - x_0\| \leq t - t_0$, where $\varpi : [t_0, +\infty) \longrightarrow \mathbb{R}$ is a continuous nondecreasing function such that $\varpi(t_0; \|x_0\|, t_0) \geq 0$. In addition, the function (3.15) is the solution of the initial value problem (3.14). Note again that the majorizing function (3.15) thus obtained is the same as for the Chebyshev and the Super-Halley methods, since this function is obtained from the convergence conditions required to the operator involved F, which are the same for the three methods. After that, a result analogous to Theorem 3.20 is established, and, taking into account the additional condition (N3), the semilocal convergence of the Halley method (3.23) in the Banach space X is followed exactly the same as in Theorem 3.21 for the Chebyshev method, as we can see in the following result.

Theorem 3.42 *Let* $F : \Omega \subseteq X \longrightarrow Y$ *be a three times continuously Fréchet differentiable operator defined on a non-empty open convex domain* Ω *of a Banach space* X *with values in a Banach space* Y *and* $f(t)$ *the scalar function (3.15). Suppose that conditions (N1)-(N2)-(N3) are satisfied. Then, the Halley sequence (3.23) starting at* x_0 *converges to a root*

3.4 The Halley Method

x^* of $F(x) = 0$. Besides, $x_n, x^* \in \overline{B(x_0, t^* - t_0)}$ and

$$\|x^* - x_n\| \leq t^* - t_n, \quad \text{for all} \quad n \geq 0,$$

where $\{t_n\}$ is the Halley sequence (3.22).

To illustrate the study of the Halley method in Banach spaces, we use the boundary value problem (3.20) given in Example 3.34.

Example 3.43 We consider the boundary value problem (3.20) and follow the process of discretization given in Example 3.34, so that $\varpi(t; \|x_0\|, t_0) = 48h^2(t - t_0 + \|x_0\|)$.

If we choose the starting vector

$$\mathbf{x}_0 = (0.3, 0.3, 0.3, 0.3, 0.3, 0.3, 0.3, 0.3, 0.3)^T$$

and $t_0 = 0$, we obviously obtain the same majorizing function as in Example 3.34. Therefore, the domain of existence of solution coincides with that given there. In addition, the Halley method converges to $\mathbf{y}^*$ after three iterations and using the stopping criterion $\|\mathbf{y}_n - \mathbf{y}_{n-1}\| < 10^{-16}$.

In Table 3.14 we show the errors $\|\mathbf{y}^* - \mathbf{y}_n\|$ using the stopping criterion $\|\mathbf{y}^* - \mathbf{y}_n\| < 10^{-16}$ and the sequence $\{\|F(\mathbf{y}_n)\|\}$. From the last, we notice that the vector shown in Table 3.11 is a good approximation of the solution of the system $F(\mathbf{y}) = 0$, since $\|F(\mathbf{y}^*)\| \leq \text{constant} \times 10^{-16}$. See the sequence $\{\|F(\mathbf{y}_n)\|\}$ in Table 3.14.

Table 3.14 Absolute errors, a priori error estimates, and $\{\|F(\mathbf{y}_n)\|\}$ from Example 3.43

n	$\|\mathbf{y}^* - \mathbf{y}_n\|$	$|t^* - t_n|$	$\|F(\mathbf{y}_n)\|$
0	$5.4539\ldots \times 10^{-1}$	$6.1564\ldots \times 10^{-1}$	$6.7016\ldots \times 10^{-1}$
1	$1.1104\ldots \times 10^{-2}$	$5.6414\ldots \times 10^{-2}$	$2.1157\ldots \times 10^{-3}$
2	$1.0476\ldots \times 10^{-7}$	$2.2562\ldots \times 10^{-4}$	$2.8113\ldots \times 10^{-8}$
3	0.0×10^{-23}	$1.8336\ldots \times 10^{-11}$	0.0×10^{-23}

4 Newton-Like Methods with High Order of Convergence

To obtain a characterization of Newton-like iterative methods with order of convergence at least three, we extend to Banach spaces the result given by Gander in [48] for scalar equations and mentioned in Chap. 3. For scalar equations, $f(t) = 0$, Gander characterizes the third-order Newton-like methods of the form

$$t_{n+1} = G(t_n) = t - H(L_f(t_n))\frac{f(t_n)}{f'(t_n)}, \quad n \geq 0, \quad \text{with } t_0 \text{ given}, \tag{4.1}$$

from the following result [48]:

> If H is any function with $H(0) = 1$, $H'(0) = \frac{1}{2}$, and $|H''(t)| < \infty$, then the iteration $t_{n+1} = G(t_n)$, $n \geq 0$, defined in (4.1) is of third order.

So, many well-known third-order iterative methods are special cases of Gander's result:

- The Chebyshev method: (4.1) with $H(L_f(t_n)) = 1 + \frac{1}{2}L_f(t_n)$
- The Super-Halley method: (4.1) with $H(L_f(t_n)) = 1 + \frac{1}{2}L_f(t_n) + \frac{1}{2}\sum_{k\geq 2} L_f(t_n)^k$
- The Halley method: (4.1) with $H(L_f(t_n)) = 1 + \frac{1}{2}L_f(t_n) + \sum_{k\geq 2}\frac{1}{2^k}L_f(t_n)^k$
- Chebyshev-like methods [4, 34]: (4.1) with $H(L_f(t_n)) = 1 + \frac{1}{2}L_f(t_n) + A_2 L_f(t_n)^2$, $A_2 \in \mathbb{R}^+$
- The Ostrowski method [76]: (4.1) with $H(L_f(t_n)) = 1+\frac{1}{2}L_f(t_n)+\sum_{k\geq 2}(-1)^k\binom{-1/2}{k}L_f(t_n)^k$
- The Exponential method [3]: (4.1) with $H(L_f(t_n)) = 1+\frac{1}{2}L_f(t_n)+\sum_{k\geq 2}\frac{1}{k+1}L_f(t_n)^k$

- The Logarithmic method [3]: (4.1) with $H\left(L_f(t_n)\right) = 1 + \frac{1}{2}L_f(t_n) + \sum_{k\geq 2} \frac{1}{(k+1)!} L_f(t_n)^k$
- The Euler method [48]: (4.1) with $H\left(L_f(t_n)\right) = 1 + \frac{1}{2}L_f(t_n) + \sum_{k\geq 2}(-1)^k 2^{k+1}\binom{1/2}{k+1} L_f(t_n)^k$

From the last methods, it is clear that we can generalize them to obtain an iterative method (4.1) by observing the sequential development of the powers of the function H that has each method. So, we can consider

$$\begin{cases} t_0 \text{ given,} \\ t_{n+1} = G(t_n) = t_n - H(L_f(t_n))\dfrac{f(t)}{f'(t)}, \quad n \geq 0, \\ H(L_f(t_n)) = 1 + \dfrac{1}{2}L_f(t_n) + \displaystyle\sum_{k\geq 2} A_k L_f(t_n)^k, \quad A_k \in \mathbb{R}^+, \end{cases} \qquad (4.2)$$

where $\{A_k\}_{k\geq 2}$ is a nonincreasing scalar sequence. To guarantee a good definition of H, we require that $|L_f(t)| < r$, where r is the radius of convergence of the power series $\sum_{k\geq 0} A_k t^k$ with $A_0 = 1$ and $A_1 = \frac{1}{2}$.

Moreover, we have seen in Chap. 3 that we obtain the well-known third-order iterative methods of Chebyshev, Super-Halley, and Halley from different accelerations of the Newton method, and the three methods have cubic convergence in the scalar case. We can also see that the three methods have R-order of convergence at least three in Banach spaces [9, 11, 12, 18, 19, 22]. We see in this chapter that all methods with R-order of convergence at least three in Banach spaces also admit a general expression.

Our aim is to obtain a theory, the most general possible, relative to these iterative methods. In order to obtain a unified theory of the convergence analysis, we extend the family of iterations (4.2) to solve the equation $F(x) = 0$, where F is an operator $F : \Omega \subseteq X \to Y$ that is at least twice Fréchet differentiable and defined on a non-empty subset of a Banach space X with values in a Banach space Y as [62]

$$\begin{cases} x_0 \text{ given in } \Omega, \\ x_{n+1} = G(x_n) = x_n - H(L_F(x_n))[F'(x_n)]^{-1}F(x_n), \quad n \geq 0, \\ H(L_F(x_n)) = I + \dfrac{1}{2}L_F(x_n) + \displaystyle\sum_{k\geq 2} A_k L_F(x_n)^k, \quad A_k \in \mathbb{R}^+, \end{cases} \qquad (4.3)$$

where $\{A_k\}_{k\geq 2}$ is a nonincreasing scalar sequence such that

$$\sum_{k\geq 0} A_k t^k < +\infty, \text{ for } |t| < r, \text{ with } A_0 = 1, A_1 = \frac{1}{2}, \qquad (4.4)$$

and r is the radius of convergence of the power series $\sum_{k\geq 0} A_k t^k$, taking into account $I = L_F(x)^0$ and that the composition $L_F(x)^k = \overset{k}{\overbrace{L_F(x) \cdots L_F(x)}}$ is a linear operator in X.

Note that the family (4.3) is well defined when the operator H exists, which is such that

$$H(L_F(\_)) : \Omega \xrightarrow{L_F} \mathcal{L}(\Omega, \Omega) \xrightarrow{H} \mathcal{L}(\Omega, \Omega),$$

and $H(L_F(x_n)) = \sum_{k\geq 0} A_k L_F(x_n)^k$, $A_0 = 1$, $A_1 = \frac{1}{2}$, and $A_k \in \mathbb{R}^+$ with $k \geq 2$. Now, from [24], we conclude that the operator H is well defined as an operator of $\mathcal{L}(X, X)$ into $\mathcal{L}(X, X)$ if the operator degree of logarithmic convexity L_F is such that $\|L_F(x)\| < r$.

As particular cases of the family of iterations (4.3) in Banach spaces, we have:

- The Chebyshev method: $A_k = 0$ for $k \geq 2$.
- The Super-Halley method: $A_k = \frac{1}{2}$ for $k \in \mathbb{N}$ and $k \geq 2$.
- The Halley method: $A_k = \frac{1}{2^k}$ for $k \in \mathbb{N}$ and $k \geq 2$.
- Chebyshev-like methods: $A_2 \in \mathbb{R}^+$, $A_k = 0$ for $k \in \mathbb{N}$ and $k > 2$.
- The Ostrowski method: $A_k = \frac{1}{2^{2k}}\binom{2k}{k}$ for $k \in \mathbb{N}$ and $k \geq 2$.
- The Exponential method: $A_k = \frac{1}{k+1}$ for $k \in \mathbb{N}$ and $k \geq 2$.
- The Logarithmic method: $A_k = \frac{1}{(k+1)!}$ for $k \in \mathbb{N}$ and $k \geq 2$.
- The Euler method: $A_k = (-1)^k 2^{k+1} \binom{1/2}{k+1}$ for $k \in \mathbb{N}$ and $k \geq 2$.

After that, we study the semilocal convergence of the family (4.3) under Kantorovich-type conditions for iterative methods with R-order of convergence at least three, where it is usually required that the second Fréchet derivative of the operator involved is bounded in some domain [62], by a technique that uses a system of recurrence relations (see [45] for the Newton method).

Next, since there are many situations in which the second derivative of the operator involved is not bounded, we relax this condition by requiring that the second derivative is only bounded in the initial point of the method to consider [63]. In this case, we use the method of majorizing sequences to prove the semilocal convergence of the family (4.3).

Finally, we consider a particular family of iterative methods included in (4.3) with the particularity that it has R-order of convergence at least four when it is applied to solve quadratic equations [39]. We analyze the semilocal convergence of this family in Banach spaces by the method of majorizing sequences.

Again, we use the theoretical significance of the methods to draw conclusions about the existence and uniqueness of solution of the equation to solve and illustrate all the theoretical results with examples.

4.1 Kantorovich-Type Convergence Conditions

In this section, we obtain a semilocal convergence result for the family of iterations (4.3) under Kantorovich-type conditions on the operator F. This type of conditions habitually allows us to use the method of majorizing sequences described in Sect. 2.2.2.1. However, it is not simple to bound the R-order of convergence of the iterations involved in (4.3). To solve this, we introduce another technique to study the semilocal convergence of an iterative method, which is based on a system of recurrence relations (see [45]), so that we can bound the R-order of convergence.

First, we write the family of iterations (4.3) as follows:

$$\begin{cases} x_0 \text{ given in } \Omega, \\ y_n = x_n - [F'(x_n)]^{-1} F(x_n), \quad n \geq 0, \\ x_{n+1} = G(x_n) = y_n + \frac{1}{2} L_F(x_n) \widetilde{H}(L_F(x_n))(y_n - x_n), \\ \widetilde{H}(L_F(x_n)) = I + 2 \sum_{k \geq 2} A_k L_F(x_n)^{k-1}, \quad A_k \in \mathbb{R}^+, \end{cases} \quad (4.5)$$

where $\{A_k\}_{k \geq 2}$ is a nonincreasing scalar sequence that satisfies (4.4).

4.1.1 Semilocal Convergence

We use the technique based on recurrence relations, which is developed in [45] for the Newton method, to prove the semilocal convergence of the family (4.5). For this, we suppose that F is a twice continuously Fréchet differentiable operator $F : \Omega \subseteq X \to Y$ defined on a non-empty subset of a Banach space X with values in a Banach space Y and the following classical Kantorovich-type conditions:

(C1) There exists $\Gamma_0 = [F'(x_0)]^{-1} \in \mathcal{L}(Y, X)$, for some $x_0 \in \Omega$, with $\|\Gamma_0\| \leq \beta$ and $\|\Gamma_0 F(x_0)\| \leq \eta$.
(C2) There exists a constant $M > 0$ such that $\|F''(x)\| \leq M$, for $x \in \Omega$.
(C3) There exists a constant $K > 0$ such that $\|F''(x) - F''(y)\| \leq K\|x - y\|$, for $x, y \in \Omega$.

4.1.1.1 Recurrence Relations

Consider $a_0 = M\beta\eta$, $b_0 = K\beta\eta^2$, and $v(t) = \sum_{k \geq 2} B_k t^{k-2}$ with $B_k = 2A_k$. From (C1)–(C2)–(C3), it follows $\|L_F(x_0)\| \leq a_0$ and $K\|\Gamma_0\|\|\Gamma_0 F(x_0)\|^2 \leq b_0$. Moreover, the operator $\widetilde{H}(L_F(x_0))$ exists if $\|L_F(x_0)\| < r$, so that $a_0 < r$ is required.

4.1 Kantorovich-Type Convergence Conditions

In addition, if $y_0 \in \Omega$, we obtain

$$\|\widetilde{H}(L_F(x_0))\| = \left\| I + \sum_{k \geq 2} 2A_k L_F(x_0)^{k-1} \right\| \leq 1 + \sum_{k \geq 2} B_k a_0^{k-1} = 1 + a_0 v(a_0),$$

$$\|x_1 - y_0\| = \left\| \frac{1}{2} L_F(x_0) \widetilde{H}(L_F(x_0)) \Gamma_0 F(x_0) \right\| \leq \frac{1}{2} a_0 (1 + a_0 v(a_0)) \|y_0 - x_0\|,$$

$$\|x_1 - x_0\| \leq \|x_1 - y_0\| + \|y_0 - x_0\| \leq \left(1 + \frac{1}{2} a_0 (1 + a_0 v(a_0))\right) \eta.$$

Besides, $\|I - \widetilde{H}(L_F(x_0))\| = \left\| \sum_{k \geq 2} 2A_k L_F(x_0)^{k-1} \right\| \leq \sum_{k \geq 2} B_k a_0^{k-1} = a_0 v(a_0)$.

Next, we try to generalize these initial bounds to any step of the iterative method. For this, we define the scalar auxiliary sequences

$$a_{n+1} = a_n f(a_n)^2 g(a_n, b_n) \quad \text{and} \quad b_{n+1} = b_n f(a_n)^3 g(a_n, b_n)^2, \tag{4.6}$$

where

$$f(t) = \frac{2}{2 - 2t - t^2 - t^3 v(t)}, \tag{4.7}$$

$$g(t, u) = \frac{t^2}{2}\left(1 + (1 + t)v(t) + \frac{t}{4}(1 + t v(t))^2\right) + \frac{u}{6}, \tag{4.8}$$

with $v(t) = \sum_{k \geq 2} B_k t^{k-2}$ and $B_k = 2A_k$. In addition, we consider the scalar auxiliary function

$$h(t) = 1 + \frac{t}{2}(1 + t v(t)). \tag{4.9}$$

From the last sequences and functions, we prove the system of recurrence relations given in the next lemma.

Lemma 4.1 *Suppose that $y_0, y_n, x_n \in \Omega$, for $n \in \mathbb{N}$. If $a_0 < r$, $a_0 h(a_0) < 1$, and the sequences $\{a_n\}$ and $\{b_n\}$ given in (4.6) are decreasing, then the following recurrence relations are satisfied:*

(i_n) $\Gamma_n = [F'(x_n)]^{-1}$ *exists and* $\|\Gamma_n\| \leq f(a_{n-1})\|\Gamma_{n-1}\|$.
(ii_n) $\|\Gamma_n F(x_n)\| = \|y_n - x_n\| \leq f(a_{n-1})g(a_{n-1}, b_{n-1})\|\Gamma_{n-1} F(x_{n-1})\|$.
(iii_n) $M\|\Gamma_n\|\|\Gamma_n F(x_n)\| \leq a_n$.

(iv_n) $\|\widetilde{H}(L_F(x_n))\| \leq 1 + a_n\, v(a_n).$
(v_n) $K\|\Gamma_n\|\|\Gamma_n F(x_n)\|^2 \leq b_n.$
(vi_n) $\|x_{n+1} - x_n\| \leq h(a_n)\|\Gamma_n F(x_n)\|.$
(vii_n) $\|x_{n+1} - x_0\| \leq h(a_0)\left(\sum_{k=0}^{n} f(a_0)^k g(a_0, c_0)^k\right)\eta.$
$(viii_n)$ $\|x_{n+1} - y_n\| \leq \frac{1}{2}a_n\,(1 + a_n\, v(a_n))\,\|\Gamma_n F(x_n)\|.$

Proof We begin by proving that conditions (i_1)–(ii_1)–(iii_1)–(iv_1)–(v_1)–(vi_1)–(vii_1)–$(viii_1)$ are satisfied. First, we see that the operator Γ_1 exists and is bounded. Indeed, from

$$F'(x_1) - F'(x_0) = \int_{x_0}^{x_1} F''(x)dx = \int_0^1 F''(x_0 + \tau(x_1 - x_0))(x_1 - x_0)d\tau,$$

it follows that $\|F'(x_1) - F'(x_0)\| \leq M\|x_1 - x_0\|$, so that

$$\|I - \Gamma_0 F'(x_1)\| \leq \left(1 + \frac{1}{2}a_0\,(1 + a_0\, v(a_0))\right)a_0 = h(a_0)a_0 < 1,$$

and, from the Banach Lemma on invertible operators, Γ_1 exists and (i_1) follows, since

$$\|\Gamma_1\| = \|\Gamma_1 F'(x_0)\Gamma_0\| \leq \left\|[\Gamma_0 F'(x_1)]^{-1}\right\|\|\Gamma_0\| \leq \frac{\|\Gamma_0\|}{1 - h(a_0)a_0} = f(a_0)\|\Gamma_0\|.$$

Next, we prove (ii_1). For this, we use the decompositions

$$F(x_{n+1}) = F(y_n) + F'(y_n)(x_{n+1} - y_n) + \int_{y_n}^{x_{n+1}} F''(x)(x_{n+1} - x)\,dx, \qquad (4.10)$$

$$F(y_n) = F(x_n) + F'(x_n)(y_n - x_n) + \int_{x_n}^{y_n} F''(x)(y_n - x)\,dx, \qquad (4.11)$$

which are obtained from Taylor's formula. Besides,

$$F'(y_n)(x_{n+1} - y_n) = \int_{x_n}^{y_n} F''(x)(x_{n+1} - y_n)\,dx - \frac{1}{2}F''(x_n)\widetilde{H}\,(L_F(x_n))\,(y_n - x_n)^2. \qquad (4.12)$$

Now, if $x = x_n + \tau(y_n - x_n)$, with $\tau \in [0, 1]$, in (4.11) and (4.12) and $x = y_n + \tau(x_{n+1} - y_n)$, with $\tau \in [0, 1]$, in (4.10), we obtain

$$F(x_{n+1}) = \int_0^1 \left(F''(x_n + \tau(y_n - x_n)) - F''(x_n)\right)(y_n - x_n)^2(1 - \tau)\,d\tau$$

4.1 Kantorovich-Type Convergence Conditions

$$+ \int_0^1 F''(x_n)\left(I - \widetilde{H}(L_F(x_n))\right)(y_n - x_n)^2(1-\tau)\,d\tau$$

$$+ \int_0^1 F''(x_n + \tau(y_n - x_n))(y_n - x_n)(x_{n+1} - y_n)\,d\tau$$

$$+ \int_0^1 F''(y_n + \tau(x_{n+1} - y_n))(x_{n+1} - y_n)^2(1-\tau)\,d\tau.$$

If $n = 0$, we have

$$\|F(x_1)\| \leq \frac{K}{6}\|y_0 - x_0\|^3 + \frac{M}{2}a_0(1 + v(a_0) + a_0\,v(a_0))\|y_0 - x_0\|^2$$

$$+ \frac{M}{8}a_0^2(1 + a_0\,v(a_0))^2\|y_0 - x_0\|^2.$$

Thus,

$$\|\Gamma_1 F(x_1)\| \leq f(a_0)\left(\frac{b_0}{6} + \frac{a_0^2}{2}(1 + (1 + a_0)v(a_0)) + \frac{a_0^3}{8}(1 + a_0\,v(a_0))^2\right)\|y_0 - x_0\|,$$

so that

$$\|\Gamma_1 F(x_1)\| \leq f(a_0) g(a_0, b_0) \|\Gamma_0 F(x_0)\|,$$

and (ii_1) is then proved.

From

$$\|L_F(x_1)\| \leq f(a_0)\|\Gamma_0\| M f(a_0) g(a_0, b_0) \|\Gamma_0 F(x_0)\| \leq a_0 f(a_0)^2 g(a_0, b_0) = a_1,$$

we obtain (iii_1), and, taking into account that $\{a_n\}$ is a decreasing sequence, it follows that $a_1 < r$. Hence,

$$\left\|\widetilde{H}(L_F(x_1))\right\| = \left\|I + \sum_{k \geq 2} 2 A_k L_F(x_1)^{k-1}\right\| \leq 1 + \sum_{k \geq 2} B_k a_1^{k-1} = 1 + a_1 v(a_1),$$

and (iv_1) follows.

As a consequence of the abovementioned definitions, (v_1) is clear, since

$$K\|\Gamma_1\|\|\Gamma_1 F(x_1)\|^2 \leq K\beta\eta^2 f(a_0)^3 g(a_0, b_0)^2 = b_0 f(a_0)^3 g(a_0, b_0)^2 = b_1.$$

For (vi_1), we have

$$\|x_2 - x_1\| = \left\| I + \frac{1}{2} L_F(x_1) \widetilde{H}(L_F(x_1)) \right\| \|\Gamma_1 F(x_1)\| \leq h(a_1) \|\Gamma_1 F(x_1)\|.$$

In addition,

$$\|x_2 - x_0\| \leq \|x_2 - x_1\| + \|x_1 - x_0\| \leq h(a_1) \|y_1 - x_1\| + h(a_0) \|y_0 - x_0\|,$$

and taking into account that $\{a_n\}$ is a decreasing sequence and the scalar function h is increasing, it follows (vii_1), since

$$\|x_2 - x_0\| \leq h(a_0) \left(1 + f(a_0) g(a_0, b_0)\right) \eta.$$

Moreover,

$$\left\| I - \widetilde{H}(L_F(x_1)) \right\| = \left\| \sum_{k \geq 2} 2 A_k L_F(x_1)^{k-1} \right\| \leq \sum_{k \geq 2} B_k a_1^{k-1} = a_1 v(a_1),$$

$$\|x_2 - y_1\| \leq \left\| \frac{1}{2} L_F(x_1) \widetilde{H}(L_F(x_1)) \right\| \|\Gamma_1 F(x_1)\|$$

$$\leq \frac{1}{2} a_1 (1 + a_1 v(a_1)) \|\Gamma_1 F(x_1)\|,$$

so that $(viii_1)$ is proved.

After that, we can easily prove (i_n)–(ii_n)–(iii_n)–(iv_n)–(v_n)–(vi_n)–(vii_n)–$(viii_n)$ by mathematical induction on n. ∎

4.1.1.2 Analysis of the Scalar Sequences

Our next aim is to study the scalar sequences given in (4.6) and guarantee the convergence of the family (4.5). First, we give two technical lemmas whose proofs are trivial.

Lemma 4.2 *Let f, g, and h be the three scalar functions given in (4.7), (4.8), and (4.9), respectively. Then:*

(a) *If $h(a_0) a_0 < 1$, f is increasing and $f(t) > 1$ in $(0, a_0)$.*
(b) *Fixed t, the function $g(t, u)$ is increasing as a function of u, and fixed u, $g(t, u)$ is increasing as a function of t.*

Lemma 4.3 *Let f, g, and h be the three scalar functions given in (4.7), (4.8), and (4.9), respectively. Suppose (C1)–(C2)–(C3). If $b_0 < \kappa_0$, where $\kappa_0 = 12 + 6 a_0 - 6 h(a_0)(1 +$*

4.1 Kantorovich-Type Convergence Conditions

$2a_0) + 3h(a_0)^2 a_0(2a_0 - 1)$, *then the sequences* $\{a_n\}$ *and* $\{b_n\}$ *given in (4.6) are strictly decreasing. Besides,* $f(a_0)g(a_0, b_0) < 1$.

4.1.1.3 Semilocal Convergence Result

Next, we establish the semilocal convergence result for the family (4.5) when it is applied to operators that satisfy conditions (C1)–(C2)–(C3).

Theorem 4.4 *Let* $F : \Omega \subseteq X \longrightarrow Y$ *be a twice continuously Fréchet differentiable operator defined on a non-empty open convex domain* Ω *of a Banach space* X *with values in a Banach space* Y. *Suppose that conditions (C1)–(C2)–(C3) are satisfied. Suppose also that* $a_0 < r$, *where* r *is the radius of convergence of the power series* $\sum_{k \geq 0} A_k t^k$ *with* A_k *as in (4.5),* $a_0 h(a_0) < 1$ *and* $b_0 < \kappa_0$ *with* $\kappa_0 = 12 + 6a_0 - 6h(a_0)(1 + 2a_0) + 3h(a_0)^2 a_0(2a_0 - 1)$ *and* h *defined in (4.9). Then, if* $B(x_0, R) \subset \Omega$, *where* $R = \frac{h(a_0)\eta}{1 - f(a_0)g(a_0, b_0)}$ *and* f *and* g *are given, respectively, in (4.7) and (4.8), the family of iterations (4.5) starting at* x_0 *converges to a solution* x^* *of* $F(x) = 0$. *Besides,* $x_n, x^* \in \overline{B(x_0, R)}$, *and* x^* *is unique in* $B\left(x_0, \frac{2}{M\beta} - R\right) \cap \Omega$.

Proof From

$$\|y_0 - x_0\| = \|\Gamma_0 F(x_0)\| \leq \eta < R \quad \text{and} \quad \|x_1 - x_0\| \leq h(a_0)\eta < R,$$

we have that $y_0, x_1 \in \Omega$. We now suppose that $y_k, x_{k+1} \in \Omega$, for $k = 1, \ldots, n-1$. Then, we prove by mathematical induction on n that $y_n, x_{n+1} \in \Omega$. From the recurrence relations (ii_n) and (vi_n) given in Lemma 4.1, it follows

$$\|y_n - x_0\| \leq \|y_n - x_n\| + \|x_n - x_0\| < h(a_0)\left(\sum_{k=0}^{n} f(a_0)^k g(a_0, b_0)^k\right)\eta < R,$$

$$\|x_{n+1} - x_0\| \leq h(a_0)\left(\sum_{k=0}^{n} f(a_0)^k g(a_0, b_0)^k\right)\eta < R.$$

Therefore, $y_n, x_{n+1} \in \Omega$.

Besides, from the recurrence relations (ii_n) and (vi_n) of Lemma 4.1, we obtain

$$\|x_{n+1} - x_n\| \leq h(a_n)\|\Gamma_n F(x_n)\| \leq h(a_0)\prod_{k=0}^{n-1} f(a_k)g(a_k, b_k)\eta.$$

Next, we prove that $\{x_n\}$ is a Cauchy sequence. Taking into account that $\{a_n\}$ and $\{b_n\}$ are decreasing sequences and $f(a_0)g(a_0, b_0) < 1$, it follows that

$$\|x_{n+m} - x_n\| \leq \sum_{i=0}^{m-1} \|x_{n+i+1} - x_{n+i}\|$$

$$\leq h(a_0) \sum_{i=n-1}^{n+m-2} \left(\prod_{j=0}^{n+m-2} f(a_j)g(a_j, b_j) \right) \eta$$

$$\leq h(a_0) \frac{1 - (f(a_0)g(a_0, b_0))^m}{1 - f(a_0)g(a_0, b_0)} \eta (f(a_0)g(a_0, b_0))^n.$$

Therefore, $\{x_n\}$ is a Cauchy sequence and convergent to a solution x^* of $F(x) = 0$. From

$$\|F(x_n)\| \leq \|F'(x_n)\Gamma_n F(x_n)\| \leq \|F'(x_n)\| \|\Gamma_n F(x_n)\| \leq \|F'(x_n)\| \prod_{i=0}^{n-1} f(a_i)g(a_i, b_i)\eta$$

and taking into account that $\|F'(x_n)\|$ is bounded, since

$$\|F'(x_n)\| \leq \|F'(x_n) - F'(x_0)\| + \|F'(x_0)\|$$

$$\leq \sup_{0<\theta<1} \|F''(x_0 + \theta(x_n - x_0))\| \|x_n - x_0\| + \|F'(x_0)\|$$

$$\leq Mh(a_0) \left(\sum_{i=0}^{n-1} f(a_0)^i g(a_0, b_0)^i \right) \eta + \|F'(x_0)\|,$$

by the Mean Value Theorem, it follows that $\lim_{n\to\infty} \|F(x_n)\| = 0$. As a consequence of this and the continuity of the operator F in $\overline{B(x_0, R)}$, we obtain that $F(x^*) = 0$, so that x^* is a solution of $F(x) = 0$.

To prove the uniqueness of the solution x^*, we suppose that y^* is another solution of $F(x) = 0$ in $B\left(x_0, \frac{2}{M\beta} - R\right) \cap \Omega$. Now, from

$$0 = \Gamma_0(F(y^*) - F(x^*)) = \int_{x^*}^{y^*} \Gamma_0 F'(x)\, dx = \int_0^1 \Gamma_0 F'(x^* + \tau(y^* - x^*))\, d\tau (y^* - x^*),$$

we obtain $y^* = x^*$ if the operator J^{-1} exists with $J = \int_0^1 \Gamma_0 F'(x^* + \tau(y^* - x^*))\, d\tau$. For this, we consider

$$F'(x^* + \tau(y^* - x^*)) - F'(x_0) = \int_{x_0}^{x^* + \tau(y^* - x^*)} F''(x) dx$$

$$= \int_0^1 F''\left(x_0 + \tau(x^* + t(y^* - x^*) - x_0)\right) \left((x^* - x_0)(1 - \tau) + \tau(y^* - x_0)\right) d\tau$$

4.1 Kantorovich-Type Convergence Conditions

and take norms to obtain

$$\left\| F'\left(x^* + \tau(y^* - x^*)\right) - F'(x_0) \right\| \leq M \left(\|x^* - x_0\|(1 - \tau) + \tau \|y^* - x_0\| \right).$$

Now, as $y^* \in B(x_0, \frac{2}{M\beta} - R) \cap \Omega$ and $x^* \in \overline{B(x_0, R)}$, we have $\|y^* - x_0\| < \frac{2}{M\beta} - R$ and $\|x^* - x_0\| \leq R$, so that

$$\|I - J\| \leq M\beta \int_0^1 \left(\|x^* - x_0\|(1 - \tau) + \tau \|y^* - x_0\| \right) d\tau < 1,$$

and, by the Banach lemma on invertible operators, the operator J^{-1} exists. ∎

4.1.2 R-Order of Convergence

Once the semilocal convergence of the family of iterations (4.5) has been studied, we see that it converges to a solution x^* of $F(x) = 0$ with R-order of convergence at least three. Before, we present the following technical lemma.

Lemma 4.5 *Let* $\gamma = \dfrac{a_1}{a_0}$. *Under the hypotheses of Lemma 4.3, we have:*

(a) $\gamma = f(a_0)^2 g(a_0, b_0) \in (0, 1)$.
(b) $f(\gamma x) < f(x)$, $g(\gamma x, \gamma^2 y) < \gamma^2 g(x, y)$ for $x, y > 0$ and $\gamma \in (0, 1)$.
(c) $a_n \leq \gamma^{3^{n-1}} a_{n-1} \leq \gamma^{\frac{3^n - 1}{2}} a_0$, $b_n \leq \left(\gamma^{3^{n-1}}\right)^2 b_{n-1} \leq \gamma^{3^n - 1} b_0$, $n \geq 1$.
(d) $f(a_n) g(a_n, b_n) \leq \gamma^{3^n} \Delta$, where $\Delta = \dfrac{1}{f(a_0)}$, $n \geq 1$.

Proof The proofs of items (a) and (b) are immediate. To prove item (c), we use mathematical induction on n. So, if $n = 1$, then $a_1 = \gamma a_0$ and $b_1 \leq \gamma^2 b_0$, since

$$b_1 = b_0 f(a_0)^3 g(a_0, b_0)^2 < b_0 \left(f(a_0)^2 g(a_0, b_0) \right)^2 = b_0 \gamma^2$$

and $f(a_0) > 1$. From Lemma 4.2 and item (b), it follows

$$a_{n+1} \leq \gamma^{3^{n-1}} a_{n-1} f(a_{n-1})^2 \left(\gamma^{3^{n-1}}\right)^2 g(a_{n-1}, b_{n-1}) = \gamma^{3^n} a_n,$$

so that

$$a_{n+1} \leq \gamma^{3^n} a_n \leq \gamma^{3^n} \gamma^{3^{n-1}} a_{n-1} \leq \cdots \leq \gamma^{\frac{3^{n+1} - 1}{2}} a_0.$$

Besides,

$$b_{n+1} \leq \left(\gamma^{3^{n-1}}\right)^2 b_{n-1} f(a_{n-1})^3 \left(\gamma^{3^{n-1}}\right)^{2^2} g(a_{n-1}, b_{n-1})^2 = \left(\gamma^{3^n}\right)^2 b_n,$$

and then

$$b_{n+1} \leq \left(\gamma^{3^n}\right)^2 b_n \leq \left(\gamma^{3^n}\right)^2 \left(\gamma^{3^{n-1}}\right)^2 b_{n-1} \leq \cdots \leq \gamma^{3^{n+1}-1} b_0.$$

Finally, we prove item (d). Taking into account item (c), we have

$$f(a_n) g(a_n, b_n) \leq f(a_0) \gamma^{3^n - 1} g(a_0, b_0) \leq \gamma^{3^n} \frac{f(a_0) g(a_0, b_0)}{\gamma} = \frac{\gamma^{3^n}}{f(a_0)} = \gamma^{3^n} \Delta.$$

The proof is complete. ∎

Remark 4.6 As a consequence of Lemma 4.5, it follows

$$\prod_{k=0}^{n-1} f(a_k) g(a_k, b_k) \leq \prod_{k=0}^{n-1} \gamma^{3^k} \Delta = \gamma^{\frac{3^n - 1}{2}} \Delta^n.$$

Then, $\Delta < 1$ and $\gamma < 1$; it is clear that $\lim_{n \to \infty} \prod_{k=0}^{n-1} f(a_k) g(a_k, b_k) = 0$.

After that, we see that the family (4.5) has R-order of convergence at least three in the next result.

Theorem 4.7 *Under the hypotheses of Theorem 4.4, the family of iterations (4.5) starting at x_0 converges to a solution x^* of $F(x) = 0$ with R-order of convergence at least three. Moreover, we have the following a priori error estimates:*

$$\|x^* - x_n\| < h\left(a_0 \gamma^{\frac{3^n-1}{2}}\right) \eta \frac{\gamma^{\frac{3^n-1}{2}} \Delta^n}{1 - \gamma^{3^n} \Delta} < \left(\gamma^{\frac{1}{2}}\right)^{3^n} \frac{R}{\gamma^{\frac{1}{2}}}, \quad n \geq 0,$$

where $R = \frac{h(a_0)\eta}{1 - f(a_0) g(a_0, b_0)}$.

Proof As a consequence of Lemma 4.5 and the recurrence relations (ii_n) and (vi_n), we have

$$\|x_{n+1} - x_n\| \leq h\left(a_0 \gamma^{\frac{3^n-1}{2}}\right) \gamma^{\frac{3^n-1}{2}} \Delta^n \eta.$$

4.1 Kantorovich-Type Convergence Conditions

Now, from Remark 4.6, we obtain

$$\|x_{n+m} - x_n\| \leq \|x_{n+m} - x_{n+m-1}\| + \|x_{n+m-1} - x_{n+m-2}\| + \cdots + \|x_{n+1} - x_n\|$$

$$\leq h\left(a_0 \gamma^{\frac{3^n-1}{2}}\right)\left(\gamma^{\frac{3^{n+m-1}-1}{2}} \Delta^{n+m-1} + \gamma^{\frac{3^{n+m-2}-1}{2}} \Delta^{n+m-2} + \cdots \gamma^{\frac{3^n-1}{2}} \Delta^n\right) \eta.$$

Now, by the Bernoulli inequality, $(1+z)^k - 1 \geq kz$ if $z > -1$, we have $3^k - 1 > 2k$, and then

$$\gamma^{\frac{3^n(3^{m-1}-1)}{2}} \Delta^{m-1} + \gamma^{\frac{3^n(3^{m-2}-1)}{2}} \Delta^{m-2} + \cdots + 1 < \frac{1 - \left(\gamma^{3^n} \Delta\right)^m}{1 - \gamma^{3^n} \Delta},$$

so that we have, for $n, m \in \mathbb{N}$,

$$\|x_{n+m} - x_n\| < h\left(a_0 \gamma^{\frac{3^n-1}{2}}\right) \gamma^{\frac{3^n-1}{2}} \Delta^n \eta \frac{1 - \left(\gamma^{3^n} \Delta\right)^m}{1 - \gamma^{3^n} \Delta}. \tag{4.13}$$

Then, if $m \to \infty$ in (4.13), we obtain

$$\|x^* - x_n\| < h\left(a_0 \gamma^{\frac{3^n-1}{2}}\right) \eta \frac{\gamma^{\frac{3^n-1}{2}} \Delta^n}{1 - \gamma^{3^n} \Delta} < \left(\gamma^{\frac{1}{2}}\right)^{3^n} \frac{R}{\gamma^{\frac{1}{2}}},$$

and then the family of iterations (4.5) has R-order of convergence at least three. ∎

4.1.3 Particular Cases

From Theorem 4.4, we obtain the semilocal convergence of the following well-known iterative methods with R-order of convergence at least three, which are included in the family of iterations (4.5):

1. *The Chebyshev method*: Since $v(x) = 0$, $h(x) = 1 + \frac{x}{2}$, and $r = +\infty$, Theorem 4.4 is satisfied if $a_0 < \frac{1}{2}$ and

$$b_0 < \frac{3}{4}(2 + a_0)(2a_0 - 1)\left(-4 + 2a_0 + a_0^2\right). \tag{4.14}$$

2. *The Super-Halley method*: Since $v(x) = \sum_{k \geq 2} x^{k-2}$, $h(x) = \frac{x-2}{2(x-1)}$, and $r = 1$, Theorem 4.4 is satisfied if $a_0 < 0.380778\ldots$ and

$$b_0 < \frac{3\left(8 - 32a_0 + 32a_0^2 - 9a_0^3 + 2a_0^4\right)}{4(a_0 - 1)^2}. \tag{4.15}$$

3. *The Halley method*: Since $v(x) = \sum_{k \geq 2} \frac{x^{k-2}}{2^{k-1}}$, $h(x) = \frac{2}{2-x}$, and $r = 2$, Theorem 4.4 is satisfied if $a_0 < 0.434624\ldots$ and

$$b_0 < \frac{6\left(4 - 12a_0 + 6a_0^2 + a_0^3\right)}{(a_0 - 2)^2}. \tag{4.16}$$

Observe that the conditions required on the parameters a_0 and b_0 in Theorem 4.4 guarantee the semilocal convergence of the family of iterations (4.5). In addition, we can then associate the following set of the real plane $\mathbb{R}^2$ with Theorem 4.4

$$D = \{(a_0, b_0) \in \mathbb{R}^2 : a_0 < r, a_0 h(a_0) < 1, b_0 < \kappa_0\},$$

where r is the radius of convergence of the power series $\sum_{k \geq 0} A_k t^k$ with A_k as in (4.4), h is defined in (4.9), and $\kappa_0 = 12 + 6a_0 - 6h(a_0)(1 + 2a_0) + 3h(a_0)^2 a_0(2a_0 - 1)$, so that D is the region of the xy-plane whose points satisfy the conditions $a_0 < r$, $a_0 h(a_0) < 1$, $b_0 < \kappa_0$, and the convergence of the family of iterations (4.5) is guaranteed from Theorem 4.4. This set is called *domain of parameters* associated with Theorem 4.4 and represents the accessibility of the family (4.3) according to Theorem 4.4.

In the graph on the left of Fig. 4.1, the dashed line represents the curve

$$b_0 = \frac{6(1 - 2a_0)\left(4 - 2a_0 - a_0^2\right)}{(2 + a_0)^2},$$

which defines the domain of parameters given by Candela and Marquina in [19] for the Chebyshev method. The solid line represents the curve (4.14) and defines the domain of parameters associated with Theorem 4.4. As a consequence, we observe that the domain of parameters obtained from Theorem 4.4 is larger than that given by Candela and Marquina, so that the region of accessibility of the Chebyshev method is improved from Theorem 4.4.

In the graphs in the center and on the right of Fig. 4.1, we compare the domains of parameters of the Super-Halley and Halley methods, respectively. The dashed lines represent the curves

$$b_0 = \frac{6(1 - 2a_0)(1 - a_0)(4 - 6a_0 + a_0^2)}{(2 - a_0)^2}, \qquad b_0 = 3(1 - a_0)(2 - a_0)(1 - 2a_0),$$

which are obtained for the Super-Halley method in [53] and for the Halley method in [19], respectively. The solid lines represent the curves that define the domains of parameters that

4.1 Kantorovich-Type Convergence Conditions

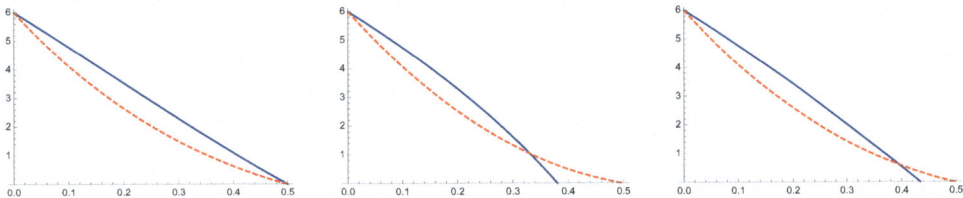

Fig. 4.1 Domains of parameters of the Chebyshev method, the Super-Halley method, and the Halley method, respectively

are obtained from (4.15) and (4.16), respectively. As a consequence, we can say that the domains of parameters are improved if $a_0 < 0.330098\ldots$ for the Super-Halley method and $a_0 < 0.393553\ldots$ for the Halley method.

Remark 4.8 Notice that the domains of parameters associated with Theorem 4.4 for the methods of Chebyshev, Super-Halley, and Halley can be improved if we analyze each of the three methods in a particular and independent way, and not in a general way, as we have previously done from the family of iterations (4.5).

Remark 4.9 Note that the bounds established for a_0 in the study of the best known iterative methods of R-order at least three are apparently more restrictive than those required in Theorem 4.4. This fact is due to the fact that b_0 has to be positive.

4.1.4 Applications

In this section, we give two examples that illustrate the previous results. We obtain a priori error estimates for the Chebyshev method in the first example and for the family of iterations

$$x_{n+1} = x_n - \left(I + \frac{1}{2}L_F(x_n) + \alpha L_F(x_n)^2\right)[F'(x_n)]^{-1}F(x_n), \ n \geq 0, \text{ with } x_0 \text{ given in } \Omega,$$
(4.17)

where $\alpha \in \mathbb{R}^+$, in the second example. In both, we improve the a priori error estimates obtained earlier by other authors.

Example 4.10 We consider $F(x) = x^3 - 10$, choose $x_0 = 2$, and denote the positive root of $F(x) = 0$ by x^*. We give an upper bound C of $10^{11}\|x^* - x_2\|$, where x_2 is the second iteration of the Chebyshev method. If the domain is the interval $(1, 3)$, we have $\beta = 1/12$, $\eta = 1/6$, $M = 18$, and $K = 6$. Then, $a_0 = 1/4$ and $b_0 = 1/72$. From the decomposition given by Candela and Marquina in [19] and calculating the smallest value of n such that $\|x^* - x_n\|$ is of order 10^{-11}, we consider

$$\|x^* - x_2\| \leq \|x^* - x_n\| + \|x_n - x_{n-1}\| + \|x_{n-1} - x_{n-2}\| + \cdots + \|x_3 - x_2\|$$

and obtain $C = 130931.5282$ for $n = 3$. Candela and Marquina obtain $C = 142360.973$ in [19], so that we improve the a priori error estimate.

Example 4.11 We consider the Chandrasekhar equation given by

$$x(s) = 1 + \frac{1}{4} s x(s) \int_0^1 \frac{x(t)}{s+t} dt, \quad s \in [0, 1]. \quad (4.18)$$

This type of integral equations appears in [20] and arises from the study of the radiative transfer theory, the transport of neutrons, and the kinetic theory of the gases [5, 6, 35].

Solving the previous Chandrasekhar equation is equivalent to solving the equation $\mathcal{F}(x) = 0$ with $\mathcal{F} : \mathcal{C}([0, 1]) \to \mathcal{C}([0, 1])$ and

$$[\mathcal{F}(x)](s) = x(s) - 1 - \frac{1}{4} s x(s) \int_0^1 \frac{x(t)}{s+t} dt, \quad s \in [0, 1].$$

Moreover,

$$[\mathcal{F}'(x)y](s) = y(s) - \frac{1}{4} s x(s) \int_0^1 \frac{y(t)}{s+t} dt - \frac{1}{4} s y(s) \int_0^1 \frac{x(t)}{s+t} dt, \quad s \in [0, 1],$$

$$[\mathcal{F}''(x)yz](s) = -\frac{1}{4} s y(s) \int_0^1 \frac{z(t)}{s+t} dt - \frac{1}{4} s z(s) \int_0^1 \frac{y(t)}{s+t} dt, \quad s \in [0, 1],$$

where $x, y, z \in \mathcal{C}([0, 1])$.

Observe that a reasonable choice for the starting point is $x_0(s) = 1$, for all $s \in [0, 1]$, since $x(0) = 1$. We then calculate β, η, M, and K for Theorem 4.4. As

$$[\mathcal{F}(x_0)](s) = -\frac{s}{4} \int_0^1 \frac{dt}{s+t} = -\frac{s}{4} \log\left(\frac{s+1}{s}\right),$$

it follows $\|\mathcal{F}(x_0)\| \leq \frac{\log 2}{4}$. Moreover,

$$\|I - \mathcal{F}'(x_0)\| \leq \frac{\log 2}{2} = 0.3465\ldots < 1,$$

and, by the Banach lemma on invertible operators, the operator $[\mathcal{F}'(x_0)]^{-1}$ exists and

$$\|[\mathcal{F}'(x_0)]^{-1}\| \leq \frac{1}{1 - \|I - \mathcal{F}'(x_0)\|} \leq \frac{2}{2 - \log 2} = 1.5303\ldots = \beta,$$

4.2 Mild Convergence Conditions

Table 4.1 A priori error estimates $\|x^* - x_n\|$ according to [34] and by Theorem 4.7 from Example 4.11

n	Ref. [34]	Theorem 4.7
0	$2.8704\ldots \times 10^{-1}$	$2.8709\ldots \times 10^{-1}$
1	$7.3599\ldots \times 10^{-2}$	$6.2292\ldots \times 10^{-2}$
2	$4.4812\ldots \times 10^{-14}$	$2.2833\ldots \times 10^{-14}$
3	$6.1643\ldots \times 10^{-55}$	$6.7586\ldots \times 10^{-56}$

$$\|[\mathcal{F}'(x_0)]^{-1}\mathcal{F}(x_0)\| \leq \frac{\log 2}{4 - 2\log 2} = 0.2651\ldots = \eta.$$

In addition,

$$\|\mathcal{F}''(x)\| \leq \frac{\log 2}{2} = 0.3465\ldots = M \quad \text{and} \quad K = 0.$$

As a consequence, $R = 0.2870\ldots$, and the Chandrasekhar equation has a solution in $\overline{B(x_0, R)} = \overline{B(x_0, 0.2870\ldots)}$, which is unique in $B\left(x_0, \frac{2}{M\beta} - R\right) = B(x_0, 3.4836\ldots)$.

In Table 4.1, we give some a priori error estimates $\|x^* - x_n\|$ for the iterative method (4.17) with $\alpha = \frac{1}{2}$, which improve those obtained in [34].

4.2 Mild Convergence Conditions

As we have indicated previously, there are situations in which the second derivative of the operator involved is not bounded, as we can see in [37], so that we cannot prove the semilocal convergence of the family (4.3) from Theorem 4.4 established in Sect. 4.1. In this section, we avoid this problem by considering that the second derivative is only bounded in x_0 instead of (C2). In addition, we prove the semilocal convergence of the family (4.3) by the method of majorizing sequences [63].

4.2.1 Semilocal Convergence

In Sect. 3.2.2, we have seen the method of majorizing sequences to prove the semilocal convergence of the Newton method in Banach spaces. In this section, we give a method to find majorizing sequences for the family of iterations (4.3). For this, we use the degree of logarithmic convexity introduced in Sect. 2.2.1, which plays a key role.

Next, we present in the following lemma the property of the degree of logarithmic convexity for a scalar function that is given in [2].

Lemma 4.12 *Let f be a scalar function such that $f(t) \geq 0$, $f'(t) \neq 0$, $f'''(t) \geq 0$ in $[0, t^*]$, where t^* is the smallest positive root of $f(t) = 0$. Then, $L_f(t) < \frac{1}{2}$ in $[0, t^*]$.*

Now, we consider the family of iterations (4.2), where f is a scalar, convex, nonincreasing, and sufficiently differentiable function in an interval $[a,b]$ such that $a \leq 0 < b$ and $f(a) > 0 > f(b)$. Besides, $f'''(t) \geq 0$ in $[a, t^*]$, where t^* is the unique root of $f(t) = 0$ in $[a,b]$. In addition, we give the following convergence result.

Theorem 4.13 *Let f be a nonincreasing and convex scalar function in $[a,b]$ such that $a \leq 0 < b$, $f(a) > 0 > f(b)$, and $f'''(t) \geq 0$ in $[a,t^*]$, where t^* is the unique root of $f(t) = 0$ in $[a,b]$. Besides, $|L_f(t)| < r$ in $[a, t^*]$, where r is the radius of convergence of the power series (4.4). Then, the sequence (4.2) is nondecreasing and convergent to t^*.*

Proof We prove that the scalar sequence (4.2) is bounded by t^*, is increasing, and converges to t^*.

Without loss of generality, we choose $t_0 = 0 \in [a,b]$. From (4.2), it follows

$$t_1 - t_0 = -H(L_f(t_0))\frac{f(t_0)}{f'(t_0)} \geq 0,$$

since $H(L_f(t_0)) \geq 0$, $f(t_0) > 0$, and $f'(t_0) < 0$. Thus, $t_1 \geq t_0$. Now, by the Mean Value Theorem, we obtain $t_1 \leq t^*$, since

$$t_1 - t^* = G(t_0) - G(t^*) = G'(\xi_0)(t_0 - t^*), \text{ with } \xi_0 \in (t_0, t^*).$$

Moreover,

$$G'(t) = 1 - H'(L_f(t))L'_f(t)\frac{f(t)}{f'(t)} - H(L_f(t))(1 - L_f(t)).$$

As

$$L'_f(t) = \frac{f''(t)}{f'(t)} + \frac{f(t)f'''(t)}{f'(t)^2} - 2\frac{f(t)f''(t)^2}{f'(t)^3},$$

we have

$$L'_f(t)\frac{f(t)}{f'(t)} = L_f(t) + L_f(t)^2 L_{f'}(t) - 2L_f(t)^2 = L_f(t)\left(1 - L_f(t)\left(2 - L_{f'}(t)\right)\right),$$

so that

$$G'(t) = 1 - \left(\frac{1}{2} + \sum_{k \geq 2} k A_k L_f(t)^{k-1}\right) L_f(t)\left(1 - L_f(t)\left(2 - L_{f'}(t)\right)\right)$$

4.2 Mild Convergence Conditions

$$-\left(1+\frac{1}{2}L_f(t)+\sum_{k\geq 2}A_k L_f(t)^k\right)(1-L_f(t)).$$

Taking now into account that $A_1 = \frac{1}{2}$, we obtain

$$G'(t) = \sum_{k\geq 2}\left(((k-1)(2-L_{f'}(t))+1)A_{k-1} - (k+1)A_k\right)L_f(t)^k.$$

From

$$L_{f'}(t) = \frac{f'(t)f'''(t)}{f''(t)^2} \leq 0, \quad t \in (t_0, r),$$

and, since $\{A_k\}_{k\geq 2}$ is a nonnegative sequence, it follows

$$(2(k-1)+1-(k+1))A_k = (k-2)A_k \geq 0, \quad k \geq 2.$$

Therefore,

$$\left((k-1)(2-L_{f'}(t))+1\right)A_{k-1} - (k+1)A_k \geq 0, \quad k \geq 2,$$

and, as a consequence, $G'(t) \geq 0$ in (t_0, t^*), and then $t_1 \leq t^*$.

Next, we suppose that $t_k \geq t_{k-1}$ and $t_k \leq t^*$, with $k = 1, 2, \ldots, n-1$. As above, from (4.2), we obtain

$$t_n - t_{n-1} = -H\left(L_f(t_{n-1})\right)\frac{f(t_{n-1})}{f'(t_{n-1})} \geq 0,$$

since $H\left(L_f(t_{n-1})\right) \geq 0$, $f(t_{n-1}) > 0$, and $f'(t_{n-1}) < 0$. Now, by the Mean Value Theorem, we have

$$t_n - t^* = G(t_{n-1}) - G(t^*) = G'(\xi_{n-1})(t_{n-1} - t^*), \text{ with } \xi_{n-1} \in (t_{n-1}, t^*).$$

From $G'(t) \geq 0$ in (t_k, t^*), since $(t_k, t^*) \subset (t_0, t^*)$, for $k = 1, 2, \ldots, n-1$, we deduce that $t_n \leq t^*$. Hence, $\{t_n\}$ is an increasing and upper bounded sequence. Then, there exists $v \in [a, b]$ such that $v \leq t^*$ and $v = \lim_{n\to\infty} t_n$. By letting $n \to \infty$ in (4.2), it follows

$$v = v - H(L_f(v))\frac{f(v)}{f'(v)},$$

and $f(v) = 0$, since $v \leq t^*$ and $H(L_f(v)) > 0$. Now, since t^* is the unique solution of $f(t) = 0$, we obtain $v = t^*$. ∎

From Theorem 4.13, we establish the convergence of the sequence (4.2) to the unique root t^* of $f(t) = 0$ in $[a, b]$. Next, we prove the convergence of the sequence (4.3) to a solution x^* of $F(x) = 0$. Before, we provide two technical lemmas, whose proofs are analogous.

Lemma 4.14 *Let $F : \Omega \subseteq X \longrightarrow Y$ be a twice continuously Fréchet differentiable operator defined on a non-empty open convex domain Ω of a Banach space X with values in a Banach space Y. Then, for the family (4.3), we have*

$$F(x_{n+1}) = \frac{1}{2} F''(x_n) \Gamma_n F(x_n) \left(\sum_{k \geq 2} 2(A_{k-1} - A_k) L_F(x_n)^{k-1} \right) \Gamma_n F(x_n)$$

$$+ \frac{1}{8} F''(x_n) L_F(x_n) \widetilde{H}(L_F(x_n)) \Gamma_n F(x_n) L_F(x_n) \widetilde{H}(L_F(x_n)) \Gamma_n F(x_n)$$

$$+ \int_{x_n}^{x_{n+1}} (F''(x) - F''(x_n))(x_{n+1} - x) \, dx,$$

where $\widetilde{H}(z) = I + 2 \sum_{k \geq 2} A_k z^{k-1}$.

Proof From Taylor's formula and (4.5), we obtain

$$F(x_{n+1}) = F(x_n) + F'(x_n)(x_{n+1} - x_n) + \frac{1}{2} F''(x_n)(x_{n+1} - x_n, x_{n+1} - x_n)$$

$$+ \int_{x_n}^{x_{n+1}} (F''(x) - F''(x_n))(x_{n+1} - x) \, dx,$$

$$F'(x_n)(x_{n+1} - x_n) = F'(x_n) \left(-\left(I + \frac{1}{2} L_F(x_n) \widetilde{H}(L_F(x_n)) \right) \right) \Gamma_n F(x_n)$$

$$= -F(x_n) - \frac{1}{2} F''(x_n) \Gamma_n F(x_n) \widetilde{H}(L_F(x_n)) \Gamma_n F(x_n).$$

If we denote $W(x_n) = -\Gamma_n F(x_n) - \frac{1}{2} L_F(x_n) \widetilde{H}(L_F(x_n)) \Gamma_n F(x_n)$, then we can write

$$F(x_{n+1}) = -\frac{1}{2} F''(x_n) \Gamma_n F(x_n) \widetilde{H}(L_F(x_n)) \Gamma_n F(x_n)$$

4.2 Mild Convergence Conditions

$$+ \frac{1}{2} F''(x_n) W(x_n)^2 + \int_{x_n}^{x_{n+1}} \left(F''(x) - F''(x_n) \right) (x_{n+1} - x) \, dx,$$

and, since F'' is a symmetric bilinear operator, it follows

$$F''(x_n) W(x_n)^2 = F''(x_n) \Gamma_n F(x_n) \left(I + L_F(x_n) \widetilde{H} \left(L_F(x_n) \right) \right) \Gamma_n F(x_n)$$
$$+ \frac{1}{4} F''(x_n) L_F(x_n) \widetilde{H} \left(L_F(x_n) \right) \Gamma_n F(x_n) L_F(x_n) \widetilde{H} \left(L_F(x_n) \right) \Gamma_n F(x_n).$$

Therefore,

$$F(x_{n+1}) = -\frac{1}{2} F''(x_n) \Gamma_n F(x_n) \widetilde{H}(L_F(x_n)) \Gamma_n F(x_n)$$
$$+ \frac{1}{2} F''(x_n) \Gamma_n F(x_n) \left(I + L_F(x_n) \widetilde{H} \left(L_F(x_n) \right) \right) \Gamma_n F(x_n)$$
$$+ \frac{1}{8} F''(x_n) L_F(x_n) \widetilde{H} \left(L_F(x_n) \right) \Gamma_n F(x_n) L_F(x_n) \widetilde{H} \left(L_F(x_n) \right) \Gamma_n F(x_n)$$
$$+ \int_{x_n}^{x_{n+1}} \left(F''(x) - F''(x_n) \right) (x_{n+1} - x) \, dx,$$

and, since

$$\frac{1}{2} F''(x_n) \Gamma_n F(x_n) \left(I + L_F(x_n) \widetilde{H} \left(L_F(x_n) \right) - \widetilde{H} \left(L_F(x_n) \right) \right) \Gamma_n F(x_n)$$
$$= \frac{1}{2} F''(x_n) \Gamma_n F(x_n) \left((1 - 2A_2) L_F(x_n) + \sum_{k \geq 2} 2(A_k - A_{k+1}) L_F(x_n)^k \right) \Gamma_n F(x_n),$$

the thesis follows. ∎

Analogously, we give a similar result for the scalar function f in the following lemma.

Lemma 4.15 *Let f be a twice differentiable scalar function. Then, for the family* (4.2), *we have*

$$f(t_{n+1}) = \frac{1}{2} f(t_n) \sum_{k \geq 2} 2 \left(A_{k-1} - A_k \right) L_f(t_n)^k + \frac{1}{8} f(t_n) L_f(t_n)^3 \widetilde{h} \left(L_f(t_n) \right)^2$$
$$+ \int_0^1 \left(f''(t_n + \tau(t_{n+1} - t_n)) - f''(t_n) \right) (1 - \tau)(t_{n+1} - t_n)^2 \, d\tau,$$

where $\widetilde{h}(t) = 1 + 2\sum_{k \geq 2} A_k t^{k-1}$.

After that, we prove that (4.2) is a majorizing sequence of the family of iterations (4.3).

Theorem 4.16 (General Semilocal Convergence) *Let f be a nonincreasing convex scalar function in $[a, b]$ such that $f(a) > 0 > f(b)$, with $a \leq 0 < b$, and $f(t_0) > 0$ with $t_0 \in [a, b]$. Suppose that $f'''(t) \geq 0$ in $[a, t^*]$, where t^* is the unique solution of $f(t) = 0$ in $[a, b]$, and $|L_f(t)| < r$ in $[a, t^*]$, where r is the radius of convergence of the power series (4.4). If:*

- *There exists $\Gamma_0 = [F'(x_0)]^{-1} \in \mathcal{L}(Y, X)$, for some $x_0 \in \Omega$, with $\|\Gamma_0\| \leq -\frac{1}{f'(t_0)}$ and $\|\Gamma_0 F(x_0)\| \leq -\frac{f(t_0)}{f'(t_0)}$, and $\|F''(x_0)\| \leq f''(t_0)$*
- *$\|F''(x) - F''(y)\| \leq |f''(u) - f''(v)|$, for $\|x - y\| \leq |u - v|$, with $x, y \in \Omega$ and $u, v \in [a, t^*]$*

and $B(x_0, t^* - t_0) \subset \Omega$, then the sequence (4.3) starting at x_0 converges to a solution x^* of $F(x) = 0$. Moreover,

$$\|x^* - x_n\| \leq t^* - t_n, \quad n \geq 0,$$

where t_n is defined by (4.2).

Proof From Theorem 4.13, it follows that the sequence (4.2) is nondecreasing and convergent to t^*. Next, we prove that the approximations x_n of the sequence (4.3) are well defined and form a convergent sequence.

Observe that the approximation x_1 is well defined, since

$$x_1 = x_0 - \left(I + \frac{1}{2} L_F(x_0) \widetilde{H}(L_F(x_0))\right) \Gamma_0 F(x_0),$$

where $\widetilde{H}(z) = I + 2\sum_{k \geq 2} A_k z^{k-1}$, and there exist Γ_0, $L_F(x_0)$ and $\widetilde{H}(L_F(x_0))$. Moreover, as

$$\left\|\widetilde{H}(L_F(x_0))\right\| \leq 1 + \sum_{k \geq 2} 2 A_k L_f(t_0)^{k-1} = \widetilde{h}(L_f(t_0)),$$

$$\|x_1 - x_0\| \leq \left(1 + \frac{1}{2} L_f(t_0) \widetilde{h}(L_f(t_0))\right) \left(\frac{-f(t_0)}{f'(t_0)}\right) = t_1 - t_0 \leq t^* - t_0,$$

4.2 Mild Convergence Conditions

where $\widetilde{h}(t) = 1 + 2\sum_{k\geq 2} A_k t^{k-1}$ and r is the radius of convergence of the power series (4.4), we have

$$x_1 = x_0 - \left(I + \frac{1}{2}L_F(x_0)\widetilde{H}\left(L_F(x_0)\right)\right)\Gamma_0 F(x_0) \in \Omega,$$

and $x_1 \in B(x_0, t^* - t_0) \subset \Omega$.

Next, we prove the existence of Γ_1. From

$$\|I - \Gamma_0 F'(x_1)\| \leq \|\Gamma_0\| \|F'(x_0) - F'(x_1)\| \leq \|\Gamma_0\| \int_0^1 \|F''(x_0 + \tau(x_1 - x_0))\| \|x_1 - x_0\| d\tau$$

and $\|x_0 + \tau(x_1 - x_0) - x_0\| \leq \tau \|x_1 - x_0\| \leq \tau(t_1 - t_0) = t_0 + \tau(t_1 - t_0) - t_0$, it follows

$$\|I - \Gamma_0 F'(x_1)\| \leq \frac{-1}{f'(t_0)} \int_0^1 \left(\|F''(x_0)\| + \|F''(x_0 + \tau(x_1 - x_0)) - F''(x_0)\|\right) \|x_1 - x_0\| d\tau$$

$$\leq \frac{-1}{f'(t_0)} \int_0^1 f''(t_0 + \tau(t_1 - t_0))(t_1 - t_0) d\tau$$

$$= \frac{-1}{f'(t_0)} \left(f'(t_1) - f'(t_0)\right)$$

$$< 1,$$

since f' is an increasing and negative function. As consequence of the Banach lemma on invertible operators, the operator $[\Gamma_0 F'(x_1)]^{-1}$ exists, and, taking into account $[\Gamma_0 F'(x_1)]^{-1} = [I - (I - \Gamma_0 F'(x_1))]^{-1}$, it follows

$$\left\|[\Gamma_0 F'(x_1)]^{-1}\right\| \leq \frac{1}{1 - \|I - \Gamma_0 F'(x_1)\|} \leq \frac{f'(t_0)}{f'(t_1)}.$$

Now, as $\Gamma_1 = [\Gamma_0 F'(x_1)]^{-1} \Gamma_0$, we have

$$\|\Gamma_1\| \leq \left\|[\Gamma_0 F'(x_1)]^{-1}\right\| \|\Gamma_0\| \leq \frac{f'(t_0)}{f'(t_1)} \frac{-1}{f'(t_0)} = -\frac{1}{f'(t_1)}.$$

Next, from Lemma 4.14 with $x = x_0 + \tau(x_1 - x_0)$ and $\tau \in [0, 1]$, it follows

$$\|F(x_1)\| \leq \frac{1}{2}\|F''(x_0)\| \|\Gamma_0 F(x_0)\|^2 \left\| \sum_{k\geq 2} 2(A_{k-1} - A_k) L_F(x_0)^{k-1} \right\|$$

$$+ \frac{1}{8}\|F''(x_0)\| \|L_F(x_0)\|^2 \left\| \widetilde{H}(L_F(x_0)) \right\|^2 \|\Gamma_0 F(x_0)\|^2$$

$$+ \left\| \int_0^1 \left(F''(x_0 + \tau(x_1 - x_0)) - F''(x_0) \right) (1-\tau)(x_1 - x_0)^2 \, d\tau \right\|.$$

As the sequence $\{A_k\}_{k\geq 2}$ is nonincreasing, we obtain

$$\|F(x_1)\| \leq \frac{1}{2} f(t_0) \sum_{k\geq 2} 2(A_{k-1} - A_k) L_f(t_0)^k + \frac{1}{8} f(t_0) L_f(t_0)^3 \widetilde{h}(L_f(t_0))^2$$

$$+ \int_0^1 \left(f''(t_0 + \tau(t_1 - t_0)) - f''(t_0) \right) (1-\tau)(t_1 - t_0)^2 \, d\tau = f(t_1),$$

and then

$$\|\Gamma_1 F(x_1)\| \leq \|\Gamma_1\| \|F(x_1)\| \leq -\frac{f(t_1)}{f'(t_1)}.$$

Moreover,

$$\|L_F(x_1)\| \leq \|\Gamma_1\| \|F''(x_1)\| \|\Gamma_1 F(x_1)\| \leq \frac{-1}{f'(t_1)} \left(\|F''(x_0)\| + \|F''(x_1) - F''(x_0)\| \right) \frac{-f(t_1)}{f'(t_1)}$$

$$\leq \frac{-1}{f'(t_1)} f''(t_1) \frac{-f(t_1)}{f'(t_1)} = L_f(t_1) < r,$$

since $\|x_1 - x_0\| \leq t_1 - t_0 \leq t^* - t_0$, and

$$\|\widetilde{H}(L_F(x_1))\| = \left\| I + \sum_{k\geq 2} 2 A_k L_F(x_1)^{k-1} \right\| \leq 1 + \sum_{k\geq 2} 2 A_k L_f(t_1)^{k-1} = \widetilde{h}(L_f(t_1)).$$

Furthermore,

$$\|x_2 - x_1\| \leq \left(1 + \frac{1}{2} L_f(t_1) \widetilde{h}(L_f(t_1)) \right) \left(\frac{-f(t_1)}{f'(t_1)} \right) = t_2 - t_1,$$

$$\|x_2 - x_0\| \leq \|x_2 - x_1\| + \|x_1 - x_0\| \leq t_2 - t_1 + t_1 - t_0 = t_2 - t_0 \leq t^* - t_0,$$

4.2 Mild Convergence Conditions

and $x_2 \in B(x_0, t^* - t_0) \subset \Omega$.

After that, for $i = 1, 2, \ldots, n$, we suppose that $x_1, x_2, \ldots, x_n \in \Omega$, and

- There exists $\Gamma_i = [F'(x_i)]^{-1}$ with $\|\Gamma_i\| \leq -\frac{1}{f'(t_i)}$ and $\|\Gamma_i F(x_i)\| \leq -\frac{f(t_i)}{f'(t_i)}$.
- $\|L_F(x_i)\| \leq L_f(t_i) < r$.
- $\|\widetilde{H}(L_F(x_i))\| \leq \widetilde{h}(L_f(t_1))$.
- $\|x_i - x_{i-1}\| \leq t_i - t_{i-1}$.

As a consequence, we have $\|x_i - x_0\| \leq t_i - t_0 \leq t^* - t_0$ and $x_i \in B(x_0, t^* - t_0)$, for $i = 1, 2, \ldots, n$.

We note that there exist Γ_n, $L_F(x_n)$, and $\widetilde{H}(L_F(x_n))$, so that

$$x_{n+1} = x_n - \left(I + \frac{1}{2} L_F(x_n) \widetilde{H}(L_F(x_n))\right) \Gamma_n F(x_n) \in \Omega,$$

and x_{n+1} is then well defined. Besides, there exist Γ_n, $L_F(x_n)$, and $\widetilde{H}(L_F(x_n))$. In addition, as

$$\|L_F(x_n)\| \leq \|\Gamma_n\| \|F''(x_n)\| \|\Gamma_n F(x_n)\| \leq \frac{-1}{f'(t_n)} f''(t_n) \frac{-f(t_n)}{f'(t_n)} = L_f(t_n) < r,$$

$$\|\widetilde{H}(L_F(x_n))\| = \left\| I + \sum_{k \geq 2} 2 A_k L_F(x_n)^{k-1} \right\| \leq 1 + \sum_{k \geq 2} 2 A_k L_f(t_n)^{k-1} = \widetilde{h}(L_f(t_n)),$$

we have

$$\|x_{n+1} - x_n\| \leq \left(1 + \frac{1}{2} L_f(t_n) \widetilde{h}(L_f(t_n))\right) \left(\frac{-f(t_n)}{f'(t_n)}\right) = t_{n+1} - t_n,$$

$$\|x_{n+1} - x_0\| \leq \|x_{n+1} - x_n\| + \|x_n - x_0\| \leq t_{n+1} - t_n + t_n - t_0 = t_{n+1} - t_0 \leq t^* - t_0,$$

and $x_{n+1} \in B(x_0, t^* - t_0) \subset \Omega$. Therefore, by mathematical induction on n, $x_n \in B(x_0, t^* - t_0) \subset \Omega$, for all $n \geq 0$, and $\|x_{n+1} - x_n\| \leq t_{n+1} - t_n$, for all positive integers n. As a consequence, the sequence (4.2) majorizes the sequence (4.3), so that there exists $\lim_{n \to +\infty} x_n = x^*$ and $\|x^* - x_n\| \leq t^* - t_n$, for all $n \geq 0$.

In addition,

$$\lim_{n \to +\infty} \|F(x_n)\| \leq \lim_{n \to +\infty} f(t_n) = f(t^*) = 0.$$

Then, $\lim_{n \to +\infty} F(x_n) = 0$; from the continuity of operator F, we obtain $F(x^*) = 0$, and x^* is a solution of $F(x) = 0$. ∎

Notice that condition (C2), given in Sect. 4.1.1, requires that F'' is bounded in Ω. But, this condition is sometimes difficult to prove in practice. There are situations in which F'' is not bounded, as we can see below in Example 4.18 (or more generally in [63]). In these situations, we can relax condition (C2) by the condition:

(C2*) $\|F''(x_0)\| \leq \delta$.

Observe that condition (C2*) is milder than (C2), since F'' is only bounded in the initial iterate x_0.

Now, we suppose conditions (C1)–(C2*)–(C3) and prove that (4.3) is convergent to a solution of $F(x) = 0$. Note that we can obtain a majorant function from conditions (C1)–(C2*)–(C3). Indeed, to find such a function, we have

$$\|F''(x) - F''(y)\| \leq K\|x - y\| \leq K(u - v) = f''(u) - f''(v)$$

if $\|x - y\| \leq u - v$, so that

$$\int_u^v K\, d\tau = \int_u^v f'''(\tau)\, d\tau.$$

Therefore, we have to solve the initial value problem:

$$\begin{cases} y'''(t) - K = 0, \\ y(t_0) = \dfrac{\eta}{\beta}, \quad y'(t_0) = -\dfrac{1}{\beta}, \quad y''(t_0) = \delta, \end{cases}$$

whose solution is the cubic polynomial

$$f(t) = \frac{K}{6}(t - t_0)^3 + \frac{\delta}{2}(t - t_0)^2 - \frac{t - t_0}{\beta} + \frac{\eta}{\beta}.$$

If $t_0 = 0$, the last polynomial is reduced to

$$p(t) = \frac{K}{6}t^3 + \frac{\delta}{2}t^2 - \frac{t}{\beta} + \frac{\eta}{\beta}. \tag{4.19}$$

Next, we prove that the polynomial (4.19) satisfies the hypotheses of Theorem 4.16. For this, we choose $[t_0, b]$ as $[0, t^*]$, where t^* is the smallest positive root of (4.19). Notice that (4.19) has a maximum and a minimum, respectively, at

$$t = \frac{-\delta - \sqrt{\delta^2 + 2K/\beta}}{K} < 0 \quad \text{and} \quad \hat{t} = \frac{-\delta + \sqrt{\delta^2 + 2K/\beta}}{K} > 0.$$

4.2 Mild Convergence Conditions

A necessary and sufficient condition for the polynomial (4.19) to have positive roots is $p(\hat{t}) \leq 0$, or, equivalently,

$$\eta \leq \frac{1}{3K^2} \left(\beta \left(\delta^2 + 2\frac{K}{\beta} \right)^{3/2} - \beta\delta^3 - 3K\delta \right). \qquad (4.20)$$

Notice that conditions (C1)–(C2*)–(C3) are satisfied by the polynomial (4.19). Hence, from Theorem 4.16, we obtain the following semilocal convergence result for the family (4.3).

Theorem 4.17 *Let $F : \Omega \subseteq X \to Y$ be a twice continuously Fréchet differentiable operator defined on a non-empty open convex domain Ω of a Banach space X with values in a Banach space Y. Suppose (C1)–(C2*)–(C3) and $B(x_0, t^*) \subset \Omega$. If (4.20) is satisfied, then the polynomial (4.19) has two positive roots t^* and t^{**}, such that $t^* \leq t^{**}$, and the sequence (4.3) starting at x_0 converges to a solution x^* of $F(x) = 0$ in $\overline{B(x_0, t^*)}$. If $t^* < t^{**}$, the solution is unique in $B(x_0, t^{**}) \cap \Omega$, and if $t^* = t^{**}$, the solution is unique in $B(x_0, t^*)$. Moreover,*

$$\|x^* - x_n\| \leq t^* - t_n, \quad n \geq 0, \qquad (4.21)$$

where t_n is defined by (4.2).

Proof The convergence of the sequence (4.3) follows immediately from Theorem 4.16. To see the uniqueness, we suppose that $t^* < t^{**}$ and y^* is another solution of $F(x) = 0$ in $B(x_0, t^{**}) \cap \Omega$. From

$$0 = \Gamma_0(F(y^*) - F(x^*)) = \left(\int_0^1 \Gamma_0 F'(x^* + \tau(y^* - x^*)) \, d\tau \right) (y^* - x^*) = J(y^* - x^*),$$

it is enough to prove that the operator $J = \int_0^1 \Gamma_0 F'(x^* + \tau(y^* - x^*)) \, d\tau$ is invertible, and as a consequence, $x^* = y^*$.

After that, from

$$\|I - J\| \leq \|\Gamma_0\| \int_0^1 \|F'(F'(x^* + \tau(y^* - x^*)) - F'(x_0)\| \, d\tau$$

$$\leq \|\Gamma_0\| \int_0^1 \|F''(x_0 + s(x^* + \tau(y^* - x^*) - x_0)) - F''(x_0)\| \|x^*$$
$$+ \tau(y^* - x^*) - x_0\| \, ds$$

$$+ \|\Gamma_0\| \int_0^1 \|F''(x_0)\| \|x^* + \tau(y^* - x^*) - x_0\| \, d\tau$$

and, from $\|y^* - x_0\| < t^{**}$ and $\|x^* - x_0\| \leq t^*$, we have

$$\|I - J\| < \frac{\beta K}{6}\left((t^*)^2 + (t^{**})^2 + t^* t^{**}\right) + \frac{\beta \delta}{2}(t^* + t^{**}) = \beta\left(\psi(t^{**}) + \frac{1}{\beta}\right),$$

where $\psi(z) = \frac{K}{6}z^2 + \left(\frac{K}{6}t^* + \frac{\delta}{2}\right)z + \frac{K}{6}(t^*)^2 + \frac{\delta}{2}t^* - \frac{1}{\beta}$.

Next, from the Cardano formulas, we obtain

$$-\frac{K}{6}(t^* + t^{**} - r_0) = \frac{\delta}{2} \quad \text{and} \quad \frac{K}{6}(t^* t^{**} - r_0(t^* + t^{**})) = -\frac{1}{\beta},$$

where r_0, t^*, and t^{**} are the three roots of the polynomial (4.19). Then, $\psi(t^{**}) = \frac{-K}{6}(t^* t^{**} - r_0(t^* + t^{**})) - \frac{1}{\beta} = 0$, and as a consequence, $\|I - J\| < 1$. Thus, the operator J^{-1} exists, and the uniqueness of solution is given in $B(x_0, t^{**}) \cap \Omega$. ∎

4.2.2 Application

Now, we illustrate the last convergence result with an application where a nonlinear integral equation of Fredholm-type is involved.

Example 4.18 We consider the following nonlinear Fredholm integral equation of second kind:

$$x(s) = 1 + \frac{3}{100}\int_0^1 e^{s+t} x(t)^3 dt, \quad s \in [0, 1]. \tag{4.22}$$

Solving the integral equation (4.22) is equivalent to solving the equation $\mathcal{F}(x) = 0$ with $\mathcal{F} : \mathcal{C}([0, 1]) \to \mathcal{C}([0, 1])$ and

$$[\mathcal{F}(x)](s) = x(s) - 1 - \frac{3}{100}\int_0^1 e^{s+t} x(t)^3 dt, \quad s \in [0, 1].$$

Moreover,

$$[\mathcal{F}'(x)y](s) = w(s) = y(s) - \frac{9}{100}e^s \int_0^1 e^t x(t)^2 y(t) \, dt, \quad s \in [0, 1],$$

4.2 Mild Convergence Conditions

$([\mathcal{F}'(x)]^{-1}w)(s) = y(s)$, and from $P = \int_0^1 e^t x(t)^2 y(t)\, dt$, it follows

$$e^s x(s)^2 w(s) = e^s x(s)^2 y(s) - \frac{9}{100} e^{2s} x(s)^2 P,$$

$$\int_0^1 e^s x(s)^2 w(s)\, ds = P\left(1 - \frac{9}{100}\int_0^1 e^{2s} x(s)^2\, ds\right),$$

so that

$$[\mathcal{F}'(x)^{-1}w](s) = w(s) + \frac{9\,e^s}{100 - 9\int_0^1 e^{2s} x(s)^2\, ds}\int_0^1 e^s x(s)^2 w(s)\, ds,$$

for all $w(s) \in \mathcal{C}[0, 1]$. In addition,

$$[\mathcal{F}''(x)yz](s) = -\frac{18}{100} e^s \int_0^1 e^t x(t) y(t) z(t)\, dt, \quad s \in [0, 1],$$

Observe that we cannot obtain an upper bound for $\mathcal{F}''(x)$ that is independent of $\|x\|$, since $\mathcal{F}''(x)$ is not bounded. Then, under the usual Kantorovich conditions (C1)–(C2)–(C3) required in Theorem 4.4, we cannot obtain domains of existence and uniqueness of solution for (4.22). Thus, if we can apply Theorem 4.4, it is necessary to locate previously a solution in a suitable domain Ω where the operator $\mathcal{F}''(x)$ is bounded. So, from (4.22), we have

$$\|x^*\| - 1 - \frac{3}{100} \max_{s\in[0,1]} \left|\int_0^1 e^{s+t} dt\right| \|x^*\|^3 \leq 0,$$

which is satisfied if $\|x^*\| \leq \theta_1 = 1.328674\ldots$ and $\|x^*\| \geq \theta_2 = 1.746584\ldots$, where θ_1 and θ_2 are the two positive roots of the scalar equation

$$t - 1 - \frac{3\,e(e-1)}{100} t^3 = 0. \tag{4.23}$$

If we first look for a solution x^* such that $\|x^*\| \leq \theta_1$, we can consider $\Omega = B(0, \theta)$, with $\theta \in (\theta_1, \theta_2)$, since $\mathcal{F}''(x)$ is bounded in Ω. But, this location of a suitable domain may not be enough, since it is possible the existence of starting points x_0 from which the convergence cannot be reached, as we can see in the following. If we choose $\Omega = B(0, \theta)$ with $\theta = 1.7$, which satisfies $\theta \in (\theta_1, \theta_2)$, and $x_0(s) = 1 \in \Omega$, then the Kantorovich conditions (C1)–(C2)–(C3) are not satisfied. Second, if we look for a solution x^* such that $\|x^*\| \geq \theta_2$, we cannot fix a domain Ω. However, the convergence conditions (C1)–(C2*)–(C3) required in Theorem 4.17 are satisfied with $x_0(s) = 1$. Observe that

$$\|[\mathcal{F}'(x_0)]^{-1}\| \leq 1 + \frac{9M}{109 - 9\,\mathrm{e}^2} = \beta \quad \text{and}$$

$$\|[\mathcal{F}'(x_0)]^{-1}\mathcal{F}(x_0)\| \leq \frac{3(9\,\mathrm{e}^2 - 209\,\mathrm{e})M}{200(9\,\mathrm{e}^2 - 109)} = \eta,$$

where

$$M = \max_{s \in [0,1]} \left| \int_0^1 \mathrm{e}^{s+t}\, dt \right| = \mathrm{e}(\mathrm{e} - 1).$$

Moreover, $\|\mathcal{F}''(x_0)\| \leq \frac{9M}{50} = \delta$,

$$\|\mathcal{F}''(x) - \mathcal{F}''(y)\| \leq \frac{9M}{50} \|x - y\|,$$

and $K = \delta$. As a consequence, from (4.19), we obtain the majorant polynomial, which has one negative root $r_0 = -3.95980\ldots$ and two positive roots $t^* = 0.347775\ldots$ and $t^{**} = 0.612024\ldots$ Therefore, the conditions of Theorem 4.17 are satisfied. Then, the integral equation (4.22) has a solution, that is,

$$x^*(s) = 1 + (0.07742925365569706\ldots)\,\mathrm{e}^s,$$

in $\overline{B(1, 0.3477\ldots)}$, which is unique in $B(1, 0.6120\ldots)$. Notice in Fig. 4.2 that the solution $x^*(s)$ lies within the domain of existence of solution and within the domain of uniqueness of solution.

After that, we compare these domains of existence and uniqueness of solution with those obtained by the technique based on recurrence relations. We show in Table 4.2 the radius of the domain of existence of solution $\overline{B(1, r_1)}$ and the radius of the domain of uniqueness of solution $B(1, r_2)$ obtained for the methods of Chebyshev, Halley, Exponential, and Logarithmic (see [62]).

Notice that the method of majorizing sequences provides a unique domain of existence of solution and a unique domain of uniqueness of solution for all the iterative methods contained in the family (4.3). However, the technique based on recurrence relations provides one for every one, as we can see in Table 4.2. Observe that the method of majorizing sequences provides a better domain of existence of solution, but the domain of uniqueness of solution is worse than any of those obtained for the indicated methods of the family that appear in Table 4.2.

Table 4.2 Radii of the domains of existence and uniqueness of solution, respectively, r_1 and r_2, from Example 4.18

Method	r_1	r_2
The Chebyshev method	0.416987...	0.737427...
The Halley method	0.462943...	0.694072...
The Exponential method	0.486724...	0.671313...
The Logarithmic method	0.444379...	0.711685...

4.2 Mild Convergence Conditions

Table 4.3 The error estimates $\|x^* - x_n\| \leq t^* - t_n$, $n \in \mathbb{N}$, from Example 4.18

$t^* - t_n$	The Chebyshev method	The Halley method	Chebyshev-like method with $A_2 = 1/2$
$t^* - t_1$	$1.47529\ldots \times 10^{-2}$	$1.08404\ldots \times 10^{-2}$	$7.95655\ldots \times 10^{-3}$
$t^* - t_2$	$1.05876\ldots \times 10^{-5}$	$2.41283\ldots \times 10^{-6}$	$2.12421\ldots \times 10^{-7}$
$t^* - t_3$	$4.30928\ldots \times 10^{-15}$	$2.79278\ldots \times 10^{-17}$	$3.30991\ldots \times 10^{-21}$
$t^* - t_4$	$2.90570\ldots \times 10^{-43}$	$4.33082\ldots \times 10^{-50}$	$1.25217\ldots \times 10^{-62}$

Table 4.4 The error estimates $\|x^* - x_n\| \leq t^* - t_n$, $n \in \mathbb{N}$, from Example 4.18

$t^* - t_n$	The Ostrowski method	The exponential method	The logarithmic method
$t^* - t_1$	$8.20185\ldots \times 10^{-3}$	$9.08750\ldots \times 10^{-3}$	$1.23303\ldots \times 10^{-2}$
$t^* - t_2$	$6.28423\ldots \times 10^{-7}$	$1.04588\ldots \times 10^{-6}$	$4.45816\ldots \times 10^{-6}$
$t^* - t_3$	$2.89557\ldots \times 10^{-19}$	$1.64811\ldots \times 10^{-18}$	$2.24685\ldots \times 10^{-16}$
$t^* - t_4$	$2.83260\ldots \times 10^{-56}$	$6.44908\ldots \times 10^{-54}$	$2.87637\ldots \times 10^{-47}$

In Tables 4.3 and 4.4, we show some error estimates $\|x^* - x_n\| \leq t^* - t_n$, $n \geq 0$, for some iterative methods of (4.3). Note that it is necessary to calculate the terms of the scalar sequence $\{t_n\}$. In addition, as we know the two real roots of the majorant polynomial, it is also possible to obtain a priori error estimates from these roots, as we can see in [63].

Finally, we choose the Chebyshev method, which is included in the family (4.3), to approximate a solution of (4.22) starting at $x_0(s) = 10$. To obtain the iterations $x_{n+1}(s)$ of the method,

$$x_{n+1} = x_n - \left(I + \frac{1}{2}L_\mathcal{F}(x_n)\right)[\mathcal{F}'(x_n)]^{-1}\mathcal{F}(x_n), \quad n \geq 0,$$

we proceed as follows. We first calculate the integrals

$$\mathcal{A}_n = \int_0^1 e^t x_n(t)^3 \, dt, \quad \mathcal{B}_n = \int_0^1 e^t x_n(t)^2 \, dt, \quad \mathcal{C}_n = \int_0^1 e^{2t} x_n(t)^2 \, dt,$$

$$\mathcal{D}_n = \int_0^1 e^t x_n(t) \, dt, \quad \mathcal{E}_n = \int_0^1 e^{2t} x_n(t) \, dt, \quad \mathcal{F}_n = \int_0^1 e^{3t} x_n(t) \, dt$$

and then define

$$\mathcal{I}_n = \frac{3(9\mathcal{C}_n - 100)(\mathcal{A}_n(100 + 27\mathcal{C}_n - 36\mathcal{E}_n) + 54\mathcal{B}_n\mathcal{E}_n - 300(\mathcal{B}_n - \mathcal{D}_n) - 27\mathcal{C}_n(\mathcal{B}_n + \mathcal{D}_n)) - 81(2\mathcal{A}_n - 3\mathcal{B}_n)^2 \mathcal{F}_n)}{(9\mathcal{C}_n - 100)^3},$$

so that

$$x_{n+1}(s) = 1 + e^s \mathcal{I}_n.$$

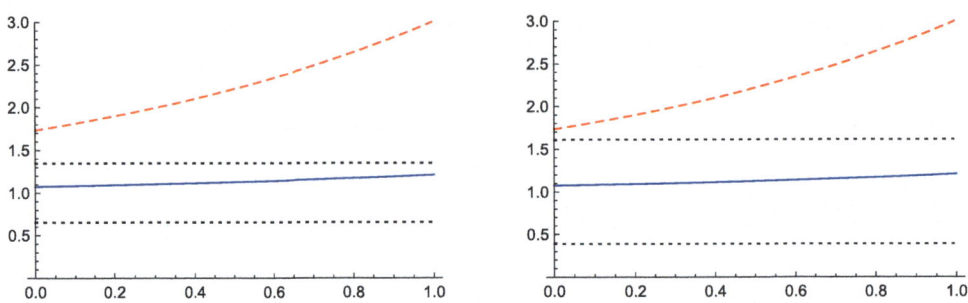

Fig. 4.2 On the left, the domain of existence of solution (dotted lines) with the solutions $x^*(s)$ (solid line) and $x^{**}(s)$ (dashed line). On the right, the domain of uniqueness of solution (dotted lines) with the solutions $x^*(s)$ (solid line) and $x^{**}(s)$ (dashed line)

Then, we obtain the following eight approximations:

$$x_0(s) = 10,$$
$$x_1(s) = 1 + (2.4754667084353140\ldots)\,e^s,$$
$$x_2(s) = 1 + (0.3879369645030384\ldots)\,e^s,$$
$$x_3(s) = 1 + (0.8948316243541690\ldots)\,e^s,$$
$$x_4(s) = 1 + (0.7488560894880802\ldots)\,e^s,$$
$$x_5(s) = 1 + (0.7390789263490810\ldots)\,e^s,$$
$$x_6(s) = 1 + (0.7390736023792825\ldots)\,e^s,$$
$$x_7(s) = 1 + (0.7390734160818100\ldots)\,e^s,$$
$$x_8(s) = 1 + (0.7390734095593162\ldots)\,e^s$$

with the stopping criterion $\|x_n(s) - x_{n-1}(s)\| < 10^{-10}$, $n \in \mathbb{N}$.

Moreover, the convergence conditions given in Theorem 4.17 are satisfied from the fourth iteration of the Chebyshev method. As a consequence, Eq. (4.22) has a solution $x^{**}(s)$ in $\overline{B(x_4(s), 0.6832\ldots)}$, which is unique in $B(x_4(s), 0.0270\ldots)$. Notice in Fig. 4.2 that the solution $x^{**}(s)$ is neither within the domain of existence of the solution nor within the domain of uniqueness of the solution.

4.3 Quadratic Equations

The study of quadratic equations is interesting, since there are problems that can be expressed by means of them, for example, equations that appear in the theory of dynamics

4.3 Quadratic Equations

of gases [79] and those related with Chandrasekhar's work [20], which arise in the theories of radiative transfer, neutron transport, and the kinetic theory of gases. An extensive literature exists on equations of this type, see [5] and the references therein.

As we have already seen at the beginning of the chapter, Gander provides an algebraic technique to obtain third-order iterative methods for solving scalar equations, which is extended to obtain iterative methods in Banach spaces with R-order of convergence at least three in Sect. 4.1. In this section, we extend this technique to obtain a family of iterative methods, which is included in (4.3), with R-order of convergence at least four when it is applied to solve quadratic equations [39].

Next, we establish in Banach spaces the semilocal convergence of the family just mentioned by using the method of majorizing sequences and prove the R-order of convergence at least four for quadratic equations by Ostrowski's technique [75].

4.3.1 A Family of Iterations with R-Order of Convergence At Least Four

According to Schröder [86], the family (4.1) has order of convergence four if the zero s^* of the function f is such that $s^* = G(s^*)$, $G'(s^*) = G''(s^*) = G'''(s^*) = 0$, and $G^{(4)}(s^*) \neq 0$. Besides, following Gander, these conditions are required to (4.1) to obtain an expression for the function H. Thus, from the conditions $s^* = G(s^*)$ and $G'(s^*) = G''(s^*) = 0$, it follows $H(0) = 1$ and $H'(0) = \frac{1}{2}$, so that (4.1) has order of convergence at least three.

Moreover, if f is a quadratic function, then $f''(x) = C \in \mathbb{R}$, $f^{(n)}(x) = 0$, for all $n \in \mathbb{N}, n \geq 3$, and it is easy to prove that

$$L'_f(x) = \frac{f''(x)}{f'(x)}(1 - 2L_f(x)), \qquad L''_f(x) = -3\frac{f''(x)^2}{f'(x)^2}(1 - 2L_f(x)).$$

In this case,

$$G'''(s) = 3(H'(0) - H''(0))L'_f(s)^2 + (H(0) - 3H'(0))L''_f(s),$$

so that $G'''(s) = 0$ if and only if $H''(0) = \frac{1}{2}$. If we now consider the analytic scalar function H, which appears in (4.1), as

$$H(L_f(x)) = 1 + \frac{1}{2}L_f(x) + \frac{1}{2}L_f(x)^2 + \sum_{k\geq 3} A_k L_f(x)^k, \qquad A_k \in \mathbb{R}^+, \qquad (4.24)$$

where $\{A_k\}_{k\geq 3}$ is a positive nonincreasing scalar sequence, then the iterations (4.1) have order of convergence four when they are applied to solve scalar quadratic equations.

Notice that (4.1)–(4.24) is reduced to the following Chebyshev-like method [3, 34]:

$$x_{n+1} = x_n - \left(1 + \frac{1}{2}L_f(x_n) + \frac{1}{2}L_f(x_n)^2\right)\frac{f(x_n)}{f'(x_n)}, \quad n \geq 0, \tag{4.25}$$

if $A_0 = 1$, $A_1 = A_2 = \frac{1}{2}$, and $A_k = 0$, for all $k \in \mathbb{N}$, $k \geq 3$, and to the Super-Halley method [11, 54]:

$$x_{n+1} = x_n - \left(1 + \frac{L_f(x_n)}{2(1 - L_f(x_n))}\right)\frac{f(x_n)}{f'(x_n)}, \quad n \geq 0, \tag{4.26}$$

if $A_0 = 1$ and $A_k = \frac{1}{2}$, for all $k \in \mathbb{N}$.

As a consequence of the last, we can then consider the family of iterative methods

$$\begin{cases} x_0 \text{ given,} \\ x_{n+1} = G(x_n) = x_n - H(L_F(x_n))[F'(x_n)]^{-1}F(x_n), \quad n \geq 0, \\ H(L_F(x_n)) = \sum_{k \geq 0} A_k L_F(x_n)^k, \end{cases} \tag{4.27}$$

where $L_F(x)^0 = I$, $A_0 = 1$, $A_1 = A_2 = \frac{1}{2}$, and $\{A_k\}_{k \geq 0}$ is a positive nonincreasing scalar sequence such that $\sum_{k \geq 0} A_k t^k < +\infty$ for $|t| < r$, to solve quadratic equations in Banach spaces.

4.3.2 Semilocal Convergence

To prove the semilocal convergence and establish the R-order of converge of (4.27) in Banach spaces, we suppose that F is a twice continuously Fréchet differentiable operator $F : \Omega \subseteq X \to Y$, defined on a non-empty subset of a Banach space X with values in a Banach space Y, and the following conditions:

(O1) There exists $\Gamma_0 = [F'(x_0)]^{-1} \in \mathcal{L}(Y, X)$, for some $x_0 \in \Omega$ with $\|\Gamma_0\| \leq \beta$ and $\|\Gamma_0 F(x_0)\| \leq \eta$.
(O2) There exists a constant $M \geq 0$ such that $\|F''(x)\| = M$ for some $x \in \Omega$.
(O3) $\mu = M\beta\eta < \min\left\{\frac{1}{2}, r\right\}$, where r is the radius of convergence of the power series $\sum_{k \geq 0} A_k t^k$, and $B(x_0, t^*) \subset \Omega$, where $t^* = \frac{1 - \sqrt{1 - 2\mu}}{\mu}\eta$.

A simple form of finding a majorizing sequence $\{t_n\}$ is to consider the sequence obtained when the iterations (4.27) are applied to the quadratic equation $p(t) = 0$, where

$$p(t) = \frac{M}{2}t^2 - \frac{t}{\beta} + \frac{\eta}{\beta} \tag{4.28}$$

4.3 Quadratic Equations

is a polynomial which satisfies conditions (O1)–(O2) with $t_0 = 0$. In this case, the majorizing sequence is

$$\begin{cases} t_0 \text{ given,} \\ t_{n+1} = P(t_n) = t_n - h(L_p(t_n))\dfrac{p(t_n)}{p'(t_n)}, & n \geq 0, \\ h(L_p(t_n)) = \sum_{k \geq 0} A_k L_p(t_n)^k, \end{cases} \quad (4.29)$$

where $\{A_k\}_{k \geq 0}$ is the sequence defined previously in (4.27). It is easy to prove that the sequence (4.29) is increasing and starting at $t_0 = 0$ converges to the smallest root t^* of $p(t) = 0$.

After that, we establish the semilocal convergence of iterations (4.27) and give domains of existence and uniqueness of solution and some a priori error estimates in the following theorem.

Theorem 4.19 *Let $F : \Omega \subseteq X \to Y$ be a twice continuously Fréchet differentiable quadratic operator defined on a non-empty open convex subset Ω of a Banach space X with values in a Banach space Y. Suppose that conditions (O1)–(O2)–(O3) are satisfied. Then, the sequence (4.27) starting at x_0 converges to a solution x^* of the equation $F(x) = 0$. Moreover, $x_n, x^* \in \overline{B(x_0, t^*)}$, and x^* is unique in $B(x_0, t^{**}) \cap \Omega$, where $t^{**} = \frac{1+\sqrt{1-2\mu}}{\mu}\eta$. Furthermore,*

$$\|x^* - x_n\| \leq t^* - t_n, \quad n \geq 0, \quad (4.30)$$

where t_n is defined in (4.29).

Proof First, we have that $\|L_F(x_0)\| \leq \mu$, and then $H(L_F(x_0))$ is well defined, since $x_0 \in \Omega$ and $\Gamma_0 F(x_0) \in \Omega$. Taking into account that the scalar sequence $\{A_k\}_{k \geq 0}$ is nonincreasing, it follows

$$\|x_1 - x_0\| \leq \left(\sum_{k \geq 0} A_k \mu^k\right)\eta = -h(L_p(t_0))\frac{p(t_0)}{p'(t_0)} = t_1 - t_0 < t^*,$$

where p is defined in (4.28), and then $x_1 \in B(x_0, t^*)$.

Next, from

$$\|I - \Gamma_0 F'(x)\| \leq \|\Gamma_0 \left(F'(x_0) - F'(x)\right)\| \leq \|\Gamma_0\| \int_{x_0}^{x} \|F''(z)\| dz \leq M\beta\|x - x_0\|$$

$$< M\beta t^* = 1 - \sqrt{1-2\mu} < 1$$

and the Banach lemma on invertible operators, we have that the operator $[F'(x)]^{-1}$ exists and

$$\left\|[F'(x)]^{-1}\right\| \leq \frac{1}{1 - \|I - F'(x)\|} < \frac{1}{1 - M\beta t^*}.$$

Moreover, as

$$F(x_1) = \frac{1}{2}F''(x_0)\Gamma_0 F(x_0)\left(\sum_{k\geq 3} 2(A_{k-1} - A_k)L_F(x_0)^{k-1}\right)\Gamma_0 F(x_0)$$

$$+ \frac{1}{8}F''(x_0)\left(L_F(x_0)\widetilde{H}(L_F(x_0))\Gamma_0 F(x_0)\right)^2,$$

where $\widetilde{H}(z) = I + 2\sum_{k\geq 2} A_k z^{k-1}$, we have

$$\|F(x_1)\| \leq \|F''(x_0)\|\|\Gamma_0 F(x_0)\|\sum_{k\geq 3}(A_{k-1} - A_k)\|L_F(x_0)\|^{k-1}$$

$$+ \frac{1}{8}\|F''(x_0)\|\|L_F(x_0)\|^2\|\widetilde{H}(L_F(x_0))\|^2\|\Gamma_0 F(x_0)\|^2$$

$$\leq M\eta^2 \left(\sum_{k\geq 3}(A_{k-1} - A_k)\mu^{k-1} + \frac{1}{2}\left(\sum_{k\geq 1} A_k \mu^k\right)^2\right)$$

$$= p(t_1),$$

$$\|\Gamma_1 F(x_1)\| \leq -\frac{p(t_1)}{p'(t_1)}.$$

Furthermore,

$$\|L_F(x_1)\| \leq \frac{M\beta}{1 - M\beta t_1}\left(-\frac{p(t_1)}{p'(t_1)}\right) = L_p(t_1) < r,$$

$$\left\|\widetilde{H}(L_F(x_1))\right\| \leq 1 + 2\sum_{k\geq 2} A_k L_p(t_1)^{k-1} = \widetilde{h}(L_p(t_1)),$$

so that x_2 is well defined

$$\|x_2 - x_1\| \leq \left(1 + \frac{1}{2}L_p(t_1)\widetilde{h}(L_p(t_1))\right)\left(-\frac{p(t_1)}{p'(t_1)}\right) = t_2 - t_1,$$

$$\|x_2 - x_0\| \leq \|x_2 - x_1\| + \|x_1 - x_0\| \leq t_2 - t_1 + t_1 - t_0 = t_2 - t_0 < t^*,$$

4.3 Quadratic Equations

and $x_2 \in B(x_0, t^*)$.

Thus, we invoke mathematical induction on n and see that the operator Γ_n exists, $\|\Gamma_n\| \leq -\frac{1}{p'(t_n)}$, $\|\Gamma_n F(x_n)\| \leq -\frac{p(t_n)}{p'(t_n)}$, and $\|F''(x_n)\| \leq p''(t_n)$, for all $x_n \in B(x_0, t^*)$. As a consequence,

$$\|L_F(x_n)\| \leq L_p(t_n) < r.$$

Then, following the same procedure as above, we prove that

$$x_{n+1} = x_n - \left(I + \frac{1}{2}L_F(x_n)\widetilde{H}(L_F(x_n))\right)\Gamma_n F(x_n)$$

is well defined, since x_n and $\Gamma_n F(x_n)$ belong to Ω and $\widetilde{H}(L_F(x_n))$ is well defined. Besides,

$$\left\|\widetilde{H}(L_F(x_n))\right\| \leq 1 + 2\sum_{k \geq 2} A_k L_p(t_n)^{k-1} = \widetilde{h}(L_p(t_n)),$$

$$\|x_{n+1} - x_n\| \leq \left(1 + \frac{1}{2}L_p(t_n)\widetilde{h}(L_p(t_n))\right)\left(-\frac{p(t_n)}{p'(t_n)}\right) = t_{n+1} - t_n,$$

$$\|x_{n+1} - x_0\| \leq \|x_{n+1} - x_n\| + \|x_n - x_0\| \leq t_{n+1} - t_n + t_n - t_0 = t_{n+1} - t_0 < t^*,$$

$x_{n+1} \in B(x_0, t^*) \subset \Omega$, and the sequence $\{t_n\}$ is a majorizing sequence of $\{x_n\}$.

In addition,

$$\|x_{n+m} - x_n\| \leq t_{n+m} - t_n, \quad n, m \in \mathbb{N}, \tag{4.31}$$

and $\{x_n\}$ is then a Cauchy sequence. Therefore, $\{x_n\}$ is convergent and $\lim\limits_{n \to +\infty} x_n = v$. Now, since $\|F(x_n)\| \leq p(t_n)$, we obtain

$$\lim_{n \to +\infty} \|F(x_n)\| \leq \lim_{n \to +\infty} p(t_n) = p(t^*) = 0.$$

Thus, $\lim\limits_{n \to +\infty} F(x_n) = 0$, and from the continuity of the operator F, we have $v = x^*$, where x^* is a solution of the equation $F(x) = 0$. Moreover, if $m \to \infty$ and we choose $n = 0$ in (4.31), we obtain $\|x^* - x_0\| < t^*$ and $x^* \in B(x_0, t^*)$.

Finally, we prove the uniqueness of the solution x^* in $B(x_0, t^{**}) \cap \Omega$. We suppose that y^* is another solution of $F(x) = 0$ in $B(x_0, t^{**}) \cap \Omega$. From

$$0 = \Gamma_0(F(y^*) - F(x^*)) = \left(\int_0^1 \Gamma_0 F'(x^* + t(y^* - x^*))\, dt\right)(y^* - x^*) = J(y^* - x^*),$$

it suffices to prove that the operator $J = \int_0^1 \Gamma_0 F'(x^* + t(y^* - x^*))\, dt$ is invertible, and then $x^* = y^*$. Thus,

$$\|I - J\| \leq \int_0^1 \left\|\Gamma_0 \left(F'(x^* + t(y^* - x^*)) - F'(x_0)\right)\right\| dt$$
$$\leq M\beta \int_0^1 \|x^* + t(y^* - x^*) - x_0\| dt,$$

and, since $\|y^* - x_0\| < t^{**}$ and $\|x^* - x_0\| \leq t^*$, it follows

$$\|I - J\| < \frac{M\beta}{2}(t^{**} + t^*) = 1.$$

Hence, x^* is the unique solution of $F(x) = 0$ in $B(x_0, t^{**}) \cap \Omega$. ∎

4.3.3 R-Order of Convergence

After that, we use Ostrowski's technique to obtain a priori error estimates for the iterations (4.27) from the polynomial (4.28), so that the scalar sequence $\{t_n\}$ does not need to be calculated. From these error estimates, if $t^* \neq t^{**}$, we conclude that the iterations (4.27) have R-order of convergence at least four when they are applied to quadratic equations.

Theorem 4.20 *If the polynomial* (4.28) *has two positive roots* t^* *and* t^{**}, *such that* $t^* \leq t^{**}$, *and* $\{t_n\}$ *is the sequence* (4.29), *then:*

(a) *If* $t^* < t^{**}$ *and* $\sqrt[3]{5}\,\phi < 1$, *where* $\phi = \frac{t^*}{t^{**}}$, *we have*

$$(t^{**} - t^*)\frac{\phi^{4^n}}{1 - \phi^{4^n}} \leq t^* - t_n \leq (t^{**} - t^*)\frac{\left(\sqrt[3]{5}\,\phi\right)^{4^n}}{\sqrt[3]{5} - \left(\sqrt[3]{5}\,\phi\right)^{4^n}}, \quad n \geq 0. \qquad (4.32)$$

(b) *If* $t^* = t^{**}$ *and* $C = \frac{5}{16} - \sum_{k \geq 3} \frac{A_k}{2^{k+1}} < 1$, *we have*

$$t^* - t_n = t^* C^n, \quad n \geq 0. \qquad (4.33)$$

Proof From (4.29), $a_n = t^* - t_n$, and $b_n = t^{**} - t_n$, for all $n \geq 0$, we obtain

$$p(t_n) = \frac{M\beta}{2} a_n b_n, \qquad p'(t_n) = -\frac{M\beta}{2}(a_n + b_n).$$

4.3 Quadratic Equations

Moreover,

$$a_{n+1} = \frac{a_n^4}{(a_n+b_n)^5}\left(a_n^2 + 4a_nb_n + 5b_n^2 - 8b_n^4 \sum_{k\geq 0} 2^k A_{k+3} \frac{a_n^k b_n^k}{(a_n+b_n)^{2k+2}}\right),$$

$$b_{n+1} = \frac{b_n^4}{(a_n+b_n)^5}\left(5a_n^2 + 4a_nb_n + b_n^2 - 8a_n^4 \sum_{k\geq 0} 2^k A_{k+3} \frac{a_n^k b_n^k}{(a_n+b_n)^{2k+2}}\right).$$

If $t^* < t^{**}$, then $\phi = \frac{t^*}{t^{**}} < 1$ and

$$\frac{a_{n+1}}{b_{n+1}} = \left(\frac{a_n}{b_n}\right)^4 h(a_n),$$

where

$$u(x) = \frac{10x^2 + 14dx + 5d^2 - 8(x+d)^4 S(x)}{10x^2 + 6dx + d^2 - 8x^4 S(x)}, \qquad S(x) = \sum_{k\geq 0} 2^k A_{k+3} \frac{x^k(x+d)^k}{(2x+d)^{2k+2}},$$

and $d = t^{**} - t^*$. Observe that $1 \leq u(x) \leq 5$, so that

$$\phi^{4^n} \leq \cdots \leq \left(\frac{a_{n-1}}{d+a_{n-1}}\right)^4 \leq \frac{a_n}{a_n+d} \leq 5\left(\frac{a_{n-1}}{a_{n-1}+d}\right)^4 \leq \cdots \leq \frac{1}{\sqrt[3]{5}}\left(\sqrt[3]{5}\phi\right)^{4^n},$$

and (4.32) is satisfied, since $\sqrt[3]{5}\,\phi < 1$.

If $t^* = t^{**}$, then $a_n = b_n$,

$$a_n = \frac{a_{n-1}}{16}\left(5 - 8\sum_{k\geq 3}\frac{A_k}{2^k}\right),$$

and (4.33) is satisfied, since $C < 1$. ∎

Remark 4.21 From Theorem 4.20, more precise a priori error estimates can be given for particular iterations of (4.27):

- If (4.27) is reduced to the Super-Halley method, we have

$$t^* - t_n = (t^{**} - t^*)\frac{\phi^{4^n}}{1 - \phi^{4^n}}, \quad n \geq 0,$$

where $\phi = \frac{t^*}{t^{**}}$ and $t^* < t^{**}$, and if $t^* = t^{**}$, then

$$t^* - t_n = t^* \left(\frac{1}{4}\right)^n, \quad n \geq 0.$$

- If (4.27) is reduced to the Chebyshev-like method (4.25), we have

$$(t^{**} - t^*) \frac{\phi^{4^n}}{1 - \phi^{4^n}} \leq t^* - t_n \leq (t^{**} - t^*) \frac{(\sqrt[3]{5}\,\phi)^{4^n}}{\sqrt[3]{5} - (\sqrt[3]{5}\,\phi)^{4^n}}, \quad n \geq 0,$$

where $\phi = \frac{t^*}{t^{**}}$ and $t^* < t^{**}$, and if $t^* = t^{**}$, then

$$t^* - t_n = t^* \left(\frac{5}{16}\right)^n, \quad n \geq 0.$$

In both cases we deduce that the R-order of convergence of the methods is at least four when they are applied to solve quadratic equations, which is well known, as we can see in [35] and [34], respectively.

4.3.4 Applications

We illustrate the previous results with two applications, where two quadratic equations are involved: an equation of molecular interaction [79] and a Chandrasekhar equation [20].

In the equation of molecular interaction, we have to solve a boundary value problem with a partial differential equation. For this, we consider a process of discretization and use two different iterations of (4.27) to solve the corresponding finite-dimensional problem. Moreover, a priori error estimates are obtained according to Theorems 4.19 and 4.20, and the speed of convergence is computationally justified.

For the Chandrasekhar equation, we first provide domains of existence and uniqueness of solution from Theorem 4.19, and second, we use a process of discretization to approximate a solution of the equation by extending the Chebyshev-like method (4.25) to Banach spaces. In this case, due to the expression of the second derivative of the operator involved in the application of the method, whose operational cost is high, we write the method so that it is applied to quadratic equations, so that the operational cost is reduced, since the second derivative does not appear in the algorithm.

Example 4.22 We consider the following equation of molecular interaction that appears in the theory of dynamics of gases [79]:

$$u_{xx} + u_{yy} = u^2,$$

with the boundary conditions

4.3 Quadratic Equations

$$\begin{cases} u(x,0) = 2x^2 - x + 1, & 0 \leq x \leq 1, \\ u(1,y) = 2, & 0 \leq y \leq 1, \\ u(x,1) = 2, & 0 \leq x \leq 1, \\ u(0,y) = 2y^2 - y + 1, & 0 \leq y \leq 1. \end{cases}$$

We first follow a process of discretization. So, we then define a uniform mesh with knots

$$P_{i,j} = (ih, jh), \quad h = \frac{1}{n+1}, \quad i, j = 0, 1, \ldots, n+1,$$

and approximate the second derivatives of u in the points $P_{i,j}$ by the following formulas:

$$u_{xx}(P_{i,j}) = \frac{u(P_{i+1,j}) - 2u(P_{i,j}) + u(P_{i-1,j})}{h^2}, \quad i, j = 1, 2, \ldots, n,$$

$$u_{yy}(P_{i,j}) = \frac{u(P_{i,j+1}) - 2u(P_{i,j}) + u(P_{i,j-1})}{h^2}, \quad i, j = 1, 2, \ldots, n.$$

We denote $x_{i,j} = u(P_{i,j})$, $i, j = 0, 1, \ldots, n+1$, and obtain the system

$$-x_{i+1,j} - x_{i-1,j} - x_{i,j+1} - x_{i,j-1} + 4x_{i,j} = -h^2 x_{i,j}^2, \quad i, j = 1, 2, \ldots, n. \quad (4.34)$$

Observe that the values $x_{0,j}$, $x_{n+1,j}$, $x_{i,0}$, and $x_{i,n+1}$ are given by the boundary conditions. Consider $m = n^2$, and order $x_{i,j}$ ($i, j = 1, 2, \ldots, n$) in the following way:

$$x_1 = x_{1,1}, \ldots, x_n = x_{n,1}, x_{n+1} = x_{1,2}, \ldots, x_m = x_{n,n}.$$

Then, the system (4.34) can be written as

$$F(\mathbf{x}) = A\mathbf{x} + \Phi(\mathbf{x}) - \mathbf{b} = \mathbf{0}, \qquad F : \mathbb{R}^m \to \mathbb{R}^m,$$

where

$$A = \begin{pmatrix} B & -I & 0 & \cdots & 0 \\ -I & B & -I & \ddots & \vdots \\ 0 & \ddots & \ddots & \ddots & 0 \\ \vdots & \ddots & \ddots & \ddots & -I \\ 0 & \cdots & 0 & -I & B \end{pmatrix}_{m \times m}, \quad B = \begin{pmatrix} 4 & -1 & 0 & \cdots & 0 \\ -1 & 4 & -1 & \ddots & \vdots \\ 0 & \ddots & \ddots & \ddots & 0 \\ \vdots & \ddots & \ddots & \ddots & -1 \\ 0 & \cdots & 0 & -1 & 4 \end{pmatrix}_{n \times n},$$

I is the identity matrix in $\mathbb{R}^n$, $\mathbf{x} = (x_1, x_2, \ldots, x_m)^T$, $\Phi(\mathbf{x}) = h^2(x_1^2, x_2^2, \ldots, x_m^2)^T$, and $\mathbf{b}$ is a vector obtained from the boundary conditions.

In addition, $F'(\mathbf{x})$ is the linear operator given by the matrix

$$A + \frac{1}{8}\begin{pmatrix} x_1 & 0 & \cdots & 0 \\ 0 & x_2 & \ddots & \vdots \\ \vdots & \ddots & \ddots & 0 \\ 0 & \cdots & 0 & x_m \end{pmatrix},$$

and $F''(\mathbf{x})$ is the constant bilinear operator:

$$F''(\mathbf{x})\mathbf{u}\,\mathbf{v} = \frac{1}{8}\mathbf{u}\,\mathbf{v}, \quad \mathbf{u}, \mathbf{v} \in \mathbb{R}^m.$$

If, for example, we consider $n = 3$ (and $m = 9$), then the vector $\mathbf{b}$ is

$$\mathbf{b} = \left(\frac{7}{4}, 1, \frac{27}{8}, 1, 0, 2, \frac{27}{8}, 2, 4\right)^T.$$

We denote the n-th iteration by $\mathbf{x}_n = \left(x_1^{(n)}, x_2^{(n)}, \ldots, x_9^{(n)}\right)^T$. If $x_i^{(0)} = 1$, for $i = 1, 2, \ldots, 9$, the hypotheses of Theorem 4.19 are satisfied, since $\mu = 0.0984\ldots < \frac{1}{2}$, so that (4.27) converges to a solution of the system (4.34). Moreover, the domains of existence and uniqueness of solution of (4.34) are, respectively, $\{\mathbf{u} \in \mathbb{R}^9 : \|\mathbf{u} - \mathbf{x}_0\|_\infty \leq 0.8233\ldots\}$ and $\{\mathbf{u} \in \mathbb{R}^9 : \|\mathbf{u} - \mathbf{x}_0\|_\infty \leq 15.0375\ldots\}$.

On the other hand, we observe that $F''(\mathbf{x})\mathbf{y}$ is a linear application, and the associated matrix is diagonal. Indeed, if $\mathbf{y} = (y_1, y_2, \ldots, y_m)$, the matrix associated with $F''(\mathbf{x})\mathbf{y}$ is

$$2h^2 \operatorname{diag}\{y_1, y_2, \ldots, y_m\}.$$

Thus, the application of $F''(\mathbf{x})\mathbf{y}$ is simple.

Although, in general, the use of iterative methods that bear the computation of the second derivative is not viable (mainly for big dimensions), they can be taken into account in this type of systems, since the second derivative of the operator involved is constant.

The application of the Chebyshev-like method in the finite-dimensional case is as follows:

Stage 1. Compute one LR-decomposition of F' by Gauss elimination.
Stage 2. Solve the linear system $F'(\mathbf{x}_k)\mathbf{c}_k = -F(\mathbf{x}_k)$.
Stage 3. Solve the linear system $F'(\mathbf{x}_k)\mathbf{z}_k = F''(\mathbf{x}_k)(\mathbf{c}_k)^2$.
Stage 4. Solve the linear system $F'(\mathbf{x}_k)\mathbf{w}_k = F''(\mathbf{x}_k)\mathbf{c}_k\mathbf{z}_k$.
Stage 5. Define $\mathbf{x}_{k+1} = \mathbf{x}_k + \mathbf{c}_k - \frac{1}{2}(\mathbf{z}_k - \mathbf{w}_k)$.

4.3 Quadratic Equations

If we apply the Super-Halley method, then the stages are:

Stage 1. Compute one LR-decomposition of F' by Gauss elimination.
Stage 2. Solve the linear system $F'(\mathbf{x}_k)\mathbf{c}_k = -F(\mathbf{x}_k)$.
Stage 3. Compute one LR-decomposition of $F'(\mathbf{x}_k) + F''(\mathbf{x}_k)\mathbf{c}_k$ by Gauss elimination.
Stage 4. Solve the linear system

$$[F'(\mathbf{x}_k) + F''(\mathbf{x}_k)\mathbf{c}_k]\mathbf{d}_k = -F(\mathbf{x}_k) + \frac{1}{2}F''(\mathbf{x}_k)\mathbf{c}_k^2.$$

Stage 5. Define $\mathbf{x}_{k+1} = \mathbf{x}_k + \mathbf{d}_k$.

In view of both algorithms, we note that the application of the Chebyshev-like method only uses one LR-decomposition and the Super-Halley method does two.

In Tables 4.5 and 4.6, it is shown the error estimates obtained when the Chebyshev-like method and the Super-Halley method are applied to solve the system (4.34). Notice that they are similar. However, observe that Ostrowski's technique used in Table 4.6 has the advantage of not having to calculate the elements of the scalar sequence $\{t_n\}$ if we know the roots of the majorizing polynomial (4.28). In Table 4.5, we obtain a priori error estimates (4.30) using the majorizing sequence. Observe that the a priori error estimates given in Tables 4.5 and 4.6 are rather sharp, and those for the Super-Halley method are the same, as we can deduce from Remark 4.21.

After four iterations applying the Chebyshev-like method and using the stopping criterion $\|\mathbf{x}_n - \mathbf{x}^*\|_\infty < 10^{-150}$, we obtain the numerical solution $\mathbf{x}^* = (x_1^*, x_2^*, \ldots, x_9^*)^T$ of (4.34) given in Table 4.7.

By taking the same stopping criterion in Table 4.8, we obtain the errors $\|\mathbf{x}_n - \mathbf{x}^*\|_\infty$. If we now consider the computational order of convergence defined in [51] as

Table 4.5 A priori error estimates (4.30)

n	The Chebyshev-like method	The Super-Halley method
1	$5.40570\ldots \times 10^{-4}$	$1.27736\ldots \times 10^{-4}$
2	$1.48624\ldots \times 10^{-16}$	$9.26994\ldots \times 10^{-20}$
3	$8.49486\ldots \times 10^{-67}$	$2.57121\ldots \times 10^{-80}$

Table 4.6 A priori error estimates (4.32)

n	The Chebyshev-like method	The Super-Halley method
1	$6.38706\ldots \times 10^{-4}$	$1.27736\ldots \times 10^{-4}$
2	$2.89685\ldots \times 10^{-16}$	$9.26994\ldots \times 10^{-20}$
3	$1.22605\ldots \times 10^{-65}$	$2.57121\ldots \times 10^{-80}$

Table 4.7 Numerical solution $\mathbf{x}^*$ of the system (4.34)

i	x_i^*	i	x_i^*	i	x_i^*
1	1.02591171169...	4	1.20971388713...	7	1.51670303095...
2	1.20971388713...	5	1.38770378643...	8	1.62587249195...
3	1.51670303095...	6	1.62587249195...	9	1.76429948544...

Table 4.8 Errors $\|\mathbf{x}_n - \mathbf{x}^*\|_\infty$ and the computational order of convergence for the Chebyshev-like method and the Super-Halley method, respectively

n	Chebyshev-like method	Super-Halley method	ρ_C	ρ_S
1	$4.4890\ldots \times 10^{-5}$	$8.8077\ldots \times 10^{-6}$	3.9106...	3.9213...
2	$1.2753\ldots \times 10^{-21}$	$3.7998\ldots \times 10^{-25}$	4.0116...	4.0124...
3	$5.3383\ldots \times 10^{-88}$	$7.5716\ldots \times 10^{-103}$		

$$\rho \approx \frac{\log\left(\frac{\|\mathbf{x}_{n+1} - \mathbf{x}^*\|_\infty}{\|\mathbf{x}_n - \mathbf{x}^*\|_\infty}\right)}{\log\left(\frac{\|\mathbf{x}_n - \mathbf{x}^*\|_\infty}{\|\mathbf{x}_{n-1} - \mathbf{x}^*\|_\infty}\right)}, \quad n \in \mathbb{N}, \tag{4.35}$$

the Chebyshev-like method and the Super-Halley method reach computationally the R-order of convergence at least four obtained in Remark 4.21. See Table 4.8, where ρ_C and ρ_S denote, respectively, the computational orders of convergence of the Chebyshev-like and the Super-Halley methods.

Example 4.23 We use the results obtained previously to obtain domains of existence and uniqueness of solution and some error estimates for the Chandrasekhar equation (4.18) presented in Example 4.11

$$x(s) = 1 + \frac{1}{4} s\, x(s) \int_0^1 \frac{x(t)}{s+t}\, dt, \quad s \in [0, 1].$$

If we again write the previous Chandrasekhar equation as $\mathcal{F}(x) = 0$ with $\mathcal{F} : \mathcal{C}([0, 1]) \to \mathcal{C}([0, 1])$ and

$$[\mathcal{F}(x)](s) = x(s) - 1 - \frac{1}{4} s\, x(s) \int_0^1 \frac{x(t)}{s+t}\, dt, \quad s \in [0, 1], \tag{4.36}$$

then

$$[\mathcal{F}'(x)y](s) = y(s) - \frac{1}{4} s\, x(s) \int_0^1 \frac{y(t)}{s+t}\, dt - \frac{1}{4} s\, y(s) \int_0^1 \frac{x(t)}{s+t}\, dt, \quad s \in [0, 1],$$

4.3 Quadratic Equations

$$[\mathcal{F}''(x)yz](s) = -\frac{1}{4}s\,y(s) \int_0^1 \frac{z(t)}{s+t}\,dt - \frac{1}{4}s\,z(s) \int_0^1 \frac{y(t)}{s+t}\,dt, \quad s \in [0,1],$$

and we observe that the operator $\mathcal{F}$ is quadratic and such that $\|\mathcal{F}''(x)\| \leq \frac{\log 2}{2} = 0.3465\ldots$

As we mention in Example 4.11, a reasonable choice for the starting point is $x_0(s) = 1$, since $x(0) = 1$, so that conditions (O1)–(O2)–(O3) of Theorem 4.19 are satisfied, since $\beta = 1.5303\ldots$, $\eta = 0.2651\ldots$, $M = 0.3465\ldots$, and $\mu = 0.1406\ldots < \frac{1}{2}$. Therefore, by Theorem 4.19, the operator (4.36) has a zero in $\{x \in \mathcal{C}([0,1]) : \|x-1\|_\infty \leq 0.2870\ldots\}$ and is unique in $\{x \in \mathcal{C}([0,1]) : \|x-1\|_\infty \leq 3.4836\ldots\}$. Notice that the domains of existence and uniqueness of solution are the same as those obtained in Example 4.11.

Finally, we approximate numerically a solution of $\mathcal{F}(x) = 0$ with $\mathcal{F}$ given in (4.36). For this, we use a process of discretization to transform $\mathcal{F}(x) = 0$ into a finite-dimensional problem. Thus, we approach the integral of (4.36) by a Gauss-Legendre quadrature formula of the form

$$\int_0^1 f(t)\,dt \simeq \sum_{j=1}^8 \varpi_j f(t_j),$$

where the nodes t_j and the weights ϖ_j are known. If we denote by x_i the approximations of $x(t_i)$, with $i = 1, 2, \ldots, 8$, we obtain the following nonlinear system:

$$\mathsf{x}_i = 1 + \frac{1}{4}\mathsf{x}_i t_i \sum_{j=1}^8 \varpi_j \frac{\mathsf{x}_j}{t_i + t_j}, \quad i = 1, 2, \ldots, 8. \tag{4.37}$$

Now, we denote $\mathbf{x} = (\mathsf{x}_1, \mathsf{x}_2, \ldots, \mathsf{x}_8)^T$, $\mathbf{1} = (1, 1, \ldots, 1)^T$, $A = (a_{ij})$ and write (4.37) in the matrix form

$$F(\mathbf{x}) = \mathbf{x} - \mathbf{1} - \mathbf{x} \odot A\mathbf{x}, \qquad F: \mathbb{R}^8 \to \mathbb{R}^8,$$

where $\odot$ denotes the inner product. In addition, $F'(\mathbf{x})$ is then the linear operator given by

$$F'(\mathbf{x})\mathbf{y} = \mathbf{y} - (\mathbf{x} \odot A\mathbf{y} + \mathbf{y} \odot A\mathbf{x}),$$

and $F''(\mathbf{x})$ is the bilinear operator defined as

$$F''(\mathbf{x})\mathbf{y}\mathbf{z} = -(\mathbf{z} \odot A\mathbf{y} + \mathbf{y} \odot A\mathbf{z}),$$

where $\mathbf{y} = (\mathsf{y}_1, \mathsf{y}_2, \ldots, \mathsf{y}_8)^T$ and $\mathbf{z} = (\mathsf{z}_1, \mathsf{z}_2, \ldots, \mathsf{z}_8)^T$.

Contrary to the previous equation of molecular interaction, we observe that it is necessary to calculate two scalar products and two matrix products for this equation, so that the operational cost is increased considerably when the Chebyshev-like method is applied. The efficiency of this method can be improved without any additional computations if F'' is replaced in each step by Taylor's formula using F', since the order of convergence of the method is preserved. So, from Taylor's formula and taking into account that F is a quadratic operator, we have

$$F'(\mathbf{y}_n) = F'(\mathbf{x}_n) + F''(\mathbf{x}_n)(\mathbf{y}_n - \mathbf{x}_n),$$

where $\mathbf{y}_n = \mathbf{x}_n - [F'(\mathbf{x}_n)]^{-1} F(\mathbf{x}_n)$. Then,

$$F''(\mathbf{x}_n)[F'(\mathbf{x}_n)]^{-1} F(\mathbf{x}_n) = F'(\mathbf{x}_n) - F'(\mathbf{y}_n), \qquad L_F(\mathbf{x}_n) = I - [F'(\mathbf{x}_n)]^{-1} F'(\mathbf{y}_n).$$

In addition, the Chebyshev-like method can be written for a quadratic operator F as

$$\mathbf{x}_{n+1} = \mathbf{x}_n - \left(2I - \frac{3}{2}[F'(\mathbf{x}_n)]^{-1} F'(\mathbf{y}_n) + \frac{1}{2}\left([F'(\mathbf{x}_n)]^{-1} F'(\mathbf{y}_n)\right)^2\right)$$
$$\times [F'(\mathbf{x}_n)]^{-1} F(\mathbf{x}_n), \quad n \geq 0, \tag{4.38}$$

and it is not necessary to consider F''.

Starting at $\mathbf{x}_n$, for the finite-dimensional case, the computation of the $(n+1)$-step of (4.38) is as follows:

Stage 1. Compute one LR-decomposition of F' by Gauss elimination.
Stage 2. Solve the linear system $F'(\mathbf{x}_n)\mathbf{z}_n = -F(\mathbf{x}_n)$.
Stage 3. Set $\mathbf{y}_n = \mathbf{x}_n + \mathbf{z}_n$.
Stage 4. Solve the linear system $F'(\mathbf{x}_n)\mathbf{u}_n = F'(\mathbf{y}_n)\mathbf{z}_n$.
Stage 5. Solve the linear system $F'(\mathbf{x}_n)\mathbf{v}_n = F'(\mathbf{y}_n)\mathbf{u}_n$.
Stage 6. Set $\mathbf{x}_{n+1} = \mathbf{x}_n + 2\mathbf{z}_n - \frac{3}{2}\mathbf{u}_n + \frac{1}{2}\mathbf{v}_n$.

Notice that the linear systems considered above have the same associated matrix, and then we only need one LR-decomposition of the matrix $F'(\mathbf{x}_n)$ in each step.

The iteration (4.38) approximates the numerical solution $\mathbf{x}^* = (x_1^*, x_2^*, \ldots, x_8^*)^T$ appearing in Table 4.9 after four iterations and using the stopping criterion $\|\mathbf{x}_{n+1} - \mathbf{x}_n\|_\infty < 10^{-150}$ and the starting point $\mathbf{x}_0 = \mathbf{1}$ (taking into account that we have previously chosen $x_0(s) = 1$, a reasonable choice for initial approximation seems to be the vector $\mathbf{x}_0 = \mathbf{1}$). Observe in Table 4.10 that, according to (4.35), the R-order of convergence computationally reached is at least four.

Finally, by interpolating the values given in Table 4.9 and taking into account that a solution of the Chandrasekhar equation satisfies $x(0) = 1$, we obtain the solution drawn in Fig. 4.3. Observe also that this interpolated solution lies within the domain of existence of solution, $\overline{B(\mathbf{1}, 0.2870\ldots)}$, provided by Theorem 4.19.

4.3 Quadratic Equations

Table 4.9 Numerical solution $\mathbf{x}^*$ of the system (4.37)

i	x_i^*	i	x_i^*
1	1.02171973146...	5	1.20307175130...
2	1.07318638173...	6	1.22649087463...
3	1.12572489365...	7	1.24152460059...
4	1.16975331216...	8	1.24944851669...

Table 4.10 Errors and computational order of convergence for the method (4.38)

n	$\|\mathbf{x}^* - \mathbf{x}_n\|$	order of convergence ρ
1	$4.2052\ldots \times 10^{-5}$	$4.1325\ldots$
2	$1.0737\ldots \times 10^{-20}$	$4.0221\ldots$
3	$2.0579\ldots \times 10^{-83}$	

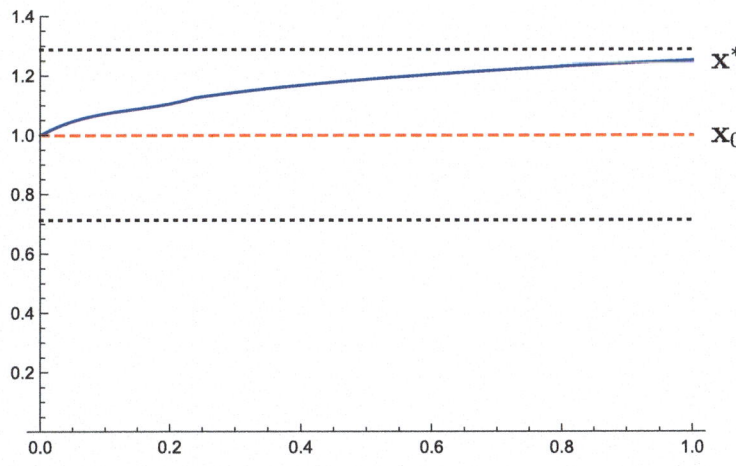

Fig. 4.3 The approximate solution of the Chandrasekhar equation (4.18) (solid line) and the domain of existence of solution (dotted lines)

Optimization of the Chebyshev Method 5

The main problem in the one-point iterative methods with R-order of convergence at least three is the evaluation of the second derivative of the operator involved. Moreover, the convergence conditions required to prove the semilocal convergence of these methods are usually more restrictive than those required to Newton-like methods [27]. In this chapter, we try to solve both problems. Thus, from the Chebyshev method, we construct families of iterative methods where the second derivative of the operator involved does not appear in the algorithms, but with the same R-order of convergence as the Chebyshev method, so that the operational cost of the new methods is lower than that of the Chebyshev method. These constructions are developed in the next sections and are called optimizations in this chapter.

It is known that there are three determining factors to measure the quality of an iterative method: the number of operator evaluations that must be done in the application of the method, the operational cost required in each iteration and the accessibility of the method to the solution of the equation to solve, which is the set of starting points from which the iterative method is convergent.

In this context, throughout this chapter, we construct four modifications of the Chebyshev method that improve this method in some sense, and all of them are free of second derivatives.

When it comes to choosing an iterative method for solving nonlinear equations, we have especially to take into account the speed of convergence and the computational cost, which are directly related to the efficiency of the method. For this, and following Traub [88], we can analyze the efficiency of an iterative method by studying the efficiency index, EI, which is defined by the order of convergence to the inverse power of the number of evaluations of the function involved and its derivatives (in particular, this index is usually considered in the analysis of scalar equations, where the computational cost of the successive derivatives is not very different), and the computational efficiency, CE, which is defined by the order of convergence to the inverse power of the number of the operations.

Then, in Sect. 5.2, we construct a family of iterative methods with a better efficiency index than the Chebyshev method and a better computational efficiency than the Newton method to solve systems of equations with more than ten equations [36, 60]. In Sect. 5.3, we construct a family of iterative methods with a better efficiency index than the Newton and the Chebyshev methods and a better computational efficiency than the Newton method to solve systems of equations with more than six equations [38]. Since we do not reach the accessibility of the Newton method with the previous two families of iterative methods, our next aim is to achieve that. So, in Sect. 5.4, we construct a family of hybrid iterative methods with R-order of convergence at least three that has a better accessibility than the Chebyshev method and the same region of accessibility as the Newton method. Finally, in Sect. 5.5, we construct two derivative-free iterative methods [40] with the aim of improving the efficiency index and the computational efficiency of the secant method, which is the most well-known and used derivative-free iterative method.

All studies are done in Banach spaces and using the technique based on recurrence relations described in Chap. 4. Finally, the established theoretical results are illustrated with applications related to nonlinear Fredholm integral equations.

5.1 Preliminary Analysis

When we analyze the efficiency of an iterative method to solve a scalar equation, in the sense defined by Traub [88], we usually use the efficiency index, which is defined as $EI = q^{1/d}$, where q is the order of convergence and d the number of new evaluations of the function involved and its derivatives per iteration. For one-point iterative methods of order q, a restrictive condition is that these methods depend explicitly on the first $q-1$ derivatives of the operator involved. Moreover, for these methods, $q = d$ and $EI = q^{1/q}$, so that the best situation is obtained for $q = 3$, namely for one-point iterative methods of third order. However, for nonlinear systems of equations, third-order methods are not considered as the most favorable, rather the Newton method is, although its efficiency index $EI = 2^{1/2}$ is worse. This is due to the fact that the efficiency index does not take into account other considerations.

For solving nonlinear systems of dimension m by iterative methods, we have to bear in mind two important points: the number of evaluations of functions per iteration and the number of operations (products and divisions) needed to apply a step of the iterative methods. Then, we can use the efficiency index and the computational efficiency given by $EI = q^{1/\nu(\mathrm{m})}$ and $CE = q^{1/\vartheta(\mathrm{m})}$, respectively, where q is the order of convergence, $\nu(\mathrm{m})$ the number of evaluations of functions per iteration, and $\vartheta(\mathrm{m})$ the number of operations (products and divisions) needed to apply a step of the method.

At first glance, the Newton method is the most efficient with the two previous efficiency indices. Thus, based on the Chebyshev method, we try to improve these efficiency indices

5.1 Preliminary Analysis

and then obtain new iterative methods as efficient as the Newton method. We then begin by studying the efficiency of the Newton and Chebyshev methods to analyze later the improvements that we obtain from the four optimizations of the Chebyshev method presented in this chapter.

If we consider the case of solving nonlinear systems of equations of dimension m, $F(x_1, x_2, \ldots, x_m) = 0$, where $F : \Omega \subseteq \mathbb{R}^m \to \mathbb{R}^m$ is a nonlinear function and $F \equiv (F_1, F_2, \ldots, F_m)$ with $F_i : \Omega \subseteq \mathbb{R}^m \to \mathbb{R}$, $i = 1, 2, \ldots, m$, it is necessary to compute the m functions F_i ($i = 1, 2, \ldots, m$) for computing F. Moreover, for $\mathbf{x} = (x_1, x_2, \ldots, x_m)$, the computation of F'

$$F'(\mathbf{x}) = \begin{pmatrix} (F_1)_1(\mathbf{x}) & (F_1)_2(\mathbf{x}) & \cdots & (F_1)_m(\mathbf{x}) \\ (F_2)_1(\mathbf{x}) & (F_2)_2(\mathbf{x}) & \cdots & (F_2)_m(\mathbf{x}) \\ \vdots & \vdots & \ddots & \vdots \\ (F_m)_1(\mathbf{x}) & (F_m)_2(\mathbf{x}) & \cdots & (F_m)_m(\mathbf{x}) \end{pmatrix}$$

requires the computations of m^2 partial derivatives of first order, and the computation of F''

$$F''(\mathbf{x}) = \left(\begin{pmatrix} (F_1)_{11}(\mathbf{x}) & (F_1)_{12}(\mathbf{x}) & \cdots & (F_1)_{1m}(\mathbf{x}) \\ (F_1)_{21}(\mathbf{x}) & (F_1)_{22}(\mathbf{x}) & \cdots & (F_1)_{2m}(\mathbf{x}) \\ \vdots & \vdots & \ddots & \vdots \\ (F_1)_{m1}(\mathbf{x}) & (F_1)_{m2}(\mathbf{x}) & \cdots & (F_1)_{mm}(\mathbf{x}) \end{pmatrix} \cdots \begin{pmatrix} (F_m)_{11}(\mathbf{x}) & (F_m)_{12}(\mathbf{x}) & \cdots & (F_m)_{1m}(\mathbf{x}) \\ (F_m)_{21}(\mathbf{x}) & (F_m)_{22}(\mathbf{x}) & \cdots & (F_m)_{2m}(\mathbf{x}) \\ \vdots & \vdots & \ddots & \vdots \\ (F_m)_{m1}(\mathbf{x}) & (F_m)_{m2}(\mathbf{x}) & \cdots & (F_m)_{mm}(\mathbf{x}) \end{pmatrix} \right),$$

requires the computations of $\frac{m^2}{2}(m+1)$ partial derivatives of second order. In addition, the application of the Newton method

$$\begin{cases} x_0 \text{ given,} \\ F'(x_n)\delta_n = -F(x_n), \quad n \geq 0, \\ x_{n+1} = x_n + \delta_n, \end{cases} \tag{5.1}$$

to solve the nonlinear system of m equations

$$\begin{cases} F_1(x_1, x_2, \ldots, x_m) = 0, \\ F_2(x_1, x_2, \ldots, x_m) = 0, \\ \quad\quad\quad\quad \vdots \\ F_m(x_1, x_2, \ldots, x_m) = 0 \end{cases} \tag{5.2}$$

requires $m^2 + m$ evaluations of functions per iteration, whereas a one-point third-order method, as, for example, the Chebyshev method,

$$\begin{cases} x_0 \text{ given,} \\ F'(x_n)\,\delta_n = -F(x_n), \quad n \geq 0, \\ F'(x_n)\,\gamma_n = (-1/2)\,F''(x_n)\,\delta_n^2, \\ x_{n+1} = x_n + \delta_n + \gamma_n \end{cases} \qquad (5.3)$$

requires $\frac{m^2}{2}(m+1)$ evaluations of functions per iteration more than the Newton method. Therefore, for solving the system (5.2) with $m \geq 2$, the use of the Newton method is more advisable than the Chebyshev method, see Fig. 5.1.

In addition, other important point to bear in mind when choosing an iterative method is the number of operations (products and divisions) needed to apply it, which we define here as the computational cost of doing an iteration of the algorithm. So, the Newton method requires $\frac{1}{3}(m^3 + 6m^2 - 4m)$ operations to do an iteration, whereas the Chebyshev method requires to do the same operations as the Newton method plus the products $(-1/2)\,F''(x_n)\,\delta_n^2$ ($m^3 + m^2 + m$ operations) and the solution of the linear system $F'(x_n)\,\gamma_n = (-1/2)\,F''(x_n)\,\delta_n^2$ ($2m^2 - m$ operations). As a consequence, the computational cost per iteration of the Chebyshev method is $\frac{1}{3}(4m^3 + 15m^2 - 4m)$, which is higher than that of the Newton method. Therefore, for solving the system (5.2), it is clear that the application of the Newton method is a better option than the Chebyshev method, as we can deduce from Table 5.1 and Fig. 5.2.

On the other hand, when third-order methods are applied to solve nonlinear equations, it is important to note that the region of accessibility, which consists of every starting point from which an iterative method is convergent, is reduced with respect to that of the Newton method. In practice, we can see this with the attraction basins (the set of points in the space such that initial conditions chosen in the set dynamically evolve to a particular attractor [69, 89]) of iterative methods when they are applied to solve a complex equation $F(z) = 0$, where $F : \mathbb{C} \to \mathbb{C}$ and $z \in \mathbb{C}$, and we are interested in identifying the attraction

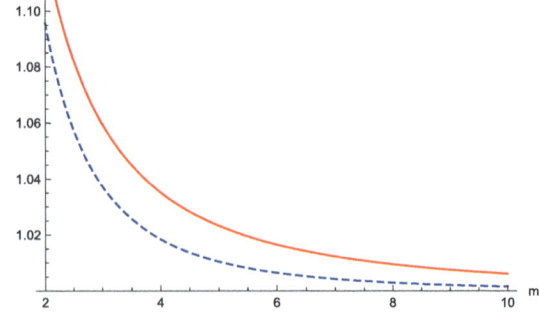

Fig. 5.1 Efficiency indices of the Newton and the Chebyshev methods for systems, respectively, $2^{1/(m^2+m)}$ (solid line) and $3^{2/(m^3+3m^2+2m)}$ (dashed line)

5.1 Preliminary Analysis

Table 5.1 Number of evaluations of functions and computational cost per iteration when the Newton and the Chebyshev methods are applied to solve nonlinear systems (10, 50, and 100 equations)

m	The Newton method		The Chebyshev method	
	$m^2 + m$	$(m^3 + 6m^2 - 4m)/3$	$(m^3 + 3m^2 + 2m)/2$	$(4m^3 + 15m^2 - 4m)/3$
10	110	520	660	1820
50	2550	46600	66300	179100
100	10100	353200	515100	1383200

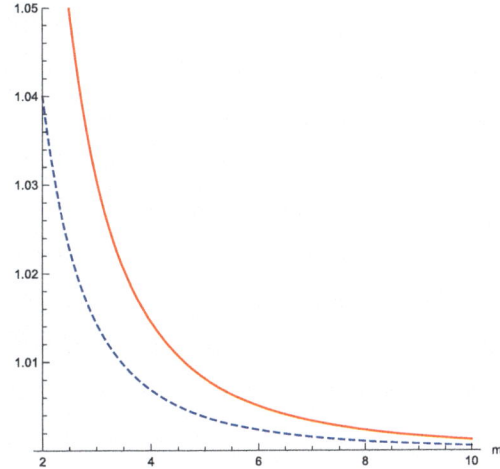

Fig. 5.2 Computational efficiencies of the Newton and the Chebyshev methods for systems, respectively, $2^{3/(m^3+6m^2-4m)}$ (solid line) and $3^{3/(4m^3+15m^2-4m)}$ (dashed line)

basin for two solutions z^* and z^{**} [89]. For this, we choose, for example, the Newton and the Chebyshev methods for solving the complex equation $F(z) = \sin z - \frac{1}{3} = 0$ and show the fractal pictures generated to approximate the solutions $z^* = \arctan(1/2\sqrt{2}) = 0.33983\ldots$ and $z^{**} = \pi - \arctan(1/2\sqrt{2}) = 2.80176\ldots$ This also allows us to compare the regions of accessibility of the methods.

We take a rectangle $D \subseteq \mathbb{C}$ and iterations starting at "every" $z_0 \in D$. In practice, a grid of 512×512 points in D is considered, and these points are chosen as z_0. The rectangle used is $[0, 3] \times [-2.5, 2.5]$, which contains the two zeros. The numerical methods starting at a point in the rectangle can converge to some of the zeros or, eventually, diverge.

In all the cases, the tolerance 10^{-3} and a maximum of 25 iterations are used. If we have not obtained the desired tolerance with 25 iterations, we do not continue and decide that the iterative method starting at z_0 does not converge to any zero.

The abovementioned rectangles corresponding to the two iterative methods when they are applied to approximate the solutions z^* and z^{**} of $F(z) = \sin z - \frac{1}{3} = 0$ are shown in Figs. 5.3 and 5.4. The strategy taken into account is the following. A color is assigned to each basin of attraction of a zero. The color is made lighter or darker according to the number of iterations needed to reach the root with the fixed precision required. Finally,

Fig. 5.3 The Newton method

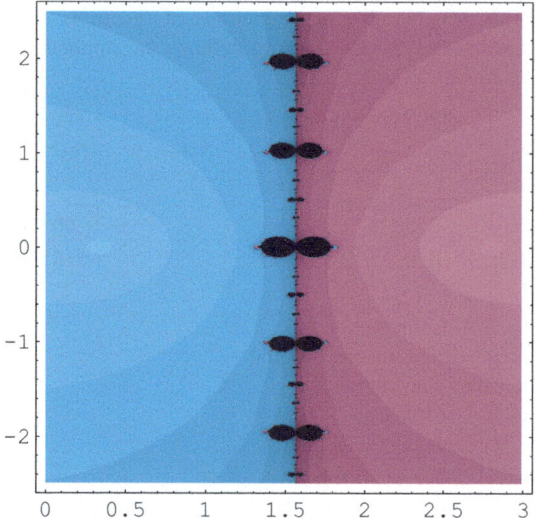

Fig. 5.4 The Chebyshev method

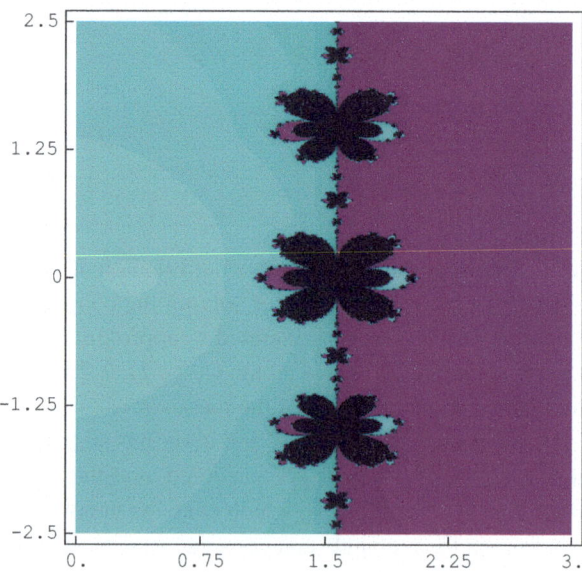

if the iteration does not converge, the color black is used. For more strategies, the reader can see [89] and the references therein. In particular, to obtain the pictures, the cyan and magenta colors have been assigned for the attraction basins of the two zeros. We mark with black the points of the rectangle for which the corresponding iterations starting at them do not reach any root with tolerance 10^{-3} in a maximum of 25 iterations. The graphics shown here have been generated with Mathematica 5.1 [91].

If we observe the behavior of the two methods, we see that the Chebyshev method is more demanding with respect to the starting point than the Newton method (see the black color). We can also observe that there exist lighter areas for the Chebyshev method. These observations are as a consequence of the higher speed of convergence of the last method (cubical convergence) as opposed to the Newton method (quadratic convergence), and therefore it is more difficult to locate starting points from which the Chebyshev method converges.

From the comments cited above, our immediate interest is focused on constructing iterations from modifications of the Chebyshev method that first reduce the number of evaluations of functions and the computational cost and then reach the region of accessibility of the Newton method.

5.2 Optimization 1

We start by constructing, from the Chebyshev method, a family of multipoint iterative methods that does not use the second derivative of the operator involved in the algorithm [60]. Multipoint methods are defined as iterations which use new information at a variable number of points (see [32] and the references therein). A very restrictive condition on one-point iterations is that they must depend explicitly on the first $q-1$ derivatives of the operator involved to obtain order of convergence q (see [88]), which implies that their efficiency index is less than or equal to one. This restriction need not hold for multipoint methods, that is, for iterations that evaluate the operator involved and its derivatives at a number of values of the independent variable.

Notice that the first family of multipoint iterative methods that we construct from a modification of the Chebyshev method provides iterative methods with a better efficiency index and a better computational efficiency than the Chebyshev method for systems of equations and a better computational efficiency than the Newton method for systems with more than ten equations.

5.2.1 Construction of the Method

If we use the Chebyshev method to solve nonlinear equations $F(x) = 0$, where $F : \Omega \subseteq X \to Y$ is a twice continuously Fréchet differentiable operator defined on a non-empty open convex subset Ω of a Banach space X with values in a Banach space Y, we have to evaluate the second derivative of F, as we can see in (5.3), that is now written as

$$\begin{cases} x_0 \text{ given in } \Omega, \\ y_n = x_n - [F'(x_n)]^{-1} F(x_n), \quad n \geq 0, \\ x_{n+1} = y_n + \tfrac{1}{2} L_F(x_n)(y_n - x_n), \end{cases}$$

so that its use is restrictive in practice as a consequence of the operational cost. In fact, the Newton method is usually applied more in practice despite its lower speed of convergence. This is due to the relation between several factors, such as the number of necessary values of the involved function and its derivatives, the computational cost, and the speed of convergence. Thus, our immediate aim is to modify the Chebyshev method so that the second derivative of F does not appear in the algorithm. Then, we derive a family of multipoint iterative methods as follows. From Taylor's formula, we have

$$F'(z_n) \approx F'(x_n) + F''(x_n)(z_n - x_n),$$

where $z_n = x_n + \theta(y_n - x_n)$ and $\theta \in (0, 1]$. Then,

$$L_F(x_n) \approx \frac{1}{\theta}[F'(x_n)]^{-1}(F'(x_n) - F'(z_n)),$$

and we derive the following family of multipoint iterations with R-order of convergence at least three:

$$\begin{cases} x_0 \text{ given in } \Omega, \\ y_n = x_n - [F'(x_n)]^{-1} F(x_n), \quad n \geq 0, \\ z_n = x_n + \theta(y_n - x_n), \quad \theta \in (0, 1], \\ P(x_n, z_n) = \frac{1}{\theta}[F'(x_n)]^{-1}(F'(x_n) - F'(z_n)), \\ x_{n+1} = y_n + \frac{1}{2} P(x_n, z_n)(y_n - x_n). \end{cases} \quad (5.4)$$

As we can see in this case, it is only necessary for the operator F to be Fréchet differentiable once.

5.2.2 Analysis of the Method

The family (5.4) can be written to solve the nonlinear system (5.2) as

$$\begin{cases} x_0 \text{ given,} \\ F'(x_n) \delta_n = -F(x_n), \quad n \geq 0, \\ z_n = x_n + \theta \delta_n, \quad \theta \in (0, 1], \\ F'(x_n) \hat{\gamma}_n = -\frac{1}{2\theta}(F'(z_n) - F'(x_n)) \delta_n, \\ x_{n+1} = x_n + \delta_n + \hat{\gamma}_n, \end{cases} \quad (5.5)$$

5.2 Optimization 1

Table 5.2 Evaluations of functions

Method	Eval. of F	Eval. of F'	LR-decomp.	Eval. of F''
The Chebyshev method	1	1	1	1
The method (5.5)	1	2	1	0
The Newton method	1	1	1	0

Table 5.3 Number of evaluations of functions and computational cost per iteration when (5.5) is applied to solve systems (10, 50, and 100 equations)

	Method (5.5)	
m	$2m^2 + m$	$(m^3 + 15m^2 - m)/3$
10	210	830
50	5050	54150
100	20100	383300

so that the two lineal systems, which we have to solve when applying (5.5), have the same associated matrix, and then we only need one LR decomposition of the matrix $F'(x_n)$ in each step (see Table 5.2).

Observe that (5.5) can be considered as a Newton-like method with a similar operational cost to that of the Newton method, including one more evaluation of F', as we see in Table 5.2. However, we can prove that (5.5) has R-order of convergence at least three. As a consequence, (5.5) is preferable to the Chebyshev method.

As the expression $F''(x_n)\delta_n^2$ of the algorithm (5.3) of the Chebyshev method is approximated by the expression $\frac{1}{\theta}(F'(z_n) - F'(x_n))\delta_n$ to obtain (5.5), then the number of evaluations of functions and the computational cost per iteration are, respectively, reduced to $2m^2 + m$ and $\frac{1}{3}(m^3 + 15m^2 - m)$; see Table 5.3. In addition, the efficiency index of the method (5.5) is better than that of the Chebyshev method, but worse than that of the Newton method (Fig. 5.5). On the other hand, the computational efficiency of (5.5) is always better than that of the Chebyshev method and better than that of the Newton method from 11 equations (Fig. 5.6).

Regarding the behavior of the Newton method and the family (5.5), from the point of view of accessibility, if, for example, we consider (5.5) with $p = 1$, we see in Figs. 5.3 and 5.7 that the last method is more demanding than the Newton method, so that it is more difficult to locate starting points from which the method (5.5) with $p = 1$ converges.

5.2.3 Convergence Analysis

Next, we study the semilocal convergence of the family of multipoint iterative methods (5.5). For this, we use the technique based on recurrence relations seen in

Fig. 5.5 Efficiency indices of the Newton and the Chebyshev methods and the method (5.5) for systems, respectively, $2^{1/(m^2+m)}$ (solid line), $3^{2/(m^3+3m^2+2m)}$ (dashed line) and $3^{1/(2m^2+m)}$ (dotted line)

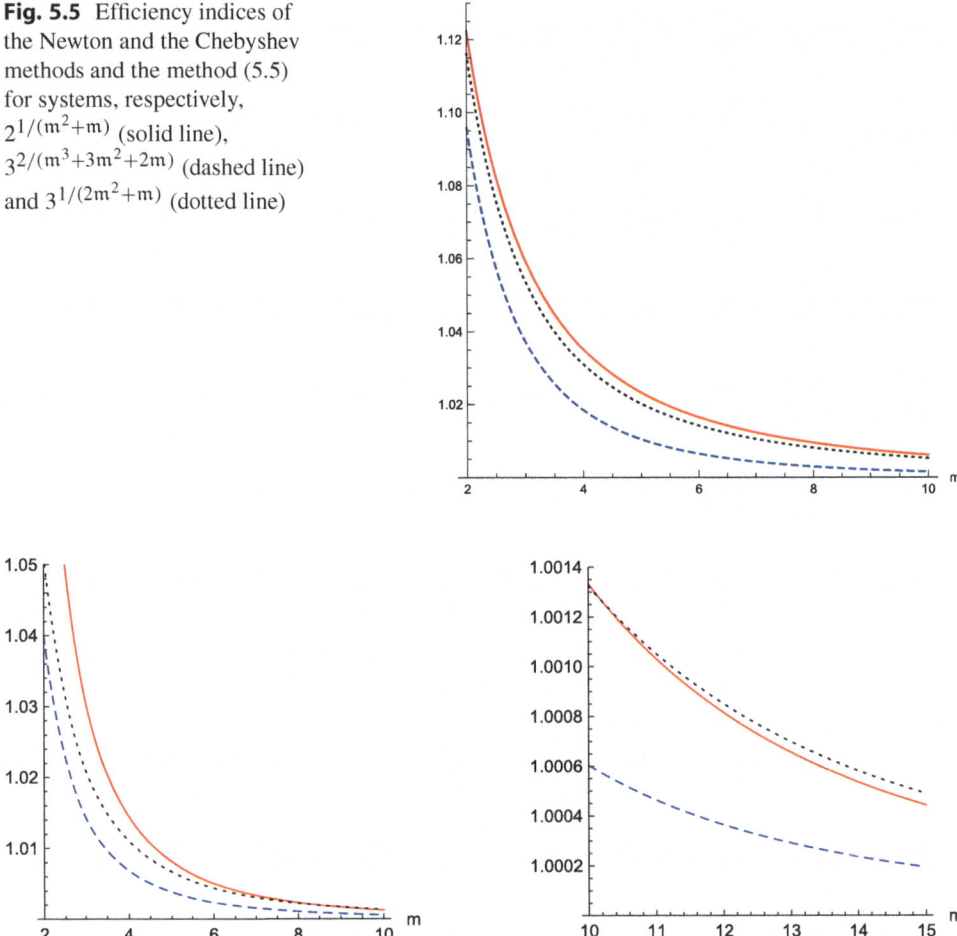

Fig. 5.6 Computational efficiencies of the Newton and the Chebyshev methods and the method (5.5) for systems, respectively, $2^{3/(m^3+6m^2-4m)}$ (solid line), $3^{3/(4m^3+15m^2-4m)}$ (dashed line), and $3^{3/(m^3+15m^2-m)}$ (dotted line)

Chap. 4. Then, we use a system of real sequences which simplifies that given by other authors [18, 53]. In order to guarantee the R-order of convergence at least three of the family, this study is applied to operators with a Hölder continuous second Fréchet derivative [8], so that the usual Lipschitz continuity of the second Fréchet derivative is thus relaxed [60]. As we have indicated previously, the family (5.5) can be applied to operators that are once Fréchet differentiable. Thus, in Sect. 5.2.4, we study the convergence of the family for once Fréchet differentiable operators.

Fig. 5.7 Method (5.5) with $p = 1$

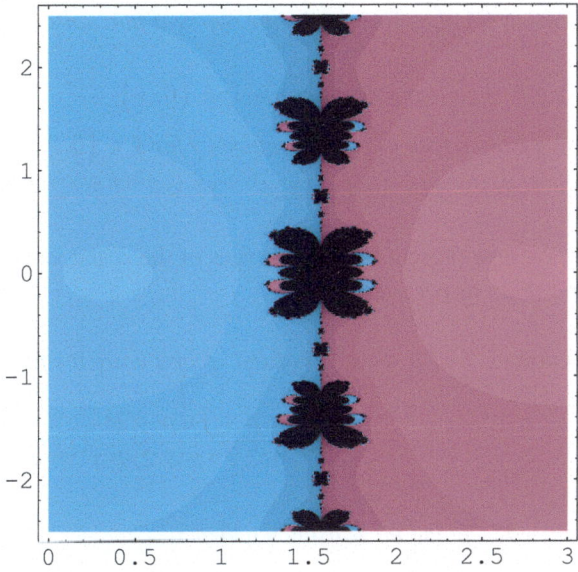

5.2.3.1 Recurrence Relations

We suppose that the operator F is twice continuously Fréchet differentiable and the following conditions:

(Q1) There exists $\Gamma_0 = [F'(x_0)]^{-1} \in \mathcal{L}(Y, X)$, for some $x_0 \in \Omega$, with $\|\Gamma_0\| \leq \beta$ and $\|\Gamma_0 F(x_0)\| \leq \eta$.

(Q2) There exists a constant $M > 0$ such that $\|F''(x)\| \leq M$, for $x \in \Omega$.

(Q3) There exist two constants $K > 0$ and $p \in [0, 1]$ such that $\|F''(x) - F''(y)\| \leq K\|x - y\|^p$, for $x, y \in \Omega$.

Denote $a_0 = M\beta\eta$, and $b_0 = K\beta\eta^{1+p}$ and define the scalar sequences

$$a_n = a_{n-1} f(a_{n-1})^2 g(a_{n-1}, b_{n-1}), \quad n \in \mathbb{N}, \tag{5.6}$$

$$b_n = b_{n-1} f(a_{n-1})^{2+p} g(a_{n-1}, b_{n-1})^{1+p}, \quad n \in \mathbb{N}, \tag{5.7}$$

where

$$f(x) = \frac{2}{2 - 2x - x^2}, \quad g(x, y) = \frac{x^3 + 4x^2}{8} + \frac{(2 + (p+2)\theta^p)y}{2(p+1)(p+2)}. \tag{5.8}$$

In order to provide some properties of the sequences (5.6) and (5.7), we first give a technical lemma whose proof is trivial.

Lemma 5.1 *Let f and g be the two real functions defined in (5.8). Then,*

(a) *f is increasing and $f(x) > 1$ in $\left(0, \frac{1}{2}\right)$.*
(b) *g is increasing in both arguments for $x \in \left(0, \frac{1}{2}\right)$ and $y > 0$.*
(c) *$f(\delta x) < f(x)$ and $g(\delta x, \delta^{1+p} y) < \delta^{1+p} g(x, y)$, for $\delta \in (0, 1)$.*

We now present some properties of the scalar sequences $\{a_n\}$ and $\{b_n\}$ defined in (5.6) and (5.7), respectively.

Lemma 5.2 *If $p \in [0, 1]$ is fixed, f and g are the two real functions defined in (5.8),*

$$h(x, \theta) = \frac{(p+1)(p+2)}{4(2+(p+2)\theta^p)}(1-2x)(8-4x^2-x^3), \tag{5.9}$$

$a_0 \in \left(0, \frac{1}{2}\right)$, and $b_0 < h(a_0, \theta)$, then

(a) *$f(a_0)^2 g(a_0, b_0) < 1$.*
(b) *The sequences $\{a_n\}$ and $\{b_n\}$ are decreasing.*
(c) *$a_n\left(1 + \frac{a_n}{2}\right) < 1$, for all $n \geq 0$.*

Proof From the hypotheses, item (a) follows immediately. We prove item (b) by mathematical induction on n. The conditions $0 < a_1 < a_0$ and $0 < b_1 < b_0$ follow by item (a) and Lemma 5.1 (a). Next, we suppose that $a_j < a_{j-1}$ and $b_j < b_{j-1}$, for $j = 1, 2, \ldots, n$. Then,

$$a_{n+1} = a_n f(a_n)^2 g(a_n, b_n) < a_n f(a_0)^2 g(a_0, b_0) < a_n,$$

since f is increasing and g is also increasing in both arguments. Analogously,

$$b_{n+1} = b_n f(a_n)^{2+p} g(a_n, b_n)^{1+p} = b_n f(a_n)^2 g(a_n, b_n)(f(a_n)g(a_n, b_n))^p < b_n,$$

by the same reasoning as above and the inequalities $f(a_n)g(a_n, b_n) < 1$ and $f(x) > 1$ in $\left(0, \frac{1}{2}\right)$.

Finally, for all $n \geq 0$, we have

$$a_n\left(1 + \frac{a_n}{2}\right) < a_0\left(1 + \frac{a_0}{2}\right) < 1,$$

since $\{a_n\}$ is decreasing and $a_0 \in \left(0, \frac{1}{2}\right)$, so that item (c) follows. ∎

5.2 Optimization 1

Lemma 5.3 *Under the hypotheses of Lemma 5.2 and $\gamma = \frac{a_1}{a_0}$, we have:*

(i_n) $a_n < \gamma^{(2+p)^{n-1}} a_{n-1} < \gamma^{\frac{(2+p)^n - 1}{1+p}} a_0$ and $b_n < \left(\gamma^{(2+p)^{n-1}}\right)^{1+p} b_{n-1} < \gamma^{(2+p)^n - 1} b_0$, *for all* $n \geq 2$.

(ii_n) $f(a_n)g(a_n, b_n) < \gamma^{(2+p)^n} \frac{f(a_0)g(a_0, b_0)}{\gamma} = \frac{\gamma^{(2+p)^n}}{f(a_0)}$, *for all* $n \geq 1$.

Proof We prove (i_n) by mathematical induction on n. Taking into account that $a_1 = \gamma a_0$ and $b_1 = b_0 f(a_0)^{2+p} g(a_0, b_0)^{1+p} < \gamma^{1+p} b_0$ is satisfied if and only if $f(a_0)^2 g(a_0, b_0)(f(a_0)g(a_0, b_0))^p < \gamma^{1+p}$, the result follows from Lemma 5.2. If we suppose that (i_n) is true, then

$$a_{n+1} = a_n f(a_n)^2 g(a_n, b_n)$$
$$< \gamma^{(2+p)^{n-1}} a_{n-1} f\left(\gamma^{(2+p)^{n-1}} a_{n-1}\right)^2 g\left(\gamma^{(2+p)^{n-1}} a_{n-1}, \left(\gamma^{(2+p)^{n-1}}\right)^{1+p} b_{n-1}\right)$$
$$< \gamma^{(2+p)^{n-1}} a_{n-1} f(a_{n-1})^2 \left(\gamma^{(2+p)^{n-1}}\right)^{1+p} g(a_{n-1}, b_{n-1})$$
$$= \gamma^{(2+p)^n} a_n.$$

In addition,

$$b_{n+1} = b_n f(a_n)^{2+p} g(a_n, b_n)^{1+p} < \left(\frac{a_{n+1}}{a_n}\right)^{1+p} b_n$$

if and only if

$$a_n^2 f(a_n)^{2+p} g(a_n, b_n)^{1+p} < a_{n+1}^2 = a_n^2 f(a_n)^4 g(a_n, b_n)^2,$$

and this is true since $f(a_n) > 1$. Now, $b_{n+1} < (\gamma^{(2+p)^n})^{1+p} b_n$, since $\frac{a_{n+1}}{a_n} < \gamma^{(2+p)^n}$. Moreover,

$$a_{n+1} < \gamma^{(2+p)^n} a_n < \gamma^{(2+p)^n} \gamma^{(2+p)^{n-1}} a_{n-1} < \cdots < \gamma^{\frac{(2+p)^{n+1} - 1}{1+p}} a_0,$$
$$b_{n+1} < \left(\gamma^{(2+p)^n}\right)^{1+p} b_n < \left(\gamma^{(2+p)^n}\right)^{1+p} \left(\gamma^{(2+p)^{n-1}}\right)^{1+p} b_{n-1} < \cdots < \gamma^{(2+p)^{n+1} - 1} b_0.$$

Furthermore, we observe that

$$f(a_n)g(a_n, b_n) < f\left(\gamma^{\frac{(2+p)^n - 1}{1+p}} a_0\right) g\left(\gamma^{\frac{(2+p)^n - 1}{1+p}} a_0, \gamma^{(2+p)^n - 1} b_0\right)$$

$$< \gamma^{(2+p)^n} \frac{f(a_0)g(a_0,b_0)}{\gamma}$$

$$= \frac{\gamma^{(2+p)^n}}{f(a_0)}.$$

The proof is complete. ∎

Now, we use the real sequences $\{a_n\}$ and $\{b_n\}$ defined in (5.6) and (5.7), respectively, to analyze, under certain conditions on the pair (F, x_0), the convergence of (5.4) to a unique solution x^* of the equation $F(x) = 0$. We also provide some a priori error estimates on the distances $\|x^* - x_n\|$ for all $n \geq 0$.

First, we have $\|P(x_0, z_0)\| \leq M\|\Gamma_0\|\|z_0 - x_0\| \leq a_0$ if $z_0 \in \Omega$, $K\|\Gamma_0\|\|y_0 - x_0\|^{1+p} \leq b_0$, and $\|x_1 - x_0\| \leq \left(1 + \frac{a_0}{2}\right)\|y_0 - x_0\|$.

Next, by mathematical induction on n, we prove the following estimates for $n \in \mathbb{N}$:

(i_n) $\|\Gamma_n\| = \|[F'(x_n)]^{-1}\| \leq f(a_{n-1})\|\Gamma_{n-1}\|$.
(ii_n) $\|y_n - x_n\| = \|\Gamma_n F(x_n)\| \leq f(a_{n-1})g(a_{n-1}, b_{n-1})\|y_{n-1} - x_{n-1}\|$.
(iii_n) $\|P(x_n, z_n)\| \leq M\|\Gamma_n\|\|y_n - x_n\| \leq a_n$.
(iv_n) $K\|\Gamma_n\|\|y_n - x_n\|^{1+p} \leq b_n$.
(v_n) $\|x_{n+1} - x_n\| \leq \left(1 + \frac{a_n}{2}\right)\|y_n - x_n\|$.

It is easy to check (i_1)–(ii_1)–(iii_1)–(iv_1)–(v_1) if conditions (Q1)–(Q2)–(Q3) are satisfied as we see in the following:

(i_1) If $x_1 \in \Omega$, we observe that

$$\|I - \Gamma_0 F'(x_1)\| \leq \|\Gamma_0\|\|F'(x_0) - F'(x_1)\| \leq M\|\Gamma_0\|\|x_1 - x_0\| \leq a_0\left(1 + \frac{a_0}{2}\right) < 1,$$

and, by the Banach lemma on invertible operators, the operator Γ_1 exists and

$$\|\Gamma_1\| \leq \frac{\|\Gamma_0\|}{1 - \|I - \Gamma_0 F'(x_1)\|} \leq f(a_0)\|\Gamma_0\|.$$

(ii_1) If $y_0 \in \Omega$ and we use Taylor's formula and (5.4), we have

$$F(x_1) = F(y_0) + F'(y_0)(x_1 - y_0) + \int_{y_0}^{x_1} F''(x)(x_1 - x)\,dx$$

$$= \int_0^1 (F''(x_0 + t(y_0 - x_0)) - F''(x_0))(y_0 - x_0)^2(1 - t)\,dt$$

$$+ \frac{1}{2}\int_0^1 (F''(x_0) - F''(x_0 + \theta t(y_0 - x_0)))(y_0 - x_0)^2\,dt$$

5.2 Optimization 1

$$+ \frac{1}{2}\int_0^1 F''(x_0 + t(y_0 - x_0))(y_0 - x_0)P(x_0, z_0)(y_0 - x_0)\,dt$$

$$+ \int_0^1 F''(y_0 + t(x_1 - y_0))(x_1 - y_0)^2(1-t)\,dt.$$

Thus,

$$\|y_1 - x_1\| \le \|\Gamma_1\|\|F(x_1)\| \le f(a_0)\|\Gamma_0\|\|F(x_1)\| \le f(a_0)g(a_0, b_0)\|y_0 - x_0\|.$$

(iii_1) If $z_1 \in \Omega$, we obtain

$$\|P(x_1, z_1)\| \le \frac{1}{\theta}\|\Gamma_1\|\|F(x_1) - F'(x_1 + \theta(y_1 - x_1))\|$$

$$\le M\|\Gamma_1\|\|y_1 - x_1\|$$

$$\le M\|\Gamma_0\|\|y_0 - x_0\|f(a_0)^2 g(a_0, b_0)$$

$$\le a_1.$$

(iv_1) Moreover,

$$K\|\Gamma_1\|\|y_1 - x_1\|^{1+p} \le K\|\Gamma_0\|\|y_0 - x_0\|^{1+p} f(a_0)^{2+p} g(a_0, b_0)^{1+p} \le b_1.$$

(v_1) Furthermore,

$$\|x_2 - x_1\| \le \left\|I + \frac{1}{2}P(x_1, z_1)\right\|\|y_1 - x_1\| \le \left(1 + \frac{a_1}{2}\right)\|y_1 - x_1\|.$$

Then, by mathematical induction on n, if $x_n, y_n \in \Omega$, then $z_n \in \Omega$, and the proof of the system of the recurrence relations is complete.

5.2.3.2 Semilocal Convergence

We are now ready to give the next semilocal convergence result.

Theorem 5.4 *Let $F : \Omega \subseteq X \to Y$ be a twice continuously Fréchet differentiable operator on a non-empty open convex domain Ω of a Banach space X with values in a Banach space Y. Suppose that (Q1)–(Q2)–(Q3) are satisfied. Denote $a_0 = M\beta\eta$ and $b_0 = K\beta\eta^{1+p}$. Suppose that $a_0 \in \left(0, \frac{1}{2}\right)$ and $b_0 < h(a_0, \theta)$, where h is defined in (5.9) and $\theta \in (0, 1]$. Then, if $\overline{B(x_0, R\eta)} \subseteq \Omega$, where $R = \left(1 + \frac{a_0}{2}\right)\frac{1}{1-\gamma\Delta}$ and $\Delta = \frac{1}{f(a_0)}$, the sequence (5.4) starting at x_0 converges with R-order of convergence at least $2 + p$ to a* solution x^* of $F(x) = 0$. In this case, the solution x^* and the iterates x_n, y_n, and z_n belong to $\overline{B(x_0, R\eta)}$. Moreover, the solution x^* is unique in $B\left(x_0, \frac{2}{M\beta} - R\eta\right) \cap \Omega$ and

$$\|x^* - x_n\| \leq \left(1 + \frac{a_0}{2}\gamma^{\frac{(2+p)^n-1}{1+p}}\right) \gamma^{\frac{(2+p)^n-1}{1+p}} \frac{\Delta^n}{1 - \gamma^{(2+p)^n}\Delta} \eta. \tag{5.10}$$

Proof First, we prove that $\{x_n\}$ is a Cauchy sequence. Observe that, for $n \in \mathbb{N}$,

$$\left(1 + \frac{a_n}{2}\right) \|y_n - x_n\| \leq \left(1 + \frac{a_0}{2}\right) f(a_{n-1})g(a_{n-1}, b_{n-1})\|y_{n-1} - x_{n-1}\|$$

$$< \cdots < \left(1 + \frac{a_0}{2}\right) \|y_0 - x_0\| \prod_{j=0}^{n-1} f(a_j)g(a_j, b_j)$$

as a consequence of (ii_n). Now, from Lemma 5.3, it follows

$$\prod_{j=0}^{n-1} f(a_j)g(a_j, b_j) < \prod_{j=0}^{n-1} (\gamma^{(2+p)^j}\Delta) = \gamma^{\frac{(2+p)^n-1}{1+p}} \Delta^n,$$

where $\gamma = \frac{a_1}{a_0} < 1$ and $\Delta = \frac{1}{f(a_0)} < 1$. So,

$$\|x_{n+m} - x_n\| \leq \|x_{n+m} - x_{n+m-1}\| + \|x_{n+m-1} - x_{n+m-2}\| + \cdots + \|x_{n+1} - x_n\|$$

$$\leq \left(1 + \frac{a_{n+m-1}}{2}\right) \eta \prod_{j=0}^{n+m-2} f(a_j)g(a_j, b_j) + \cdots$$

$$+ \left(1 + \frac{a_n}{2}\right) \eta \prod_{j=0}^{n-1} f(a_j)g(a_j, b_j)$$

$$< \left(1 + \frac{a_n}{2}\right) \left(\gamma^{\frac{(2+p)^{n+m-1}-1}{1+p}} \Delta^{n+m-1} + \cdots + \gamma^{\frac{(2+p)^n-1}{1+p}} \Delta^n\right) \eta$$

$$< \left(1 + \frac{a_0}{2}\gamma^{\frac{(p+2)^n-1}{1+p}}\right) \gamma^{\frac{(2+p)^n-1}{1+p}} \Delta^n \left(\gamma^{\frac{(2+p)^n[(2+p)^{m-1}-1]}{1+p}} \Delta^{m-1}\right.$$

$$\left. + \gamma^{\frac{(2+p)^n[(2+p)^{m-2}-1]}{1+p}} \Delta^{m-2} + \cdots + \gamma^{\frac{(2+p)^n[(2+p)-1]}{1+p}} \Delta + 1\right) \eta.$$

Then, by the Bernoulli inequality, $(1+z)^k - 1 > kz$ if $z > -1$, we have

$$\|x_{n+m} - x_n\| < \left(1 + \frac{a_0}{2}\gamma^{\frac{(2+p)^n-1}{1+p}}\right) \gamma^{\frac{(2+p)^n-1}{1+p}} \Delta^n \frac{1 - \gamma^{(2+p)^n m}\Delta^m}{1 - \gamma^{(2+p)^n}\Delta} \eta \tag{5.11}$$

5.2 Optimization 1

and, for $n = 0$,

$$\|x_m - x_0\| < \left(1 + \frac{a_0}{2}\right) \frac{(1 - \gamma^m \Delta^m)}{1 - \gamma \Delta} \eta < R\eta.$$

By letting $m \to \infty$ in (5.11), we obtain (5.10). Similarly, we have $y_n \in B(x_0, R\eta)$, for all $n \geq 0$, and therefore $z_n \in B(x_0, R\eta)$.

To see that x^* is a solution of $F(x) = 0$, we have $\|\Gamma_n F(x_n)\| \to 0$ as $n \to \infty$. Taking into account that $\|F(x_n)\| \leq \|F'(x_n)\| \|\Gamma_n F(x_n)\|$ and the sequence $\{\|F'(x_n)\|\}$ is bounded, we infer that $\|F(x_n)\| \to 0$ by letting $n \to \infty$. As a consequence, we obtain $F(x^*) = 0$ by the continuity of F.

To prove the uniqueness of the solution, we suppose some other solution $y^* \neq x^*$ of $F(x) = 0$ in $B\left(x_0, \frac{2}{M\beta} - R\eta\right) \cap \Omega$. From the approximation

$$0 = \Gamma_0(F(y^*) - F(x^*)) = \int_0^1 \Gamma_0 F'(x^* + t(y^* - x^*))\,dt\,(y^* - x^*),$$

we have to prove that the operator $\int_0^1 \Gamma_0 F'(x^* + t(y^* - x^*))\,dt$ is invertible and then $y^* = x^*$. Indeed, from

$$\|\Gamma_0\| \int_0^1 \|F'(x^* + t(y^* - x^*)) - F'(x_0)\|\,dt$$

$$\leq M\beta \int_0^1 \|x^* + t(y^* - x^*) - x_0\|\,dt$$

$$\leq M\beta \int_0^1 \left((1-t)\|x^* - x_0\| + t\|y^* - x_0\|\right)dt$$

$$< 1,$$

it follows that the operator $\left(\int_0^1 \Gamma_0 F'(x^* + t(y^* - x^*))\,dt\right)^{-1}$ exists.

Finally, we deduce that the R-order of convergence of the sequence (5.4) is at least $2 + p$. Finally, from (5.10), it follows that

$$\|x^* - x_n\| \leq \left(1 + \frac{a_0}{2}\right) \frac{\eta}{\gamma^{\frac{1}{1+p}}(1 - \gamma \Delta)} \left(\gamma^{\frac{1}{1+p}}\right)^{(2+p)^n}.$$

The proof is complete. ∎

Remark 5.5 Note that when the operator F has a Lipschitz-continuous second Fréchet derivative, we obtain R-order of convergence at least three, as occurs with the Chebyshev method.

5.2.3.3 Applications

We study two particular nonlinear Fredholm integral equations of the form

$$x(s) = \zeta(s) + \lambda \int_a^b \mathcal{K}(s,t)\mathcal{N}(x)(t)\,dt, \quad s \in [a,b], \qquad (5.12)$$

where $\zeta \in \mathcal{C}([a,b])$, the kernel $\mathcal{K}(s,t)$ is a known function in $[a,b] \times [a,b]$, $\mathcal{N}$ is a twice continuously Fréchet differentiable operator with Hölder continuous second derivative, and $x(s)$ is the unknown function to be determined. This type of integral equations has been treated by other authors [32, 79], and, for instance, if the kernel $\mathcal{K}(s,t)$ is the Green function in $[a,b] \times [a,b]$, Eq. (5.12) is equivalent to the following boundary value problem:

$$\begin{cases} x'' + \lambda \mathcal{N}(x) = 0, \\ x(a) = x_a, \quad x(b) = x_b. \end{cases}$$

Special situations of these kinds of problems are studied in [68, 84].

Solving Eq. (5.12) is equivalent to solving the equation $\mathcal{F}(x) = 0$ with $\mathcal{F} : \Omega \subseteq \mathcal{C}([a,b]) \to \mathcal{C}([a,b])$ and

$$[\mathcal{F}(x)](s) = x(s) - \zeta(s) - \lambda \int_a^b \mathcal{K}(s,t)\mathcal{N}(x)(t)\,dt, \quad s \in [a,b]. \qquad (5.13)$$

We consider two situations. First, a solution is located under the condition that the operator $\mathcal{N}$ satisfies a Lipschitz condition. Then, the solution is approximated by the methods proposed, and some a priori error bounds are obtained which improve those given by other authors for the Chebyshev method. Second, from the new family of iterative methods, we obtain a result of existence and uniqueness of solution when the operator $\mathcal{N}$ is Hölder continuous. And we obtain an approximation of the solution by a process of discretization.

First, we consider a particular case of the integral equation (5.12) that has been discussed by other authors [18, 19, 29] as a test function. A solution is located and approximated. We note that the direct approximation is not usual when iterative methods with R-order of convergence at least three are applied, where a process of discretization is generally used, as we see in the second application. In addition, we show that the a priori error bounds obtained for (5.4) have a good behavior, since those obtained by other authors for the Chebyshev method are improved.

Example 5.6 We consider the particular case of (5.13) given by

$$[\mathcal{F}(x)](s) = x(s) - s + \frac{1}{2}\int_0^1 s\cos(x(t))\,dt, \quad s \in [0,1], \qquad (5.14)$$

5.2 Optimization 1

where $\mathcal{F}: \Omega \subseteq \mathcal{C}([a,b]) \to \mathcal{C}([a,b])$. In addition,

$$[\mathcal{F}'(x)y](s) = y(s) - \frac{s}{2}\int_0^1 y(t)\sin(x(t))\,dt,$$

$$[\mathcal{F}''(x)yz](s) = -\frac{s}{2}\int_0^1 y(t)z(t)\cos(x(t))\,dt.$$

If we choose $x_0(s) = s$, then $[\mathcal{F}(x_0)](s) = \frac{\sin 1}{2}s$. To calculate β and η, we first obtain $[\mathcal{F}'(x)]^{-1}$. For this, we do $y(s) = [\mathcal{F}'(x)]^{-1}z(s)$, so that

$$y(s) = z(s) + \frac{s}{2}\int_0^1 y(t)\sin(x(t))\,dt.$$

We now solve the integral appearing in the last formula. Hence,

$$\int_0^1 z(s)\sin(x(s))\,ds = \int_0^1 y(s)\sin(x(s))\,ds - \int_0^1 \frac{s}{2}\sin(x(s))\left(\int_0^1 y(t)\sin(x(t))\,dt\right)ds,$$

and then

$$\int_0^1 y(s)\sin(x(s))\,ds = \frac{\int_0^1 z(s)\sin(x(s))\,ds}{1 - \int_0^1 \frac{s}{2}\sin(x(s))\,ds}.$$

As a consequence,

$$y(s) = [\mathcal{F}'(x)]^{-1}z(s) = z(s) + \frac{s}{2}\frac{\int_0^1 z(s)\sin(x(s))\,ds}{1 - \int_0^1 \frac{s}{2}\sin(x(s))\,ds}.$$

Moreover,

$$\Gamma_0 z(s) = [\mathcal{F}'(x_0)]^{-1}z(s) = z(s) + s\frac{\int_0^1 z(s)\sin s\,ds}{2 - \sin 1 + \cos 1}$$

(see [18, 29] for details).

Therefore, the parameters appearing in Theorem 5.4 are

$$\|\Gamma_0\| \leq \frac{3 - \sin 1}{2 - \sin 1 + \cos 1} = 1.2705\ldots = \beta,$$

$$\|\Gamma_0 \mathcal{F}(x_0)\| = \frac{\sin 1}{2 - \sin 1 + \cos 1} = 0.4953\ldots = \eta,$$

$$\|\mathcal{F}''(x)\| \leq \frac{1}{2} = M.$$

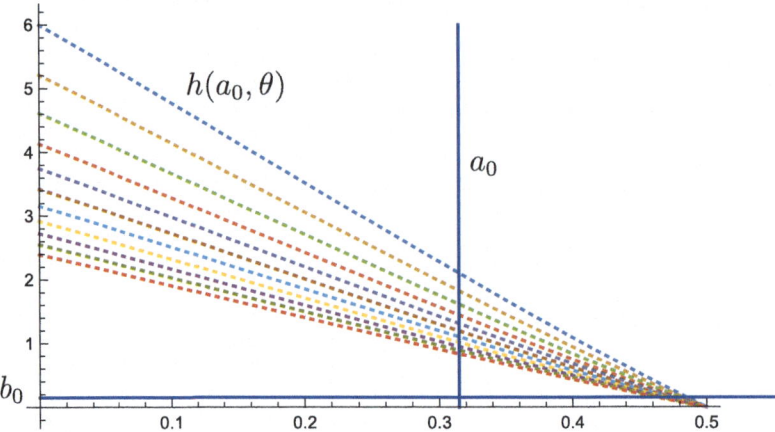

Fig. 5.8 θ-domain from initial conditions for $x_0(s) = s$ and (5.14)

In addition,

$$[(\mathcal{F}''(y) - \mathcal{F}''(x))(zw)](s) = -\frac{s}{2}\int_0^1 \cos(x(t))z(t)w(t)\,dt + \frac{s}{2}\int_0^1 \cos(y(t))z(t)w(t)\,dt$$

$$= \frac{s}{2}\int_0^1 (\cos(y(t)) - \cos(x(t)))z(t)w(t)\,dt,$$

and $\|\mathcal{F}''(y) - \mathcal{F}''(x)\| \leq \frac{1}{2}\|y - x\|$, so that $K = \frac{1}{2}$ and $p = 1$. Moreover, $a_0 = 0.3146\ldots < \frac{1}{2}$ and $b_0 = 0.1558\ldots < h(a_0, \theta) = \frac{3}{2(2+3\theta)}(1 - 2a_0)(8 - 4a_0^2 - a_0^3)$, for all $\theta \in (0, 1]$, see Fig. 5.8.

The conditions of Theorem 5.4 are therefore satisfied, and we then confirm that a solution of (5.14) exists in $\overline{B(x_0, 0.7043\ldots)}$ and is unique in $B(x_0, 4.3780\ldots)$. Another interesting feature of the new iterative methods is that the algorithm is simple, since the operator F'' is not required. So, we can approximate directly the solution $x^*(s) = ks$, where $k = 0.5224\ldots$ of (5.14), see Fig. 5.9.

Furthermore, we give an upper bound C to the number $10^{11}\|x^* - x_2\|$, where x_2 is the second iterate of (5.4). Carrying out the same decomposition as Candela and Marquina in [19] and calculating the smallest value of n such that $\|x^* - x_n\|$ is of order 10^{-11}, we obtain $n = 4$, and we can then consider

$$\|x^* - x_2\| \leq \|x^* - x_4\| + \|x_4 - x_3\| + \|x_3 - x_2\|.$$

Note that the previous value depends on the parameter $\theta \in (0, 1]$. Taking into account that the convergence of (5.4) is guaranteed for all $\theta \in (0, 1]$, we analyze its variation from the value $C = 37022683.427694$, obtained by Candela and Marquina

5.2 Optimization 1

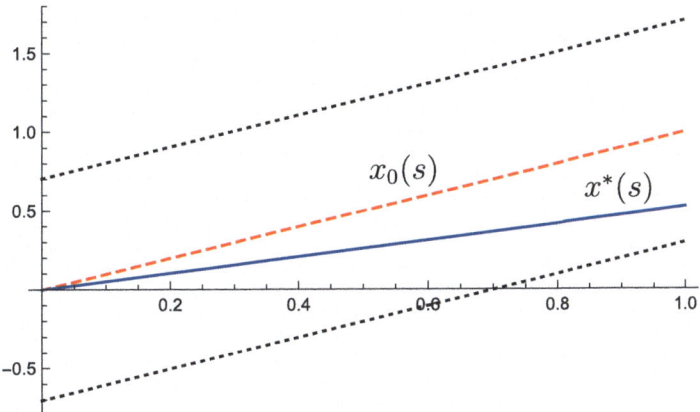

Fig. 5.9 The approximate solution of (5.14) (solid line) and the domain of existence of solution (dotted lines)

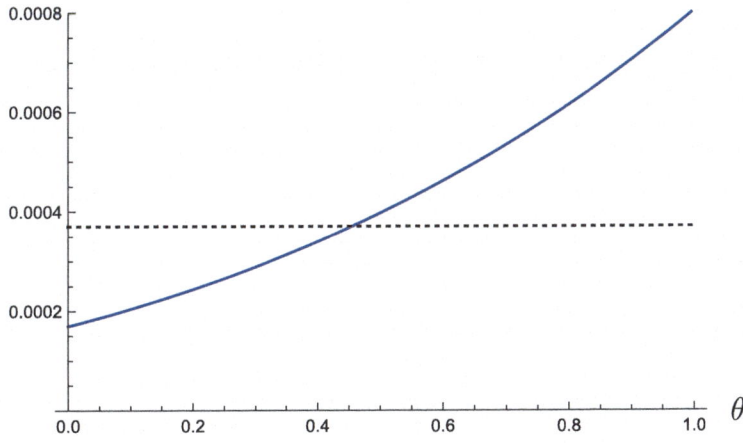

Fig. 5.10 Variation of the a priori error bounds (5.10) obtained from θ (solid line) and the a priori error bound obtained in [19] (dotted line)

in [19] for the Chebyshev method (see Fig. 5.10). Notice that this value is improved if $\theta \in (0, 0.4541\ldots)$. For example, if $\theta = 0.1$, we obtain $C = 20454900$.

In the second example, we apply Theorem 5.4 to study the existence and uniqueness of solution for a nonlinear integral equation of form (5.12). In this case, we consider an operator whose second Fréchet derivative is Hölder continuous. Moreover, we apply a process of discretization to approximate a solution.

Example 5.7 We consider the operator $\mathcal{F}: \Omega \subseteq \mathcal{C}([a, b]) \to \mathcal{C}([a, b])$ such that

$$[\mathcal{F}(x)](s) = x(s) - \zeta(s) - \lambda \int_0^1 \frac{s}{s+t} x(t)^{2+\frac{1}{n}} dt, \quad s \in [0,1], \quad \lambda \geq 0, \quad n \in \mathbb{N}.$$
(5.15)

Notice that the operator (5.15) is a generalization of the operator $\mathcal{G} : \Omega \subseteq \mathcal{C}([a,b]) \to \mathcal{C}([a,b])$ given by

$$[\mathcal{G}(x)](s) = x(s) - \lambda \int_0^1 K(s,t) x(t)^m dt, \quad s \in [0,1], \quad \lambda \geq 0, \quad m \in \mathbb{N},$$

which is studied by Davis in [26].

From (5.15), it follows

$$[\mathcal{F}'(x)y](s) = y(s) - \lambda \left(2 + \frac{1}{n}\right) \int_0^1 \frac{s}{s+t} x(t)^{1+\frac{1}{n}} y(t) dt, \quad y \in \Omega,$$

$$[\mathcal{F}''(x)yz](s) = -\lambda \left(2 + \frac{1}{n}\right)\left(1 + \frac{1}{n}\right) \int_0^1 \frac{s}{s+t} x(t)^{\frac{1}{n}} z(t) y(t) dt, \quad y, z \in \Omega.$$

In this case, $\mathcal{F}''$ is not Lipschitz continuous, but it is Hölder continuous. Indeed,

$$\|(\mathcal{F}''(x) - \mathcal{F}''(y))zw\| \leq \lambda \left(2 + \frac{1}{n}\right)\left(1 + \frac{1}{n}\right) \max_{s \in [a,b]} \left| \int_0^1 \frac{s}{s+t}(x(t)^{\frac{1}{n}} - y(t)^{\frac{1}{n}}) w(t) z(t) \, dt \right|$$

$$\leq \lambda \left(2 + \frac{1}{n}\right)\left(1 + \frac{1}{n}\right) \log 2 \, \|x - y\|^{\frac{1}{n}} \|w\| \|z\|,$$

so that $K = |\lambda| \left(2 + \frac{1}{n}\right)\left(1 + \frac{1}{n}\right) \log 2$ and $p = \frac{1}{n}$. Moreover,

$$\|[\mathcal{F}''(x)]\| \leq \lambda \left(2 + \frac{1}{n}\right)\left(1 + \frac{1}{n}\right) \log 2 \, \|x\|^{\frac{1}{n}}.$$

On the other hand,

$$\|\mathcal{F}(x_0)\| \leq \|x_0 - \zeta\| + |\lambda| \log 2 \, \|x_0\|^{2+\frac{1}{n}},$$

$$\|I - \mathcal{F}'(x_0)\| \leq |\lambda| \left(2 + \frac{1}{n}\right) \log 2 \, \|x_0\|^{1+\frac{1}{n}}.$$

If $|\lambda| \left(2 + \frac{1}{n}\right) \log 2 \, \|x_0\|^{1+\frac{1}{n}} < 1$, from the Banach lemma on invertible operators, we obtain

5.2 Optimization 1

$$\|\Gamma_0\| \leq \frac{1}{1 - |\lambda|\left(2 + \frac{1}{n}\right)\log 2 \|x_0\|^{1+\frac{1}{n}}} = \beta,$$

$$\|\Gamma_0 \mathcal{F}(x_0)\| \leq \frac{\|x_0 - \zeta\| + |\lambda|\log 2 \|x_0\|^{2+\frac{1}{n}}}{1 - |\lambda|\left(2 + \frac{1}{n}\right)\log 2 \|x_0\|^{1+\frac{1}{n}}} = \eta.$$

If we consider $n = 5$, $\zeta(s) = 1$, and $\lambda = \frac{1}{4}$, then

$$[\mathcal{F}(x)](s) = x(s) - 1 - \frac{1}{4}\int_0^1 \frac{s}{s+t} x(t)^{\frac{11}{5}} dt. \tag{5.16}$$

In addition, we choose $x_0(s) = 1$, $\theta = 1$ and look for a domain such that $\Omega = B(x_0, r) \subseteq \mathcal{C}([a, b])$.

Then,

$$\|\Gamma_0\| \leq \frac{20}{20 - 11\log 2} = 1.6161\ldots = \beta, \quad \|\Gamma_0 \mathcal{F}(x_0)\| = \frac{5\log 2}{20 - 11\log 2} = 0.2800\ldots = \eta,$$

$b_0 = 0.1605\ldots$, $K = 0.4574\ldots$, and $p = \frac{1}{5}$. To calculate the radius r, we consider $M = M(r) = Kr^{\frac{1}{5}}$ and $a_0 = a_0(r) = M(r)\beta\eta$. In this situation, from Theorem 5.4, it is necessary that $\overline{B(x_0, R\eta)} \subseteq \Omega$. Then, if $R = R(r)$, we need to see that $r - (R(r)\eta + 1) > 0$. For this, it is necessary that $r \in (1.1987\ldots, 3.3247\ldots)$, see Fig. 5.11.

Observe that $a_0(r) < \frac{1}{2}$ if and only if $r < 13.1398\ldots$ Then, if, for example, we choose $r = 2$, we obtain $b_0 < h(a_0(2), 1)$, and the conditions of Theorem 5.4 are satisfied. Therefore, there exists a solution of (5.16) in $\overline{B(1, 0.9273\ldots)} \subseteq \Omega$, which is unique in $B(1, 5.2233\ldots) \cap \Omega$.

Finally, we follow a process of discretization, and Eq. (5.16) is transformed into a finite-dimensional problem. Thus, the integral of (5.16) is approximated by a Gauss-Legendre formula of the form

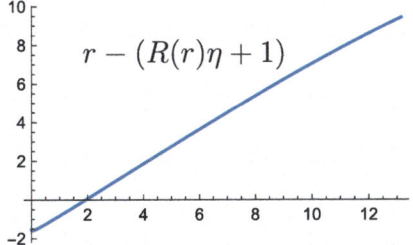
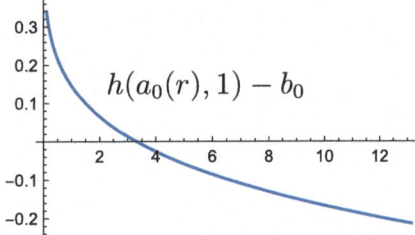

Fig. 5.11 Conditions for the parameter r

$$\int_0^1 f(t)\,dt \approx \frac{1}{2}\sum_{j=1}^{8}\beta_j f(t_j),$$

where the nodes t_j and the weights β_j are known.

If we denote the approximations of $x(t_i)$, $i = 1, 2, \ldots, 8$, by x_i, we obtain the following system of nonlinear equations:

$$\mathsf{x}_i = 1 + \frac{t_i}{8}\sum_{j=1}^{8}\beta_j \frac{\mathsf{x}_j^{\frac{11}{5}}}{t_i + t_j}, \quad i = 1, 2, \ldots, 8,$$

which can be written as

$$\mathsf{x}_i = 1 + \sum_{j=1}^{8} d_{ij}\, \mathsf{x}_j^{\frac{11}{5}}, \quad i = 1, 2, \ldots, 8, \tag{5.17}$$

where $d_{ij} = \dfrac{t_i \beta_j}{8(t_i + t_j)}$. The approximate solution of (5.17) obtained by (5.4) starting at $\mathsf{x}_0 = (1, 1, \ldots, 1)^T$ is shown in Table 5.4.

Finally, we interpolate the points (t_i, x_i), $i = 1, 2, \ldots, 8$, and taking into account that $x(0) = 1$, we obtain the approximation shown in the left graph of Fig. 5.12. The right graph in Fig. 5.12 shows that the approximation obtained lies within the domain of existence of solution.

Table 5.4 Numerical solution of the system (5.17)

i	x_i^*	i	x_i^*
1	1.02415468657...	5	1.20186439201...
2	1.07954327274...	6	1.22141388151...
3	1.13201750646...	7	1.23360029059...
4	1.17280585507...	8	1.23991266325...

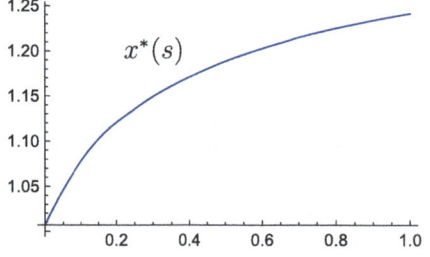
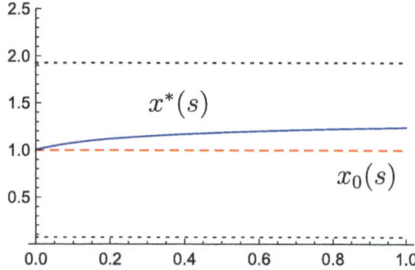

Fig. 5.12 The approximate solution of (5.16) (solid line) and the domain of existence (dotted lines)

5.2.4 Relaxing Convergence Conditions

In Sect. 5.2.1, we have seen that there exists a two-point method with R-order of convergence at least three which does not require evaluations of the second derivative, the iterative method (5.4). The second derivative of the operator F, which appears in the Chebyshev method, has been approximated by a difference of first derivatives of F in different points. It is proved that (5.4) is convergent and has R-order of convergence at least three under usual Kantorovich-type conditions for iterative methods with R-order of convergence at least three [9, 21, 54]: The second derivative of the operator F exists and is bounded and Hölder continuous. But, this result is partial, since the second derivative of F appears in the condition, although it is not used in the algorithm of the method (5.4).

In this section, we complete the study done in Sect. 5.2.3, since the convergence conditions are relaxed, so that we do not need the existence of F'' to prove the semilocal convergence of the method (5.4). For this, we require that the operator F has a ω-conditioned first derivative of the form [36]

$$\|F'(x) - F'(y)\| \leq \omega(\|x - y\|), \quad x, y \in \Omega, \tag{5.18}$$

where $\omega(z)$ is a continuous real function and is nondecreasing for $z > 0$. Observe that the condition (5.18) generalizes the cases in which F' is Lipschitz continuous ($\omega(z) = Kz$, $K \in \mathbb{R}_+$) and Hölder continuous ($\omega(z) = Kz^p$, $K \in \mathbb{R}_+$, $p \in [0, 1]$).

Notice that the fact of relaxing the convergence conditions is important, since a condition on F'' is not required, which is an advantage, as we can see in the following simple example.

Example 5.8 Let $F(x) = 0$ be the equation where $F : [0, A] \to \mathbb{R}$, with $A > 0$, and $F(x) = ax^{1+q} + bx$, with $a, b \in \mathbb{R}$ and $q \in [0, 1)$. For this F, the result given in Sect. 5.2.3 cannot be applied, since it is required that F'' exists and is bounded and Hölder continuous. Observe that

$$\|F''(x)\|_\infty = \max_{[0,A]} \left| a(1 + q) q x^{q-1} \right| = \infty,$$

so that F'' is not bounded, but a condition of type (5.18) is satisfied. Then, as we see in Sect. 5.2.3, we cannot guarantee the convergence of (5.4) to a solution of the last equation.

5.2.4.1 Recurrence Relations
Under certain conditions on the pair (F, x_0), we now study the convergence of (5.4) to a unique solution of $F(x) = 0$. To establish the semilocal convergence of (5.4) under the condition (5.18), we use the technique based on recurrence relations.

We suppose that the operator F is once continuously Fréchet differentiable and the following conditions:

(U1) There exists $\Gamma_0 = [F'(x_0)]^{-1} \in \mathcal{L}(Y, X)$, for some $x_0 \in \Omega$, with $\|\Gamma_0\| \leq \beta$ and $\|\Gamma_0 F(x_0)\| \leq \eta$.
(U2) $\|F'(x) - F'(y)\| \leq \omega(\|x - y\|)$, $x, y \in \Omega$, where $\omega : \mathbb{R}_+ \to \mathbb{R}_+$ is a continuous and nondecreasing function.
(U3) $\omega(tz) \leq t^p \omega(z)$, with $p \in (0, 1]$, $t \in [0, 1]$ and $z \in [0, +\infty)$.

Note that the condition (U3) does not involve any restriction since it is enough to choose $p = 0$, since ω is a nondecreasing function.

Denote $a_0 = \theta^{p-1} \beta \omega(\eta)$, $b_0 = \theta^{1-p} a_0 \left(1 + \frac{a_0}{2}\right)^p$, and define the scalar sequences:

$$a_n = a_{n-1} f(b_{n-1})^{1+p} g(a_{n-1})^p, \quad n \in \mathbb{N}, \tag{5.19}$$

$$b_n = \theta^{1-p} a_n \left(1 + \frac{a_n}{2}\right)^p, \quad n \in \mathbb{N}, \tag{5.20}$$

where

$$f(t) = \frac{1}{1-t}, \quad g(t) = t \left(\frac{1}{2} + \frac{\theta^{1-p}}{1+p} \left(1 + \frac{t}{2}\right)^{1+p}\right). \tag{5.21}$$

Note that the sequences (5.4), (5.19), and (5.20) satisfy the following recurrence relations:

$$\|\Gamma_1\| = \left\|[F'(x_1)]^{-1}\right\| \leq f(b_0)\|\Gamma_0\|, \tag{5.22}$$

$$\|y_1 - x_1\| \leq f(b_0)g(a_0)\|y_0 - x_0\|, \tag{5.23}$$

$$\|P(x_1, z_1)\| \leq a_1, \tag{5.24}$$

$$\|x_2 - x_1\| \leq \left(1 + \frac{a_1}{2}\right)\|y_0 - x_0\|. \tag{5.25}$$

To prove the previous relations, we suppose first that

$$x_1 \in \Omega \quad \text{and} \quad b_0 < 1.$$

By the hypotheses, the operator Γ_0 exists, and by the Banach lemma on invertible operators, the operator Γ_1 is well defined

$$\|\Gamma_1\| \leq \frac{\|\Gamma_0\|}{1 - \|I - \Gamma_0 F'(x_1)\|} \leq f(b_0)\|\Gamma_0\|,$$

5.2 Optimization 1

since

$$\|I - \Gamma_0 F'(x_1)\| \leq \|\Gamma_0\| \|F'(x_0) - F'(x_1)\| \leq \beta \omega(\|x_1 - x_0\|)$$
$$\leq \beta \left(1 + \frac{a_0}{2}\right)^p \omega(\eta) = b_0 < 1.$$

From Taylor's formula and (5.4), it follows

$$F(x_1) = \frac{1}{2\theta}(F'(x_0) - F'(z_0))(y_0 - x_0) + \int_0^1 (F'(x_0 + t(x_1 - x_0)) - F'(x_0))(x_1 - x_0)\,dt.$$

Thus,

$$\|F(x_1)\| = \left(\frac{\theta^{p-1}}{2} + \frac{\left(1 + \frac{a_0}{2}\right)^{1+p}}{1+p}\right) \omega(\eta)\|y_0 - x_0\|.$$

Hence,

$$\|y_1 - x_1\| = \|\Gamma_1 F(x_1)\| \leq \|\Gamma_1\| \|F(x_1)\| \leq f(b_0)g(a_0)\|y_0 - x_0\|,$$

$$\|P(x_1, z_1)\| \leq \theta^{p-1}\|\Gamma_1\|\omega(\|y_1 - x_1\|) \leq a_1,$$

$$\|x_2 - x_1\| \leq \|x_2 - y_1\| + \|y_1 - x_1\| \leq \left(1 + \frac{a_1}{2}\right)\|y_1 - x_1\|.$$

Next, we see that (5.22), (5.23), (5.24), and (5.25) are satisfied for every point of (5.4), and then (5.4) is a Cauchy sequence. For this, the scalar sequences (5.19) and (5.20) are first studied.

To guarantee the convergence of (5.4), we provide some properties of the sequences $\{a_n\}$ and $\{b_n\}$ given in (5.19) and (5.20), respectively. First, it is enough that $x_{n+1} \in \Omega$ and $b_n < 1$, for all $n \in \mathbb{N}$.

Lemma 5.9 *Let f and g be the two scalar functions defined in (5.21), and suppose $b_0 < 1$.*

(a) *If $f(b_0)^{1+p}g(a_0)^p < 1$, then $\{a_n\}$ and $\{b_n\}$ are both strictly decreasing sequences.*
(b) *If $f(b_0)^{1+p}g(a_0)^p = 1$, then $a_n = a_0$ and $b_n = b_0 < 1$, for all $n \in \mathbb{N}$.*

Proof Item (a) is proved by mathematical induction on n. As $f(b_0)^{1+p}g(a_0)^p < 1$, then $a_1 < a_0$ and $b_1 = h(a_0) < b_0$, since $h(x) = \theta^{1-p}x\left(1 + \frac{x}{2}\right)^p$ is increasing if $x > 0$. If we suppose that $a_i < a_{i-1}$ and $b_i < b_{i-1}$, for $i = 1, 2, \ldots, n$, then

$$a_{n+1} = a_n f(b_n)^{1+p}g(a_n)^p < a_{n-1}f(b_{n-1})^{1+p}g(a_{n-1})^p = a_n,$$

$$b_{n+1} = h(a_{n+1}) < h(a_n) = b_n,$$

as a consequence of f and h are increasing in $[0, 1)$ and g is increasing in $[0, \infty)$. Item (b) is then proved immediately. ∎

5.2.4.2 Semilocal Convergence

Now, we are ready to prove the semilocal convergence of the method (5.4).

Theorem 5.10 *Let $F : \Omega \subseteq X \to Y$ be a once continuously Fréchet differentiable operator on a non-empty open convex domain Ω of a Banach space X with values in a Banach space Y. Suppose that (U1)–(U2)–(U3) are satisfied. If $b_0 < 1$, $f(b_0)^{1+p} g(a_0)^p < 1$, where f and g are defined in (5.21), and $\overline{B(x_0, R\eta)} \subseteq \Omega$, where $R = \frac{1+a_0/2}{1-\Delta}$, $\Delta = f(b_0)^{-1/p}$, then the method (5.4) starting at x_0 converges to a solution x^* of $F(x) = 0$, and the solution x^* and the iterates x_n, y_n, z_n belong to $\overline{B(x_0, R\eta)}$. Moreover, if there exists a positive root r of the equation*

$$\beta \omega(R\eta + z) = 2^p, \quad (5.26)$$

then the solution x^* is unique in $B(x_0, r) \cap \Omega$.

Proof First, for $n \in \mathbb{N}$, we prove the following recurrence relations:

(i_n) $\Gamma_n = [F'(x_n)]^{-1}$ exists and $\|\Gamma_n\| \leq f(b_{n-1}) \|\Gamma_{n-1}\|$.
(ii_n) $\|y_n - x_n\| \leq f(b_{n-1}) g(a_{n-1}) \|y_{n-1} - x_{n-1}\|$.
(iii_n) $\|P(x_n, z_n)\| \leq \theta^{p-1} \|\Gamma_n\| \omega(\|y_n - x_n\|) \leq a_n$.
(iv_n) $\|x_{n+1} - x_n\| \leq \left(1 + \frac{a_n}{2}\right) \|y_n - x_n\|$.

Now, we suppose that $x_n, y_n, z_n \in B(x_0, R\eta)$, for all $n \in \mathbb{N}$, which is proved later.

Note that $x_1 \in \Omega$, since $1 + \frac{a_0}{2} < R$. From (5.22), (5.23), (5.24), and (5.25), it follows (i_1)–(ii_1)–(iii_1)–(iv_1). After that, we suppose that (i_n)–(ii_n)–(iii_n)–(iv_n) hold, for $n = 1, 2, \ldots, j$, and see that they are satisfied for $n = j + 1$.

(i_{j+1}) Observe that

$$\begin{aligned}
\|I - \Gamma_j F'(x_{j+1})\| &\leq \|\Gamma_j\| \omega(\|x_{j+1} - x_j\|) \\
&\leq f(b_{j-1}) \|\Gamma_{j-1}\| \omega(\|x_{j+1} - x_j\|) \\
&= f(b_{j-1}) \|\Gamma_{j-1}\| \left(1 + \frac{a_j}{2}\right)^p \omega(\|y_j - x_j\|) \\
&\leq f(b_{j-1})^{1+p} g(a_{j-1})^p \left(1 + \frac{a_j}{2}\right)^p \|\Gamma_{j-1}\| \omega(\|y_{j-1} - x_{j-1}\|) \\
&\leq b_j \\
&< 1,
\end{aligned}$$

since $\{b_n\}$ is a decreasing sequence and $b_0 < 1$. Hence, by the Banach lemma on invertible operators, the operator Γ_{j+1} exists and is such that

5.2 Optimization 1

$$\|\Gamma_{i+1}\| \leq \frac{\|\Gamma_i\|}{1-b_i} = f(b_i)\|\Gamma_i\|.$$

(ii_{j+1}) Taking into account Taylor's formula, (5.4), and (5.23), it follows

$$\|F(x_{j+1})\| = \left\| \frac{1}{2\theta}(F'(x_j) - F'(z_j))(y_j - x_j) + \int_0^1 (F'(x_j + t(x_{j+1} - x_j)) - F'(x_j))(x_{j+1} - x_j)\, dt \right\|$$

$$\leq \left(\frac{\theta^{p-1}}{2} + \frac{\left(1 + \frac{a_j}{2}\right)^{1+p}}{1+p} \right) \omega(\|y_j - x_j\|)\|y_j - x_j\|.$$

Therefore,

$$\|y_{j+1} - x_{j+1}\| \leq f(b_j)g(a_j)\|y_j - x_j\|.$$

(iii_{j+1}) Moreover,

$$\|H(x_{j+1}, z_{j+1})\| \leq \frac{1}{\theta} f(b_j) \|\Gamma_j\| \omega(\|z_{j+1} - x_{j+1}\|) \leq a_{j+1}$$

is easy to prove.

(iv_{j+1}) Furthermore,

$$\|x_{j+2} - x_{i+1}\| \leq \|x_{j+2} - y_{j+1}\| + \|y_{j+1} - x_{j+1}\| \leq \left(1 + \frac{a_{j+1}}{2}\right) \|y_{j+1} - x_{j+1}\|.$$

The induction is then complete.

Second, we prove that (5.4) is a Cauchy sequence. If $m, n \in \mathbb{N}$, $\gamma = \frac{a_1}{a_0} < 1$, and $\Delta = f(b_0)^{-1/p} < 1$, then

$$\|x_{n+m} - x_n\| \leq \sum_{j=1}^{m} \|x_{n+j} - x_{n+j-1}\|$$

$$< \sum_{i=n}^{n+m-1} \left(1 + \frac{a_i}{2}\right) \left(\prod_{j=0}^{i-1} f(b_j)g(a_j) \right) \|y_0 - x_0\|$$

$$< \sum_{i=n}^{n+m-1} \left(1 + \frac{a_0}{2}\right) \left(\prod_{j=0}^{i-1} f(b_0)g(a_0) \right) \|y_0 - x_0\|$$

$$< \left(1 + \frac{a_0}{2}\right)^{n+m-1} \sum_{i=n} \Delta^i \|y_0 - x_0\|$$

$$= \left(1 + \frac{a_0}{2}\right) \Delta^n \frac{1 - \Delta^m}{1 - \Delta} \|y_0 - x_0\|,$$

and $\{x_n\}$ is a Cauchy sequence. For $n = 0$, we obtain $x_m \in B(x_0, R\eta)$. In addition, we prove similarly that $y_n, z_n \in B(x_0, R\eta)$, for all $n \in \mathbb{N}$.

To see that x^* is a solution of $F(x) = 0$, we have that $\|\Gamma_n F(x_n)\| \to 0$, e.g., by noting that

$$\|\Gamma_n F(x_n)\| = \|y_n - x_n\| \leq \left(\prod_{j=0}^{n-1} f(b_j) g(a_j)\right) \|y_0 - x_0\| \leq \Delta^n \|y_0 - x_0\|.$$

Taking into account

$$\|F(x_n)\| \leq \|F'(x_n)\| \|\Gamma_n F(x_n)\| = \|F'(x_n)\| \|y_n - x_n\|,$$

$\{\|F'(x_n)\|\}$ is bounded, (U3), and

$$\|F'(x_n)\| \leq \|F'(x_0)\| + \omega(\|x_n - x_0\|) < \|F'(x_0)\| + \omega(R\eta),$$

it follows that $\|F(x_n)\| \to 0$ by letting $n \to \infty$. As a consequence, by the continuity of F in $\overline{B(x_0, R\eta)}$, we obtain $F(x^*) = 0$.

Finally, we prove the uniqueness of the solution. Then, we suppose that x^* and y^* are two different solutions of $F(x) = 0$ in $B(x_0, r) \cap \Omega$. From

$$0 = \Gamma_0(F(y^*) - F(x^*))$$

$$= \Gamma_0 \left(\int_0^1 F'(x^* + t(y^* - x^*)) \, dt\right)(y^* - x^*) = J(y^* - x^*),$$

it follows that the operator $J = \Gamma_0 \int_0^1 F'(x^* + t(y^* - x^*)) \, dt$ is invertible, and then $y^* = x^*$. According to this, we only have to prove that $\|I - J\| < 1$ and apply the Banach lemma on invertible operators. Indeed, if $\sigma = \frac{R\eta}{R\eta + r}$, then

$$\|I - J\| \leq \|\Gamma_0\| \int_0^1 \|F'(x^* + t(y^* - x^*)) - F'(x_0)\| \, dt$$

$$\leq \beta \int_0^1 \omega(\|x_0 - x^* - t(y^* - x^*)\|) \, dt$$

5.2 Optimization 1

$$\leq \beta \int_0^1 \omega((1-t)\|x^* - x_0\| + t\|y^* - x_0\|)\,dt$$

$$< \beta \int_0^1 \omega((1-t)R\eta + tr)\,dt$$

$$\leq \beta \int_0^1 \omega((R\eta + r)(\sigma(1-t) + (1-\sigma)t))\,dt$$

$$\leq \beta \int_0^1 (\sigma(1-t) + (1-\sigma)t)^p \omega(R\eta + r)\,dt$$

$$= \beta \frac{(1-\sigma)^{1+p} - \sigma^{1+p}}{(1+p)(1-2\sigma)} \omega(R\eta + r)$$

$$\leq \frac{\beta}{2^p} \omega(R\eta + r)$$

$$= 1.$$

The proof is now complete. ∎

Remark 5.11 If $b_0 < 1$ and $f(b_0)^{1+p} g(a_0)^p = 1$, it is clear that the sequence (5.4) is also convergent.

Remark 5.12 Observe that the value of r in Theorem 5.10, which satisfies (5.26), exists if $\omega(R\eta) < \frac{2^p}{\beta}$, since ω is a nondecreasing function. Moreover, the value of r is unique. On the other hand, the uniqueness of the solution is guaranteed in $B(x_0, R\eta)$ if $\omega(R\eta) = \frac{1}{\beta}$.

5.2.4.3 Application

The second derivative of the operator involved in the next nonlinear integral equation is not bounded, and Theorem 5.4 of Sect. 5.2.3 is then not applicable, whereas Theorem 5.10 is. For this nonlinear integral equation, the theoretical significance of Theorem 5.10 is used to obtain a domain of existence of solution of the equation, so that the solution is located in a region.

Example 5.13 Consider

$$x(s) = 1 + \int_0^1 G(s,t)\left(x(t)^{\frac{3}{2}} + \frac{1}{2}x(t)^2\right)dt, \quad s \in [0,1], \tag{5.27}$$

where $G(s,t)$ is the Green function in $[0,1] \times [0,1]$.

We consider $\mathcal{F}(x) = 0$ such that

$$\mathcal{F}: \Omega \subseteq \mathcal{C}([a,b]) \to \mathcal{C}([a,b]), \qquad \Omega = \{x \in \mathcal{C}([a,b]) : x(s) > 0,\ s \in [a,b]\},$$

$$[\mathcal{F}(x)](s) = x(s) - 1 - \int_0^1 G(s,t)\left(x(t)^{\frac{3}{2}} + \frac{1}{2}x(t)^2\right) dt, \quad s \in [0,1].$$

Then,

$$[\mathcal{F}'(x)y](s) = y(s) - \int_0^1 G(s,t)\left(\frac{3}{2}x(t)^{\frac{1}{2}} + x(t)\right) y(t)\, dt.$$

If $x_0(s) = 1$, it follows that $\|I - \mathcal{F}'(x_0)\| \leq \frac{5}{16}$, and, by the Banach lemma on invertible operators, $[\mathcal{F}'(x_0)]^{-1}$ exists and $\|[\mathcal{F}'(x_0)]^{-1}\| \leq \frac{16}{11} = \beta$. Besides, $\|y_0 - x_0\| \leq \frac{3}{11} = \eta$ and $\omega(z) = \frac{3\sqrt{z}+2z}{16}$. As $\omega(tz) \leq \sqrt{t}\omega(z)$, then $p = \frac{1}{2}$. The condition $b_0 < 1$ is satisfied if $\theta^{1/2} > 0.0036\ldots$ Thus, we choose $\theta = 1$, so that

$$a_0 = 0.1920\ldots, \qquad b_0 = 0.2010\ldots < 1, \qquad f(b_0)^{1+p} g(a_0)^p = 0.6900\ldots < 1.$$

Therefore, the conditions of Theorem 5.10 are satisfied, and Eq. (5.27) has a solution in $\{\varphi \in \mathcal{C}([0,1]) : \|\varphi - 1\| \leq 0.8265\ldots\}$, which is unique in $\{\varphi \in \mathcal{C}([0,1]) : \|\varphi - 1\| \leq 3.7445\ldots\}$.

5.3 Optimization 2

In this section, we are interested in constructing, from the Chebyshev method, some multipoint iterations with cubical convergence and a better efficiency index than the Newton method in the scalar case, taking into account that their extension to nonlinear systems or Banach spaces does not have the negative effect of one-point iterative methods of R-order of convergence at least three: the increase of computational cost. Thus, the methods that we construct in this section have a better efficiency index than the Newton and the Chebyshev methods and a better computational efficiency than the Chebyshev method for systems of equations and a better computational efficiency than the Newton method for systems with more than six equations.

Then, from the Chebyshev method and a modification of the technique presented in Sect. 5.2.1, we construct in Sect. 5.3.1 multipoint iterative methods with R-order of convergence at least three and efficiency close to that of the Newton method [38]. In Sect. 5.3.3, we establish in Banach spaces the convergence of the iterations constructed previously, so that a further generalization is then given. We give a local convergence result, where the cubical convergence of the iterations is proved, and a semilocal convergence result under Kantorovich-type conditions.

5.3 Optimization 2

5.3.1 Construction of the Method

By using a slight modification of the technique used in Sect. 5.2.1 to obtain (5.5), we obtain a family of iterations with R-order of convergence at least three, which reduces the number of evaluations of functions and the computational cost, so that these values are close to those of the Newton method.

The idea is now to approximate the expression $F''(x_n)\delta_n^2$ in the Chebyshev algorithm (5.3) by only combinations of F in different points, so that F'' is not used and F' is only evaluated in x_n. For this, we consider

$$y_n = x_n - [F'(x_n)]^{-1} F(x_n), \quad z_n = x_n + p(y_n - x_n), \quad p \in (0, 1]$$

and Taylor's formula as follows:

$$F(z_n) = F(x_n) + p F'(x_n)(y_n - x_n) + \frac{p^2}{2} F''(x_n)(y_n - x_n)^2 + \frac{1}{2} \int_{x_n}^{z_n} F'''(x)(z_n - x)^2 \, dx,$$

so that

$$F(z_n) - F(x_n) - p F'(x_n)(y_n - x_n) = \frac{p^2}{2} F''(x_n)(y_n - x_n)^2 + \frac{1}{2} \int_{x_n}^{z_n} F'''(x)(z_n - x)^2 \, dx.$$

As a consequence, since $y_n = x_n - [F'(x_n)]^{-1} F(x_n)$, we can consider the following approximation:

$$F''(x_n)(y_n - x_n)^2 \approx \frac{2}{p^2}((p-1)F(x_n) + F(z_n)),$$

and the Chebyshev method is then transformed into the method

$$\begin{cases} x_0 \text{ given}, \\ F'(x_n)\delta_n = -F(x_n), \quad n \geq 0, \\ z_n = x_n + p\delta_n, \quad p \in (0, 1], \\ F'(x_n)\tilde{\gamma}_n = -\frac{1}{p^2}((p-1)F(x_n) + F(z_n)), \\ x_{n+1} = x_n + \delta_n + \tilde{\gamma}_n. \end{cases} \quad (5.28)$$

5.3.2 Analysis of the Method

With the last modification of the Chebyshev method, when solving systems of $\mathfrak{m}$ equations, we have reduced the computational cost from $\mathfrak{m}^3 + \mathfrak{m}^2 + \mathfrak{m}$ operations to do

Table 5.5 Number of evaluations of functions and computational cost per iteration when the method (5.28) is applied to solve systems (10, 50, and 100 equations)

	Method (5.28)	
m	$m^2 + 2m$	$(m^3 + 12m^2 + 2m)/3$
10	120	740
50	2600	51700
100	10200	373400

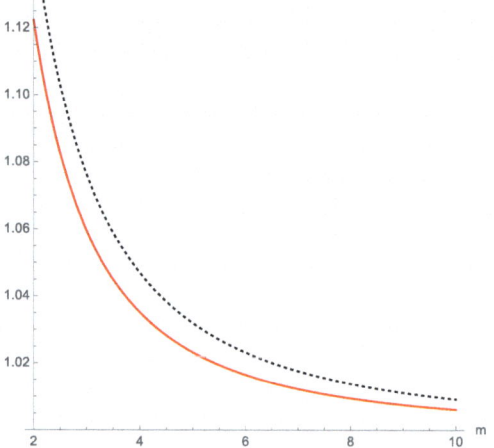

Fig. 5.13 Efficiency indices of the Newton method and the method (5.28) for nonlinear systems, respectively, $2^{1/(m^2+m)}$ (solid line) and $3^{1/(m^2+2m)}$ (dotted line)

$(-1/2) F''(x_n) \delta_n^2$ to $2m$ operations to do $(-1/p^2)((p-1)F(x_n) + F(z_n))$, which is a considerable reduction. Moreover, we observe that the efficiency is also improved, since the number of evaluations of functions per iteration is also reduced from $\frac{1}{2}(m^3 + 3m^2 + 2m)$ to $m^2 + 2m$, see Table 5.5.

Observe in Figs. 5.13 and 5.14 that the efficiency index of the Newton method is now improved by the method (5.28), even for high values of m, and the computational efficiency of the Newton method is also improved from seven equations. Taking into account the last, we can say that the method (5.28) is a better choice than the Newton method to solve the system (5.2).

On the other hand, if we now observe the problem of the region of accessibility of the method (5.28), we can see in Figs. 5.3 and 5.15 that (5.28) with $p = \frac{1}{2}(\sqrt{5} - 1)$ is still more demanding with respect to the starting points than the Newton method when we apply it to approximate the solutions $z^* = \arctan(1/2\sqrt{2}) = 0.33983\ldots$ and $z^{**} = \pi - \arctan(1/2\sqrt{2}) = 2.80176\ldots$ of the complex equation $F(z) = \sin z - \frac{1}{3} = 0$. This problem is studied and solved in Sect. 5.4.

5.3 Optimization 2

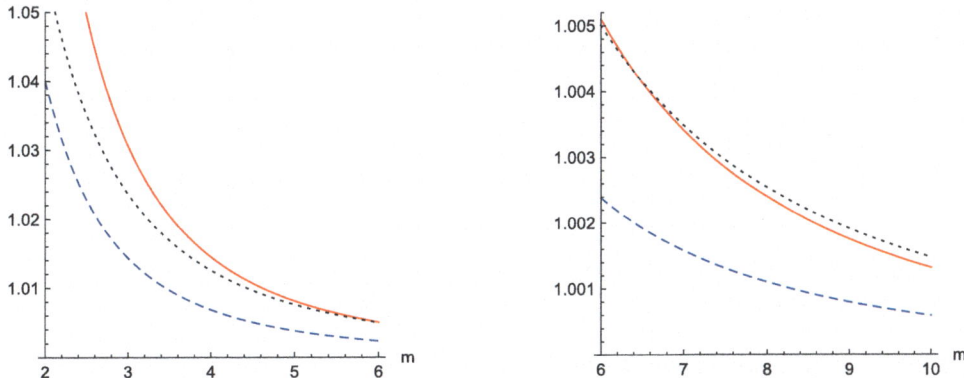

Fig. 5.14 Computational efficiencies of the Newton and the Chebyshev methods and the method (5.28) for systems, respectively, $2^{3/(m^3+6m^2-4m)}$ (solid line), $3^{3/(4m^3+15m^2-4m)}$ (dashed line), and $3^{3/(m^3+15m^2-m)}$ (dotted line)

Fig. 5.15 Method (5.28) with $p = \frac{1}{2}(\sqrt{5}-1)$

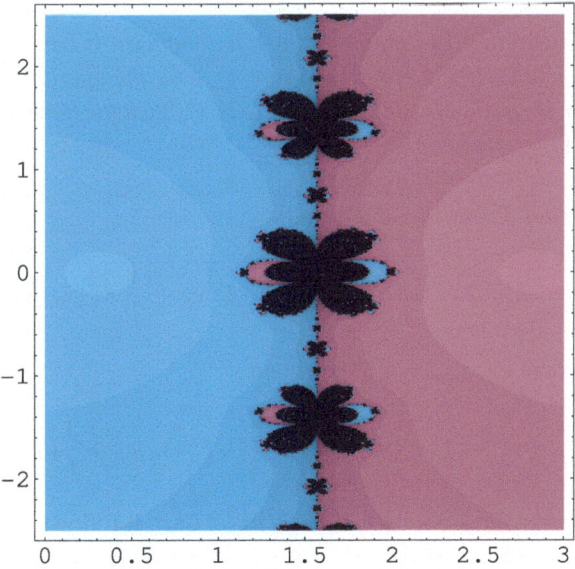

5.3.3 Convergence Analysis

We establish in this section the convergence of the method (5.28). For this, we consider $F(x) = 0$, where $F : \Omega \subseteq X \to Y$ is a nonlinear operator defined on a non-empty open convex subset Ω of a Banach space X with values in a Banach space Y.

We begin with a local convergence result, where we prove that the R-order of convergence is at least three. Next, we analyze the semilocal convergence of (5.28), which is now written as

$$\begin{cases} x_0 \text{ given in } \Omega, \\ y_n = x_n - [F'(x_n)]^{-1}(x_n), \quad n \geq 0, \\ z_n = x_n + p(y_n - x_n), \quad p \in (0, 1], \\ x_{n+1} = x_n - \frac{1}{p^2}[F'(x_n)]^{-1}\left((p^2 + p - 1)F(x_n) + F(z_n)\right) \end{cases} \quad (5.29)$$

under mild differentiability conditions. In particular, we prove that the method (5.29) converges under the same conditions as the Newton method.

5.3.3.1 Local Convergence Order

First, we see that (5.29) has R-order of convergence at least three. If $e_n = x_n - x^*$ is the error of (5.29) in the n-th iterate, the relation $e_{n+1} = C e_n^q + \mathcal{O}\left(\|e_n\|^{q+1}\right)$, where $C \in \mathbb{R}$, is called the error equation and q the order of the method (5.29) [90]. If we substitute $e_n = x_n - x^*$ in (5.29) and simplify, we obtain the error equation for (5.29).

Theorem 5.14 *Suppose that F is a sufficiently continuously Fréchet differentiable operator in Ω. If F has a simple root $x^* \in \Omega$, the operator $[F'(x)]^{-1}$ exists in a neighborhood of x^*, and x_0 is sufficiently close to x^*, then the method (5.29) has R-order of convergence at least three.*

Proof From Taylor's formula

$$0 = F(x^*) = F(x_n) - F'(x_n)e_n + \frac{1}{2!}F''(x_n)e_n^2 - \frac{1}{3!}F'''(x_n)e_n^3 + \mathcal{O}\left(\|e_n\|^4\right),$$

where $e_n = x_n - x^*$, we obtain

$$\Gamma_n F(x_n) = e_n - \frac{1}{2}\Gamma_n F''(x_n)e_n^2 + \frac{1}{6}\Gamma_n F'''(x_n)e_n^3 + \mathcal{O}\left(\|e_n\|^4\right).$$

Moreover, since $z_n - x_n = -p\Gamma_n F(x_n)$, it follows

$$z_n - x_n = -p\, e_n + \frac{p}{2}\Gamma_n F''(x_n)e_n^2 - \frac{p}{6}\Gamma_n F'''(x_n)e_n^3 + \mathcal{O}\left(\|e_n\|^4\right),$$

and, taking again into account Taylor's formula, we have

$$F(z_n) = F(x_n) + F'(x_n)(z_n - x_n) + \frac{1}{2}F''(x_n)(z_n - x_n)^2$$
$$+ \frac{1}{6}F'''(x_n)(z_n - x_n)^3 + \mathcal{O}\left(\|e_n\|^4\right)$$
$$= (1-p)F'(x_n)e_n + \frac{1}{2}(p^2 + p - 1)F''(x_n)e_n^2 + \frac{1}{6}(1 - p - p^3)F'''(x_n)e_n^3$$

5.3 Optimization 2

$$-\frac{p^2}{2}F''(x_n)\Gamma_n F''(x_n)e_n^3 + \mathcal{O}\left(\|e_n\|^4\right).$$

Therefore,

$$\Gamma_n F(z_n) = (1-p)e_n + \frac{1}{2}(p^2+p-1)\Gamma_n F''(x_n)e_n^2$$
$$+ \frac{1}{2}\left(\frac{1-p-p^3}{3}\Gamma_n F'''(x_n) - p^2(\Gamma_n F''(x_n))^2\right)e_n^3 + \mathcal{O}\left(\|e_n\|^4\right).$$

As a consequence, from (5.29), it follows

$$e_{n+1} = x_{n+1} - x^* = e_n - \Gamma_n\left(\frac{p^2+p-1}{p^2}F(x_n) + \frac{1}{p^2}F(z_n)\right)$$
$$= \frac{1}{6}\left((p-1)\Gamma_n F'''(x_n) + 3(\Gamma_n F''(x_n))^2\right)e_n^3 + \mathcal{O}\left(\|e_n\|^4\right),$$

so that (5.29) has R-order of convergence at least three. ∎

5.3.3.2 Semilocal Convergence

Now, we give a semilocal convergence result for (5.29) where mild differentiability conditions are required. In particular, we study the semilocal convergence of (5.29) under usual conditions for the Newton method [74], where F' is Lipschitz continuous in Ω. Thus, we suppose that the operator F is once continuously Fréchet differentiable and the following conditions:

(V1) There exists $\Gamma_0 = [F'(x_0)]^{-1} \in \mathcal{L}(Y, X)$, for some $x_0 \in \Omega$, with $\|\Gamma_0\| \leq \beta$ and $\|\Gamma_0 F(x_0)\| \leq \eta$.
(V2) There exists a constant $K > 0$ such that $\|F'(x) - F'(y)\| \leq K\|x - y\|$, for $x, y \in \Omega$.

Under conditions (V1)–(V2), the semilocal convergence of the Newton method is guaranteed in the next result.

Theorem 5.15 (See [45, 61]) *Let $F : \Omega \subseteq X \to Y$ be a once continuously Fréchet differentiable operator on a non-empty open convex domain Ω of a Banach space X with values in a Banach space Y. Suppose (V1)–(V2). If $B(x_0, r) \subseteq \Omega$, where $r = \frac{2(1-a)}{2-3a}\eta$ and $a = K\beta\eta < \frac{1}{2}$, then the equation $F(x) = 0$ has a solution x^*, and the Newton method converges to x^* and has R-order of convergence at least two.*

Next, we are interested in proving the semilocal convergence of (5.29) under the same conditions as for the Newton method, conditions (V1)–(V2). In view of Theorem 5.15, we consider $R = \frac{(1+a_0/2)\eta}{1-f(a_0)g(a_0)}$, where

$$f(t) = \frac{2}{2 - 2t - t^2}, \qquad g(t) = \frac{t}{8}(8 + 4t + t^2), \tag{5.30}$$

such that $B(x_0, R) \subseteq \Omega$, and denote $K\beta\eta = a_0$. Then $\|y_0 - x_0\| \leq \eta$ and $\|z_0 - x_0\| \leq p\eta$, so that $y_0, z_0 \in \Omega$. Since

$$F(z_0) = (1-p)F(x_0) + p \int_0^1 (F'(x_0 + pt(y_0 - x_0)) - F'(x_0))(y_0 - x_0)\,dt,$$

as a consequence of Taylor's formula, we have, provided that $x_1 \in \Omega$ and $a_0 < \sigma_1 = 0.4111\ldots$, where σ_1 is the root of the real equation $f(a_0)g(a_0) - 1 = 0$, the following:

$$\|x_1 - y_0\| \leq \frac{K}{2}\|\Gamma_0\|\|y_0 - x_0\|^2 \leq \frac{a_0}{2}\|y_0 - x_0\|,$$

$$\|x_1 - x_0\| \leq \|x_1 - y_0\| + \|y_0 - x_0\| \leq \left(1 + \frac{a_0}{2}\right)\|y_0 - x_0\| < R.$$

Therefore, $x_1 \in B(x_0, R)$ if $a_0 < \sigma_1$. Note that the value of R is deduced later.

On the other hand, if $a_0 < \sqrt{3} - 1$, it follows $\|I - \Gamma_0 F'(x_1)\| < 1$, and the operator $\Gamma_1 = [F'(x_1)]^{-1}$ exists by the Banach lemma on invertible operators and is such that $\|\Gamma_1\| \leq f(a_0)\|\Gamma_0\|$. Therefore, y_1 and z_1 are well defined. Moreover, $y_1, z_1 \in \Omega$.

Furthermore, from the next Taylor's formulas,

$$F(x_1) = \frac{1}{p} \int_0^1 \left(F'(x_0) - F'(x_0 + pt(y_0 - x_0))\right)(y_0 - x_0)\,dt$$

$$+ \int_0^1 \left(F'(x_0 + pt(x_1 - x_0)) - F'(x_0)\right)(x_1 - x_0)\,dt,$$

$$F(z_1) = (1-p)F(x_1) + p \int_0^1 \left(F'(x_1 + pt(y_1 - x_1)) - F'(x_1)\right)(y_1 - x_1)\,dt,$$

we obtain

$$\|F(x_1)\| \leq \frac{K}{8}(8 + 4a_0 + a_0^2)\|y_0 - x_0\|^2,$$

$$\|y_1 - x_1\| \leq f(a_0)g(a_0)\|y_0 - x_0\|,$$

$$K\|\Gamma_1\|\|y_1 - x_1\| \leq a_0 f(a_0)^2 g(a_0),$$

$$\|x_2 - y_1\| \leq \frac{a_0}{2} f(a_0)^2 g(a_0)\|y_1 - x_1\|,$$

$$\|x_2 - x_1\| \leq \|x_2 - y_1\| + \|y_1 - x_1\| \leq \left(1 + \frac{a_0}{2} f(a_0)^2 g(a_0)\right) \|y_1 - x_1\|,$$

$$\|x_2 - x_0\| \leq \|x_2 - x_1\| + \|x_1 - x_0\| \leq \left(1 + \frac{a_0}{2}\right)(1 + f(a_0)g(a_0))\|y_0 - x_0\| < R$$

5.3 Optimization 2

and $x_2 \in \Omega$, provided that $a_0 < \sigma_2 = 0.3266\ldots$, where σ_2 is the smallest positive root of the real equation $f(a_0)^2 g(a_0) - 1 = 0$.

Besides, if $a_0 < \sigma_2$, then $\|I - \Gamma_1 F'(x_2)\| < 1$, the operator $\Gamma_2 = [F'(x_2)]^{-1}$ exists by the Banach lemma on invertible operators and is such that $\|\Gamma_2\| \leq f(a_0 f(a_0)^2 g(a_0)) \|\Gamma_1\|$. After that, we can deduce in an analogous way that $y_2, z_2, x_3 \in \Omega$.

Now, we denote $a_0 f(a_0)^2 g(a_0) = a_1$ and define the real sequence:

$$a_n = a_{n-1} f(a_{n-1})^2 g(a_{n-1}), \quad n \in \mathbb{N}, \tag{5.31}$$

which is decreasing and such that $a_n \left(1 + \frac{a_2}{2}\right) < 1$, for all $n \geq 0$, provided that $a_0 < \sigma_2$. Moreover, if $y_n, z_n, x_{n+1} \in \Omega$, this real sequence satisfies a system of recurrence relations, from which we can guarantee that the sequence (5.29) is well defined. To prove them, we follow a similar reasoning to the above and then invoke the induction hypothesis.

Lemma 5.16 *Let f and g be the two real functions defined in (5.30). Suppose (V1)–(V2). If $a_0 = K\beta\eta < \sigma_2 = 0.3266\ldots$, the following items are satisfied for all $n \in \mathbb{N}$:*

(i_n) $\Gamma_n = [F'(x_n)]^{-1}$ exists and $\|\Gamma_n\| \leq f(a_{n-1}) \|\Gamma_{n-1}\|$.

(ii_n) $\|y_n - x_n\| \leq f(a_{n-1}) g(a_{n-1}) \|y_{n-1} - x_{n-1}\| \leq (f(a_0) g(a_0))^n \|y_0 - x_0\| < \eta$.

(iii_n) $K \|\Gamma_n\| \|y_n - x_n\| \leq a_n$.

(iv_n) $\|x_{n+1} - y_n\| \leq \dfrac{a_n}{2} \|y_n - x_n\|$.

(v_n) $\|x_{n+1} - x_n\| \leq \left(1 + \dfrac{a_n}{2}\right) \|y_n - x_n\|$.

(vi_n) $\|x_{n+1} - x_0\| \leq \left(1 + \dfrac{a_0}{2}\right) \dfrac{1 - (f(a_0) g(a_0))^{n+1}}{1 - f(a_0) g(a_0)} \|y_0 - x_0\| < R$, where $R = \dfrac{(1 + a_0/2)\eta}{1 - f(a_0) g(a_0)}$.

Next, the convergence of (5.29) is easily guaranteed from (V1)–(V2), as we can see in the following theorem.

Theorem 5.17 *Let $F : \Omega \subseteq X \to Y$ be a once continuously Fréchet differentiable operator on a non-empty open convex domain Ω of a Banach space X with values in a Banach space Y. Suppose (V1)–(V2). If $a_0 = K\beta\eta < \sigma_2 = 0.3266\ldots$ and $B(x_0, R) \subseteq \Omega$, where $R = \dfrac{(1 + a_0/2)\eta}{1 - f(a_0) g(a_0)}$, then the sequence (5.29) is well defined, lies in $B(x_0, R)$, and starting at x_0 converges to a solution x^* of $F(x) = 0$ in the ball $\overline{B(x_0, R)}$. Besides, the solution x^* is unique in the region $B\left(x_0, \dfrac{2}{K\beta} - R\right) \cap \Omega$ if $R < \dfrac{2}{K\beta}$.*

Proof First, we see that the sequence (5.29) is well defined. Observe

$$\|y_n - x_n\| \leq f(a_{n-1}) g(a_{n-1}) \|y_{n-1} - x_{n-1}\|$$

$$\leq \left(\prod_{i=0}^{n-1} f(a_i)g(a_i) \right) \|y_0 - x_0\|$$

$$\leq (f(a_0)g(a_0))^n \|y_0 - x_0\|$$

as a consequence of the recurrence relation (ii_n) of Lemma 5.16. Therefore, for $m \in \mathbb{N}$, we have

$$\|x_{n+m} - x_n\| \leq \sum_{j=0}^{m-1} \|x_{n+j+1} - x_{n+j}\|$$

$$\leq \sum_{j=0}^{m-1} \left(1 + \frac{a_{n+j}}{2}\right) \|y_{n+j} - x_{n+j}\|$$

$$\leq \left(1 + \frac{a_n}{2}\right) \sum_{j=n}^{n+m-1} \left(\prod_{i=0}^{j-1} f(a_i)g(a_i) \right) \|y_0 - x_0\|$$

$$\leq \left(1 + \frac{a_0}{2}\right) (f(a_0)g(a_0))^n \frac{1 - (f(a_0)g(a_0))^m}{1 - f(a_0)g(a_0)} \eta. \tag{5.32}$$

If $n = 0$ in (5.32), it follows

$$\|x_m - x_0\| \leq \left(1 + \frac{a_0}{2}\right) \frac{1 - (f(a_0)g(a_0))^m}{1 - f(a_0)g(a_0)} \eta < R.$$

Then, $x_m \in B(x_0, R)$, for all $m \in \mathbb{N}$. Analogously, $y_m, z_m \in B(x_0, R)$, for all $m \geq 0$. Therefore, $x_m, y_m, z_m \in \Omega$, for $m \in \mathbb{N}$.

Note that $\{x_n\}$ is a Cauchy sequence as a consequence of (5.32) and $a_0 < \sigma_2$. Then, $\{x_n\}$ converges to x^*, which is a solution of $F(x) = 0$. Indeed, by letting $n \to \infty$, we have $\|\Gamma_n F(x_n)\| \to 0$, and, since $\|F(x_n)\| \leq \|F'(x_n)\| \|\Gamma_n F(x_n)\|$ and the sequence $\{\|F'(x_n)\|\}$ is bounded, we obtain $\|F(x_n)\| \to 0$; by the continuity of F, it follows that $F(x^*) = 0$.

Finally, if we suppose that there exists a solution y^* of $F(x) = 0$ in $B\left(x_0, \frac{2}{K\beta} - R\right) \cap \Omega$ such that $y^* \neq x^*$, we have

$$0 = \Gamma_0(F(y^*) - F(x^*)) = \Gamma_0 \int_0^1 F'(x^* + t(y^* - x^*)) \, dt \, (y^* - x^*) = J(y^* - x^*),$$

where $J = \Gamma_0 \int_0^1 F'(x^* + t(y^* - x^*)) \, dt$. But, since

$$\|I - J\| \leq \|\Gamma_0\| \int_0^1 \|F'(x^* + t(y^* - x^*)) - F'(x_0)\| \, dt$$

$$\leq K\beta \int_0^1 \|x_0 - (x^* + t(y^* - x^*))\| \, dt$$

$$\leq K\beta \int_0^1 \left(t\|y^* - x_0\| + (1-t)\|x^* - x_0\|\right) dt$$

$$< 1,$$

we obtain that the operator J is invertible, and then $y^* = x^*$. ∎

Note that we apply the method (5.29) in the next section.

5.4 Optimization 3

Once we have obtained iterations with efficiency close to the Newton method, we now pay our attention to the region of accessibility. So, in this section, we provide a family of hybrid iterative methods [71], which combines the Newton method as a predictor method with the multipoint iteration constructed in the last section as a corrector method, with the same region of accessibility as the Newton method.

Thus, we pay attention to the conditions that the starting points of the method (5.29) must satisfy to guarantee the convergence of (5.29) from Theorem 5.17. We then observe the region of accessibility of (5.29). If we consider the complex equation $F(z) = \sin z - \frac{1}{3} = 0$ and the particular method (5.29) given for $p = \frac{1}{2}(\sqrt{5} - 1)$,

$$\begin{cases} x_0 \text{ given in } \Omega, \\ y_n = x_n - \Gamma_n F(x_n), \quad n \geq 0, \\ x_{n+1} = x_n - \dfrac{3 + \sqrt{5}}{2} \Gamma_n F\left(x_n + \dfrac{\sqrt{5} - 1}{2}(y_n - x_n)\right), \end{cases} \quad (5.33)$$

which is also presented in [25] for the scalar case, we can see the behavior of (5.33) in Fig. 5.15. Observe that the method (5.33) is also more demanding than the Newton method with respect to the starting points (see Figs. 5.3 and 5.15) as a consequence of its higher speed of convergence. In addition, it is more difficult to locate starting points for the method (5.33) than for the Newton method.

On the other hand, it is clear that the condition $a_0 < \sigma_2 = 0.3266\ldots$ required to guarantee the convergence of (5.29) in Theorem 5.17 is more demanding than that required for the Newton method, $a_0 < \frac{1}{2}$ in Theorem 5.15 under the same conditions (V1)–(V2). Therefore, the application of (5.29) is more restrictive than the application of the Newton method. To illustrate this, we can see the region of accessibility of the Newton method associated with Theorem 5.15 and the region of accessibility of the method (5.33) associated with Theorem 5.17 in Figs. 5.16 and 5.17, respectively, when they are applied to approximate the solutions $z^* = \arctan(1/2\sqrt{2}) = 0.33983\ldots$ and $z^{**} = \pi - \arctan(1/2\sqrt{2}) = 2.80176\ldots$ of $F(z) = \sin z - \frac{1}{3} = 0$ under the same conditions (V1)–(V2). Observe that the domain of starting points for the Newton method is a little bigger than that of the method (5.33) (see the size of the regions of convergence).

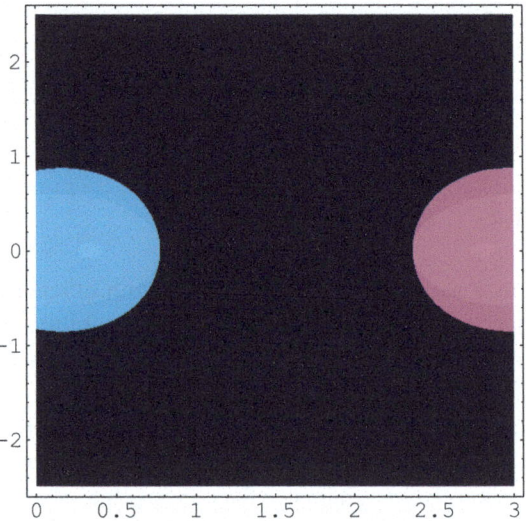

Fig. 5.16 The Newton method

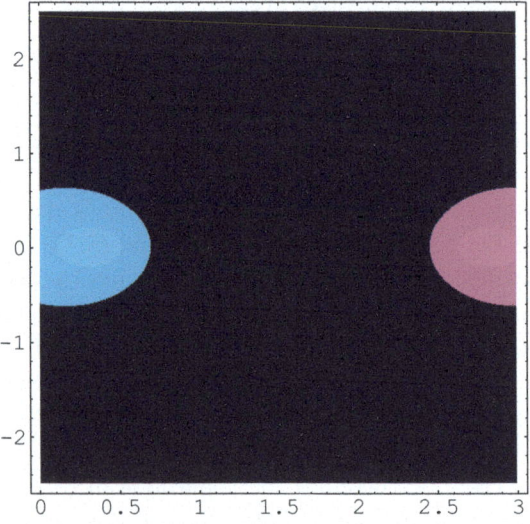

Fig. 5.17 Method (5.33)

5.4 Optimization 3

Since the main aim is to construct iterative methods from (5.29) that converge when they start at the same points as the Newton method, we define the following method:

$$\begin{cases} x_0 \in \Omega, \\ x_n = x_{n-1} - \Gamma_{n-1} F(x_{n-1}), \quad n = 1, 2, \ldots, N_0, \\ \bar{x}_0 = x_{N_0}, \\ \bar{y}_{k-1} = \bar{x}_{k-1} - [F'(\bar{x}_{k-1})]^{-1} F(\bar{x}_{k-1}), \quad k \geq 1, \\ \bar{z}_{k-1} = \bar{x}_{k-1} + p(\bar{y}_{k-1} - \bar{x}_{k-1}), \quad p \in (0, 1], \\ \bar{x}_k = \bar{x}_{k-1} - \frac{1}{p^2}[F'(\bar{x}_{k-1})]^{-1}\left((p^2 + p - 1)F(\bar{x}_{k-1}) + F(\bar{z}_{k-1})\right), \end{cases} \quad (5.34)$$

where x_0 is such that $a = K\beta\eta < \frac{1}{2}$ and $\bar{x}_0 = x_{N_0}$ such that $a_0 = K\tilde{\beta}\tilde{\eta} < \sigma_2 = 0.3266\ldots$ with $\tilde{\beta} \geq \|[F'(\bar{x}_0)]^{-1}\|$ and $\tilde{\eta} \geq \|[F'(\bar{x}_0)]^{-1} F(\bar{x}_0)\|$. In this case, we can apply the Newton method for N_0 steps, provided that the condition $a < \frac{1}{2}$ is satisfied, until the condition $a_0 < \sigma_2$ is satisfied for $\bar{x}_0 = x_{N_0}$, and then apply (5.29) to accelerate the convergence. For this, we have to guarantee the existence of N_0.

5.4.1 Semilocal Convergence

We have seen that the Newton method and the method (5.29) converge under the same conditions (V1)–(V2). The convergence of both methods is guaranteed as a consequence of the fact that both sequences are Cauchy sequences. We use the same argument to prove the semilocal convergence of the method (5.34).

We suppose that the initial iterate x_0 is such that $a = K\beta\eta \in [\sigma_2, 1/2)$, and we look for the existence of $x_{N_0} = \bar{x}_0$, $N_0 \in \mathbb{N}$, such that $a_0 \in (0, \sigma_2)$, where $a_0 = K\tilde{\beta}\tilde{\eta}$, $\tilde{\beta} \geq \|[F'(\bar{x}_0)]^{-1}\|$ and $\tilde{\eta} \geq \|[F'(\bar{x}_0)]^{-1} F(\bar{x}_0)\|$.

We define the scalar sequence

$$\alpha_0 = a, \quad \alpha_n = \frac{\alpha_{n-1}^2}{2(1 - \alpha_{n-1})^2}, \quad n \in \mathbb{N},$$

and, for $n \in \mathbb{N}$, we construct the system of the following three recurrence relations (see [45, 61]):

$$\|\Gamma_n\| \leq \frac{1}{1 - \alpha_{n-1}} \|\Gamma_{n-1}\|,$$

$$\|x_{n+1} - x_n\| \leq \frac{\alpha_{n-1}}{2(1 - \alpha_{n-1})} \|x_n - x_{n-1}\| \leq \delta^n \|\Gamma_0 F(x_0)\|,$$

$$\|x_{n+1} - x_0\| \leq \frac{1 - \delta^{n+1}}{1 - \delta} \|\Gamma_0 F(x_0)\| < r,$$

where $\delta = \frac{a}{2(1-a)}$ and $r = \frac{2(1-a)}{2-3a}\eta$.

The strict decreasing of the positive scalar sequence $\{\alpha_n\}$ guarantees the existence of the element α_{N_0} such that $\alpha_{N_0} < \sigma_2$.

Now, from $\bar{x}_0 = x_{N_0}$,

$$K\|[F'(\bar{x}_0)]^{-1}\|\|[F'(\bar{x}_0)]^{-1}F(\bar{x}_0)\| \leq K\tilde{\beta}\tilde{\eta},$$

we define the initial parameter $a_0 = K\tilde{\beta}\tilde{\eta}$ for (5.29), which starts at $\bar{x}_0 = x_{N_0}$, where x_{N_0} is the last iteration obtained by the Newton method. Next, we define the scalar sequence (5.31) and construct the corresponding system of recurrence relations given in Lemma 5.16, so that the convergence of the sequence (5.29) is guaranteed from the strict decreasing of (5.31). In addition, we can apply (5.34) to approximate a solution of the equation $F(x) = 0$, starting at the same iterate x_0 as the Newton method, since

$$K\|\Gamma_{N_0}\|\|\Gamma_{N_0}F(x_{N_0})\| \leq \alpha_{N_0} = a_0 < \sigma_2,$$

so that $x_{N_0} = \bar{x}_0$ can be chosen to start (5.29) and the convergence of (5.34) is then guaranteed from Theorem 5.17.

Since the sequence (5.34) is well defined, we only have to prove that (5.34) is a Cauchy sequence. For this, we rewrite (5.34) as

$$w_n = \begin{cases} x_n & \text{if } n \leq N_0, \\ \bar{x}_{n-N_0} & \text{if } n > N_0. \end{cases}$$

Theorem 5.18 *Let $F : \Omega \subseteq X \to Y$ be a once continuously Fréchet differentiable operator on a non-empty open convex domain Ω of a Banach space X with values in a Banach space Y. Suppose (V1)–(V2), $K\beta\eta < \frac{1}{2}$, and $B(x_0, r + R) \subseteq \Omega$. Then, the sequence $\{w_n\}$ starting at w_0 converges to a solution x^* of $F(x) = 0$. Moreover, $w_n, x^* \in \overline{B(x_0, r + R)}$. Furthermore, x^* is unique in $B\left(x_0, \frac{2}{K\beta} - (r + R)\right) \cap \Omega$ if $r + R < \frac{2}{K\beta}$.*

Proof From the abovementioned definitions, it is clear that there exists N_0. Besides, $w_i \in \Omega$ for $i = 0, 1, \ldots, N_0$. Indeed, since $w_i = x_i$, for $i = 0, 1, \ldots, N_0$, are iterates of the Newton method, then $\|w_i - x_0\| < r \leq r + R$ ($i = 1, 2, \ldots, N_0$) and $w_i \in \Omega$, for $i = 0, 1, \ldots, N_0$.

After that, we have that $w_{N_0} = \bar{x}_0 = x_{N_0}$ and w_i, for $i > N_0$, are iterates of (5.29), so that $\|w_i - w_{N_0}\| < R$, for $i > N_0$. Thus, $\|w_i - x_0\| \leq \|w_i - w_{N_0}\| + \|w_{N_0} - x_0\| < r + R$, for $i > N_0$, and $w_i \in \Omega$, for $i > N_0$. Therefore, the sequence $\{w_n\}$ is well defined.

5.4 Optimization 3

The fact that $\{w_n\}$ is a Cauchy sequence in Ω follows immediately, since $\{w_n\}_{n \geq N_0}$ is given by (5.29), which is a Cauchy sequence (see Theorem 5.17). Then, $\lim_n w_n = x^*$, $x^* \in \overline{B(x_0, R)} \subset \overline{B(x_0, r + R)}$, and $F(x^*) = 0$.

Finally, the uniqueness of the solution x^* in $B\left(x_0, \frac{2}{K\beta} - (r + R)\right) \cap \Omega$ follows as in Theorem 5.17. ∎

Remark 5.19 Observe that the domain of starting points is extended in Theorem 5.18 compared to Theorem 5.17, so that domains of existence and uniqueness of solution can be given by Theorem 5.18, which cannot be given by Theorem 5.17.

Remark 5.20 Notice that the method (5.34) has R-order of convergence at least two until iteration N_0 and R-order of convergence at least three from iteration $N_0 + 1$.

If we consider again the previous complex equation, $F(z) = \sin z - \frac{1}{3} = 0$, we can see in Figs. 5.18 and 5.19 the regions of accessibility of the method (5.34) with $p = \frac{1}{2}(\sqrt{5} - 1)$ when the condition $K\beta\eta < \frac{1}{2}$ required in Theorem 5.18 is satisfied (Fig. 5.19) or not (Fig. 5.18). Observe that the domain of starting points is the same as that of the Newton method (Fig. 5.16), but the color intensity is different, lighter or darker, according to the number of iterations needed to reach the roots. There are lighter areas for the method (5.34) with $p = \frac{1}{2}(\sqrt{5} - 1)$ as a consequence of the faster speed of convergence.

Fig. 5.18 Method (5.34) with $p = \frac{\sqrt{5}-1}{2}$

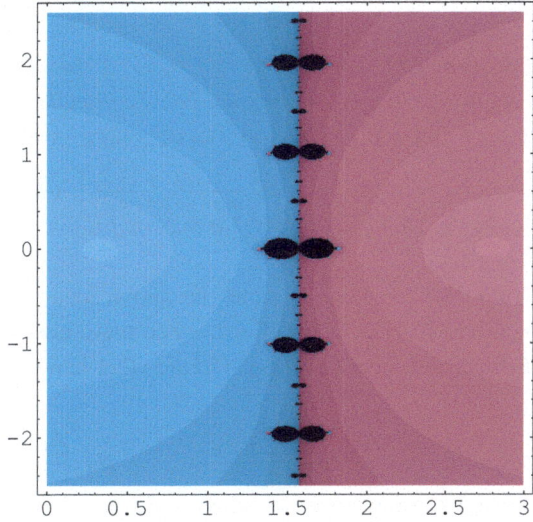

Fig. 5.19 Method (5.34) with $p = \frac{\sqrt{5}-1}{2}$

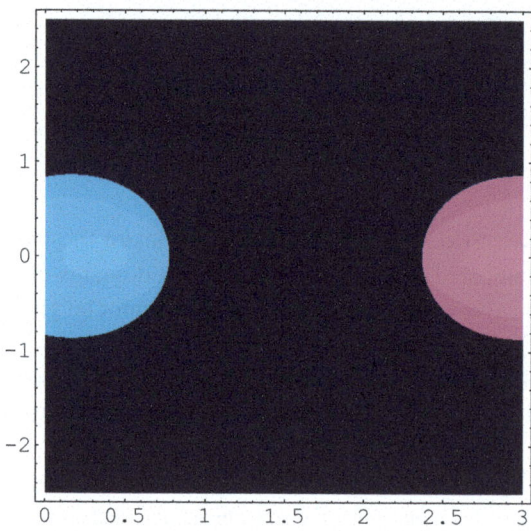

5.4.2 Application

Note that if the conditions of Theorem 5.18 are satisfied, then the method (5.34) can be applied, since N_0 always exists. The aim is now to estimate a priori the value of N_0, which improves the use of (5.34), since the verification of $a_0 < \sigma_2 = 0.3266\ldots$ is saved in every step.

Theorem 5.21 *Let $F : \Omega \subseteq X \to Y$ be a once continuously Fréchet differentiable operator on a non-empty open convex domain Ω of a Banach space X with values in a Banach space Y. Suppose (V1)–(V2) and $a = K\beta\eta \in \left[\sigma_2, \frac{1}{2}\right)$, where $\sigma_2 = 0.3266\ldots$ Set $\bar{x}_0 = x_{N_0}$ with $N_0 = 1 + \left[\frac{\ln \sigma_2 - \ln a}{\ln a - \ln(2(1-a)^2)}\right]$, where $[t]$ denotes the integer part of the real number t. Then, $\bar{x}_0$ is such that the condition $a < \sigma_2$ is satisfied.*

Proof We take into account the above ideas carried out where we guarantee that the method (5.34) is well defined, since there always exists $N_0 \in \mathbb{N}$, so that (5.29) can be applied starting at $\bar{x}_0 = x_{N_0}$. On the other hand,

$$\alpha_{N_0} = \frac{\alpha_{N_0-1}^2}{2(1-\alpha_{N_0-1})^2} = \cdots = \alpha_0 \prod_{i=0}^{N_0-1} \frac{\alpha_i}{2(1-\alpha_i)^2} < a \left(\frac{a}{2(1-a)^2}\right)^{N_0},$$

since the sequence $\{\alpha_n\}$ is decreasing and $\alpha_0 = a$. If $a\left(\frac{a}{2(1-a)^2}\right)^{N_0} < \sigma_2$, then x_{N_0} is a good starting point for (5.29). Thus, if

$$N_0 > \frac{\ln \sigma_2 - \ln a}{\ln a - \ln(2(1-a)^2)},$$

the thesis follows. ∎

Finally, we illustrate how the new iterations can be used to solve a Fredholm integral equation.

Example 5.22 Consider

$$x(s) = 1 + \frac{1}{2} \int_0^1 G(s,t) x(t)^2 \, dt, \quad s \in [0,1], \tag{5.35}$$

where $x \in \mathcal{C}([0,1])$, $t \in [0,1]$, and the kernel G is the Green function in $[0,1] \times [0,1]$. Solving nonlinear integral equation of this type is illustrated by the dynamic model of a chemical reactor (see [16]).

We see in the following that the method (5.5) cannot be applied to solve Eq. (5.35), but the method (5.34) can.

First, we follow a process of discretization to transform (5.35) into a finite-dimensional problem. If we denote the approximation of $x(t_j)$ by x_j, with $j = 1, 2, \ldots, m$, (5.35) is now equivalent to the following nonlinear system of equations:

$$x_j = 1 + \frac{1}{2} \sum_{k=1}^m d_{jk} x_k^2, \quad j = 1, 2, \ldots, m, \tag{5.36}$$

where

$$d_{jk} = \begin{cases} w_k t_k (1 - t_j) & \text{if } k \leq j, \\ w_k t_j (1 - t_k) & \text{if } k < j, \end{cases}$$

and w_k are the weights and t_j, t_k the nodes ($j, k = 1, 2, \ldots, m$) used in the Gauss-Legendre formula to approximate the integral involved in the problem. The system (5.36) can be written in the matrix form as $\mathbf{x} = \mathbf{1} + \frac{1}{2} D \mathbf{y}$, or

$$F(\mathbf{x}) \equiv \mathbf{x} - \mathbf{1} - \frac{1}{2} D \mathbf{y} = 0, \quad F : \mathbb{R}^m \to \mathbb{R}^m, \tag{5.37}$$

where

$$\mathbf{x} = (x_1, x_2, \ldots, x_m)^T, \quad \mathbf{1} = (1, 1, \ldots, 1)^T, \quad D = (d_{jk})_{j,k=1}^m, \quad \mathbf{y} = \left(x_1^2, x_2^2, \ldots, x_m^2\right)^T.$$

Table 5.6 Numerical solution $\mathbf{x}^*$ of the system (5.37)

i	x_i^*	i	x_i^*	i	x_i^*	i	x_i^*
1	1.005450...	3	1.051629...	5	1.069365...	7	1.025815...
2	1.025815...	4	1.069365...	6	1.051629...	8	1.005450...

In addition, we have

$$F'(\mathbf{x})(\mathbf{u}) = \begin{pmatrix} 1-a_{11}x_1 & -a_{12}x_2 & \cdots & -a_{1m}x_m \\ -a_{21}x_1 & 1-a_{22}x_2 & \cdots & -a_{2m}x_m \\ \vdots & \vdots & \ddots & \vdots \\ -a_{m1}x_1 & -a_{m2}x_2 & \cdots & 1-a_{mm}x_m \end{pmatrix} \begin{pmatrix} u_1 \\ u_2 \\ \vdots \\ u_m \end{pmatrix},$$

where $\mathbf{u} = (u_1, u_2, \ldots, u_m)^T$.

If we choose $m = 8$ and $\mathbf{x}_0 = (1.4, 1.4, \ldots, 1.4)^T$, then

$$K = 1, \quad \beta = 1.1382\ldots, \quad \eta = 0.0691\ldots, \quad a = K\beta\eta = 0.4755\ldots < \frac{1}{2}.$$

Observe that we can apply the Newton method to solve (5.37), but we cannot use (5.28) because $a = K\beta\eta \geq \sigma_2 = 0.3266\ldots$ However, by Theorem 5.21, we can use the method (5.33) after the third approximation given by the Newton method, since $N_0 = 3$, and obtain the numerical solution $\mathbf{x}^* = (x_1^*, x_2^*, \ldots, x_8^*)^T$ given in Table 5.6 after four more approximations.

Moreover, by Theorem 5.18, the existence of the solution is guaranteed in the ball $\overline{B(\mathbf{x}_0, 1.1182\ldots)}$ and the uniqueness in $B(\mathbf{x}_0, 0.5440\ldots)$.

Finally, we interpolate the points of Table 5.6, and taking into account that the solution of (5.35) satisfies $x(0) = x(1) = 1$, we obtain the approximation $\tilde{\mathbf{x}}$ of the numerical solution $\mathbf{x}^*$ shown in Fig. 5.20. Notice that the interpolated approximation $\tilde{\mathbf{x}}$ lies within the existence domain of the solution obtained above.

5.5 Optimization 4

The main disadvantage of the Newton method is that the first derivative of the operator involved is used in the algorithm. To solve this problem, we often apply the secant method [7, 10], instead of the Newton method, to approximate a solution of an equation, because the secant method does not use derivatives in the algorithm, since it uses a divided difference of first order as an approximation of the first derivative that appears in the Newton method.

5.5 Optimization 4

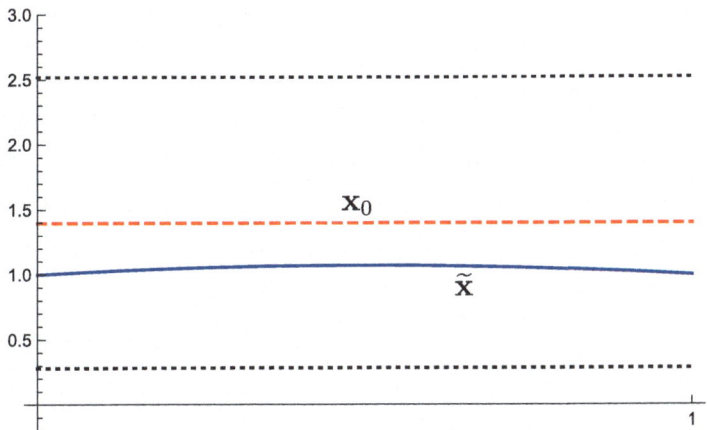

Fig. 5.20 The approximate solution $\widetilde{x}$ of (5.35) (solid line) and the domain of existence of solution (dotted lines)

In this section, we construct two iterative methods that do not use derivatives in their algorithms [40] and see that one of them is more efficient than the secant method, which is the classical method with this feature. The strategy to follow is the following. From iterative methods more efficient than the Newton method, we construct two iterative methods in a similar way that the secant method is done from the Newton method. Then, we analyze the efficiency of these two new methods from the efficiency index and the computational efficiency. After that, we propose the most efficient way to solve an equation.

From an optimization of the Chebyshev method, we construct in Sect. 5.3 the family of iterative methods (5.28) with R-order of convergence at least three, which is more efficient than the Newton method, since the efficiency index and the computational efficiency are higher than those of the Newton method. Taking then into account the abovementioned methods, we construct in Sect. 5.5.1 iterative methods from (5.28) that do not use derivatives in their algorithms, and in Sect. 5.5.2 we find an iterative method more efficient than the secant method. Next, in Sect. 5.5.3, we give a semilocal result in Banach spaces for the iterative method found previously, and we apply this method to approximate a solution of a Fredholm integral equation.

Note that the operator $[x, y; F] \in \mathcal{L}(X, Y)$ is called divided difference of first order for the operator F on the points x and y ($x \neq y$) if

$$[x, y; F](x - y) = F(x) - F(y).$$

We suppose that, for each pair of distinct points $x, y \in \Omega$, there exists a first-order divided difference of the operator F at these points. Moreover, we also suppose

$$\|[x, y; F] - [u, v; F]\| \leq K (\|x - u\| + \|y - v\|), \quad K \geq 0, \quad x, y, u, v \in \Omega,$$

$$x \neq y, u \neq v.$$

In this case, we say that F has a Lipschitz-continuous divided difference in Ω. In addition, under this situation, we have $[x, x; F] = F'(x)$, for all $x \in \Omega$ (see [10,77]). Besides, in this case, when the operator F is Fréchet differentiable, we can define the divided difference as

$$[x, y; F](x - y) = F(x) - F(y) = \int_y^x F'(z)\,dz.$$

5.5.1 Construction of Two Methods

First, we note that the efficiency indices and the computational efficiencies of the methods included in (5.28) have similar order, since the family (5.28) has R-order of convergence at least three. But we can choose (5.28) with $p = 1$ because this method has the least operational cost between the methods of (5.28). It is easy to see the last if we consider $F(x) = 0$ as a nonlinear system of dimension m, since 3m operations (products and divisions) and one evaluation of F are deducted when $p = 1$ in (5.28), although the operational cost of the algorithm is of order m^3 operations. Then, for simplification, we choose the method (5.28) with $p = 1$.

Second, we observe that (5.28) with $p = 1$ is a method of two steps, but the same first derivative of the operator F, $F'(x_n)$, is involved in the two steps. To approximate $F'(x_n)$ by a first-order divided difference, we have two alternatives (take into account in both that $F'(w)$ is approximated by $[u, v; F]$ with u and v close to w in (5.28)). The first alternative consists of approximating $F'(x_n)$ by $[x_{n-1}, x_n; F]$ in the two steps, so that (5.28) with $p = 1$ turns into

$$\begin{cases} x_0, x_{-1} \text{ given in } \Omega, \\ [x_{n-1}, x_n; F]\gamma_n = -F(x_n), \quad n \geq 0, \\ y_n = x_n + \gamma_n, \\ [x_{n-1}, x_n; F]\xi_n = -F(y_n), \\ x_{n+1} = x_n + \gamma_n + \xi_n. \end{cases} \quad (5.38)$$

The second alternative consists of approximating $F'(x_n)$ by $[x_{n-1}, x_n; F]$ in the first step and by $[y_n, x_n; F]$ in the second step, taking into account that y_n has already been computed in the first step of the method. In this case, we obtain

5.5 Optimization 4

$$\begin{cases} u_0, u_{-1} \text{ given in } \Omega, \\ [u_{n-1}, u_n; F]\gamma_n = -F(u_n), \quad n \geq 0, \\ v_n = u_n + \gamma_n, \\ [v_n, u_n; F]\varepsilon_n = -F(v_n), \\ u_{n+1} = u_n + \gamma_n + \varepsilon_n. \end{cases} \quad (5.39)$$

5.5.2 Analysis of the Methods

In this section, we analyze the efficiencies of the methods (5.38) and (5.39) with regard to the secant method

$$\begin{cases} x_0, x_{-1} \text{ given in } \Omega, \\ [x_{n-1}, x_n; F]\gamma_n = -F(x_n), \quad n \geq 0, \\ x_{n+1} = x_n + \gamma_n, \end{cases} \quad (5.40)$$

in the sense defined by Traub [88]. For this, we use the efficiency index and the computational efficiency.

We begin by introducing a result on the local convergence of the sequences defined in (5.38) and (5.39), which, following the ideas given in [87], leads to the R-order of convergence of the sequences (5.38) and (5.39). Remember that the R-order of convergence of the secant method (5.40) is at least $\frac{1+\sqrt{5}}{2}$, see [10,77].

Theorem 5.23 *Let x^* be a solution of $F(x) = 0$ such that the operator $[F'(x^*)]^{-1}$ exists. Suppose that there exists the first-order divided difference $[x, y; F] \in \mathcal{L}(\Omega, Y)$, for all $x, y \in \Omega$, which satisfies*

$$\left\| [F'(x^*)]^{-1} ([x, y; F] - [u, v; F]) \right\| \leq K (\|x - u\| + \|y - v\|), \quad K \geq 0,$$

$$x, y, u, v \in \Omega,$$

and $B(x^*, r) \subset \Omega$, where $r = \frac{1}{5K}$. Then, for $x_{-1}, x_0, u_{-1}, u_0 \in B(x^*, r)$, the sequences $\{x_n\}, \{y_n\}, \{u_n\}, \{v_n\}$ given in (5.38) and (5.39) are well defined, belong to $B(x^*, r)$, and converge to x^*. Moreover,

$$\left. \begin{aligned} \|y_n - x^*\| &= \frac{K (\|x_n - x_{n-1}\| + \|x_n - x^*\|)}{1 - K (\|x_{n-1} - x^*\| + \|x_n - x^*\|)} \|x_n - x^*\|, \\ \|x_{n+1} - x^*\| &= \frac{K (\|y_n - x_{n-1}\| + \|x_n - x^*\|)}{1 - K (\|x_{n-1} - x^*\| + \|x_n - x^*\|)} \|y_n - x^*\|, \end{aligned} \right\} \quad (5.41)$$

$$\|v_n - x^*\| = \frac{K\left(\|u_n - u_{n-1}\| + \|u_n - x^*\|\right)}{1 - K\left(\|u_{n-1} - x^*\| + \|u_n - x^*\|\right)} \|u_n - x^*\|,$$

$$\|u_{n+1} - x^*\| = \frac{K\|u_n - x^*\|}{1 - K\left(\|v_n - x^*\| + \|u_n - x^*\|\right)} \|v_n - x^*\|.$$

Proof Let $x_{-1}, x_0 \in B(x^*, r)$. As

$$\left\| I - [F'(x^*)]^{-1}[x_{-1}, x_0; F] \right\| = \left\| [F'(x^*)]^{-1} \left([x^*, x^*; F] - [x_{-1}, x_0; F] \right) \right\| < \frac{2}{5} < 1,$$

from the Banach lemma on invertible operators, we obtain that the operator $[x_{-1}, x_0; F]^{-1}$ exists and

$$\left\| [x_{-1}, x_0; F]^{-1} F'(x^*) \right\| \leq \frac{1}{1 - K\left(\|x_{-1} - x^*\| + \|x_0 - x^*\|\right)}.$$

Next, we observe that

$$y_0 - x^* = x_0 - [x_{-1}, x_0; F]^{-1} F(x_0) - x^*$$
$$= x_0 - x^* - [x_{-1}, x_0; F]^{-1}[x_0, x^*; F](x_0 - x^*)$$
$$= -[x_{-1}, x_0; F]^{-1} F'(x^*) \left(F'(x^*)^{-1} \left([x_0, x^*; F] - [x_{-1}, x_0; F] \right) \right) (x_0 - x^*),$$

and then $y_0 \in B(x^*, r)$, since

$$\|y_0 - x^*\| \leq \frac{K\left(\|x_0 - x_{-1}\| + \|x_0 - x^*\|\right)}{1 - K\left(\|x_{-1} - x^*\| + \|x_0 - x^*\|\right)} \|x_0 - x^*\| < r.$$

After that, from

$$x_1 - x^* = y_0 - [x_{-1}, x_0; F]^{-1} F(y_0) - x^*$$
$$= -[x_{-1}, x_0; F]^{-1} F'(x^*) \left(F'(x^*)^{-1} \left([y_0, x^*; F] - [x_{-1}, x_0; F] \right) \right) (y_0 - x^*),$$

we obtain

$$\|x_1 - x^*\| \leq \frac{K\left(\|y_0 - x_{-1}\| + \|x_0 - x^*\|\right)}{1 - K\left(\|x_{-1} - x^*\| + \|x_0 - x^*\|\right)} \|y_0 - x^*\| < r,$$

and $x_1 \in B(x^*, r)$.

If we now suppose that $x_n, y_n \in B(x^*, r)$, we can prove by mathematical induction on n that $x_{n+1}, y_{n+1} \in B(x^*, r)$, so that x_n and y_n belong to $B(x^*, r)$ for all positive integers

5.5 Optimization 4

n. Moreover, (5.41) is satisfied for all $n \geq 0$. Furthermore,

$$\|x_{n+1} - x^*\| \leq \frac{K\left(\|y_n - x_{-1}\| + \|x_{n-1} - x^*\| + \|x_n - x^*\|\right)}{1 - K\left(\|x_{n-1} - x^*\| + \|x_n - x^*\|\right)} \|y_n - x^*\| < \|x_n - x^*\|,$$

and then $\lim_n x_n = x^*$.

The results concerning the sequences $\{u_n\}$, $\{v_n\}$ defined in (5.39) follow in a similar way to the abovementioned results. ∎

Corollary 5.24 *Under the same hypotheses of Theorem 5.23, we have:*

(a) *The iterative method (5.38) is locally convergent to x^* with R-order of convergence at least two.*
(b) *The iterative method (5.39) is locally convergent to x^* with R-order of convergence at least $1 + \sqrt{2}$.*

Proof We prove item (a), and the proof of item (b) is left to the reader. If we denote $\|y_n - x^*\| = a_n$ and $\|x_{n+1} - x^*\| = b_{n+1}$, we have, from (5.41), that

$$a_n < \frac{3K}{1 - 2Kb_0} b_{n-1} b_n \quad \text{and} \quad b_{n+1} < \frac{3K^2(2 - Kb_0)}{(1 - 2Kb_0)^3} b_{n-1}^2 b_n,$$

since $\{b_n\}$ is a strictly decreasing sequence. In addition, it follows $b_{n+1} < C b_{n-1}^2 b_n$, where $C \in \mathbb{R}$. Based on this, we can write the associated equation of the R-order of convergence of the method (5.38), $t^2 - t - 2 = 0$, whose unique positive root is two. ∎

Once we know the R-orders of convergence of the methods (5.38), (5.39), and (5.40) we analyze the efficiency index and the computational efficiency. For this, we consider that the equation $F(\mathbf{x}) = 0$ represents a nonlinear system of dimension $\mathfrak{m}$, namely $F(x_1, x_2, \ldots, x_\mathfrak{m}) = 0$, where $F : \Omega \subseteq \mathbb{R}^\mathfrak{m} \to \mathbb{R}^\mathfrak{m}$ is a nonlinear function and $F \equiv (F_1, F_2, \ldots, F_\mathfrak{m})$ with $F_i : \Omega \subseteq \mathbb{R}^\mathfrak{m} \to \mathbb{R}$, $i = 1, 2, \ldots, \mathfrak{m}$, count the evaluations of operators involved, and analyze the operational cost needed to apply the three methods.

For solving nonlinear systems of $\mathfrak{m}$ equations, we note that the secant method (5.40) requires the $\mathfrak{m}$ functions F_i, $i = 1, 2, \ldots, \mathfrak{m}$, and the $\mathfrak{m}(\mathfrak{m} - 1)$ evaluations of functions in the divided difference matrix $[\mathbf{x}, \mathbf{y}; F] = ([\mathbf{x}, \mathbf{y}; F]_{ij})_{i,j=1}^{\mathfrak{m}} \in \mathcal{L}(\mathbb{R}^\mathfrak{m}, \mathbb{R}^\mathfrak{m})$, where

$$[\mathbf{x}, \mathbf{y}; F]_{ij} = \frac{1}{x_j - y_j} \big(F_i(x_1, \ldots, x_{j-1}, x_j, y_{j+1}, \ldots, y_\mathfrak{m})$$
$$- F_i(x_1, \ldots, x_{j-1}, y_j, y_{j+1}, \ldots, y_\mathfrak{m})\big),$$

$1 \le i, j \le m$, $\mathbf{x} = (x_1, x_2, \ldots, x_m)^T$ and $\mathbf{y} = (y_1, y_2, \ldots, y_m)^T$ to be evaluated per iteration; namely, m^2 evaluations of functions are required in total. The secant method (5.40) also requires m^2 operations to compute $[x_{n-1}, x_n; F]$, $\frac{1}{3}(m^3 - m)$ operations in the decomposition LU and m^2 operations more for solving two triangular linear systems. Therefore, $\frac{1}{3}(m^3 + 6m^2 - m)$ operations are required in total. In consequence, we have $EI = \left(\frac{1+\sqrt{5}}{2}\right)^{1/m^2}$ and $CE = \left(\frac{1+\sqrt{5}}{2}\right)^{3/(m^3 + 6m^2 - m)}$ for the secant method.

If we now consider the method (5.38), we can easily see that the total number of evaluations of functions required is $m^2 + m$, since a new evaluation of the operator F is needed, $F(y_n)$. Taking into account that the two linear systems to solve have the same matrix of coefficients, the decomposition LU is made only once, but two more triangular linear systems have to be solved (m^2 operations), so that $\frac{1}{3}(m^3 + 9m^2 - m)$ is the total number of operations required. Therefore, we obtain $EI = 2^{1/(m^2 + m)}$ and $CE = 2^{3/(m^3 + 9m^2 - m)}$ for the method (5.38).

Finally, for the method (5.39), we obtain that the total number of evaluations of functions required is $2m^2$, since a new more divided difference operator is evaluated with regard to (5.38), and the total number of operations required is $\frac{2}{3}(m^3 + 6m^2 - m)$, twice that of the method (5.40). Therefore, we have $EI = (1+\sqrt{2})^{1/(2m^2)}$ and $CE = (1+\sqrt{2})^{3/(2(m^3 + 6m^2 - m))}$ for the method (5.39).

To sum up, we observe in Figs. 5.21 and 5.22 that the method (5.38) is the most efficient of the three methods.

Fig. 5.21 Efficiency indices of the methods (5.38) (dashed line), (5.39) (dotted line), and (5.40) (solid line)

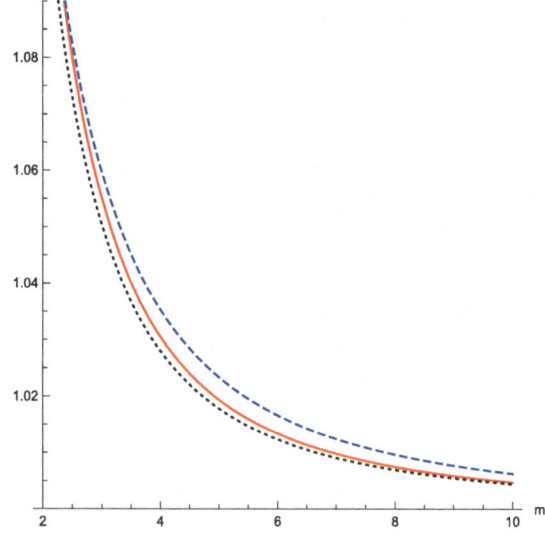

5.5 Optimization 4

Fig. 5.22 Computational efficiencies of the methods (5.38) (dashed line), (5.39) (dotted line), and (5.40) (solid line)

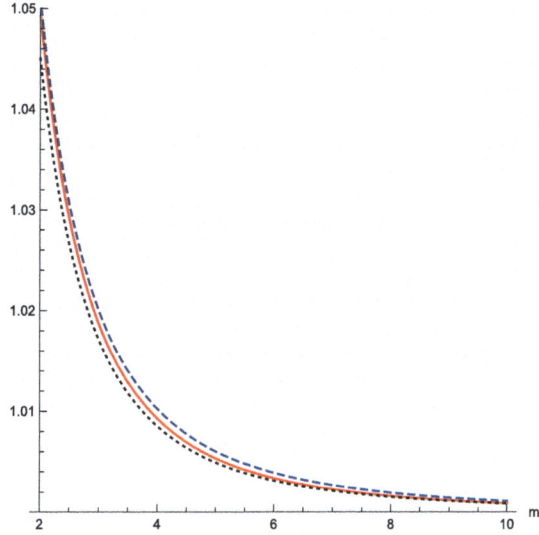

5.5.3 Convergence Analysis

In this section, we analyze the semilocal convergence of the method (5.38). Suppose:

- (W1) There exists a constant $\alpha > 0$ such that $\|x_0 - x_{-1}\| \leq \alpha$.
- (W2) There exists $[x_{-1}, x_0; F]^{-1} = A_0^{-1}$, for some $x_{-1}, x_0 \in \Omega$, with $\|A_0^{-1}\| \leq \beta$ and $\|A_0^{-1} F(x_0)\| \leq \eta$.
- (W3) There exists a constant $K \geq 0$ such that $\|[x, y; F] - [u, v; F]\| \leq K(\|x - u\| + \|y - v\|)$, for $x, y, u, v \in \Omega$.

5.5.3.1 Recurrence Relations

First, we consider the next two scalar functions

$$f(t) = \frac{1}{1-t}, \qquad g(s,t) = s(1+t) + t \tag{5.42}$$

and denote $c_0 = K\beta\eta$, $b_0 = K\beta(\alpha + \eta)$ and $a_0 = b_0(1 + c_0)$. Observe that the two scalar functions are increasing in all the arguments when $t > 0, s > 0$.

Second, we give two approximations of the operator F that we need later.

Lemma 5.25 *If $A_n = [x_{n-1}, x_n; F]$ and $B_n = [x_n, y_n; F]$, then, for all $n \in \mathbb{N}$, we have*

$$F(y_n) = (B_n - A_n)(y_n - x_n),$$
$$F(x_{n+1}) = (A_{n+1} - A_n)(x_{n+1} - x_n) + (A_n - B_n)(y_n - x_n).$$

Third, we observe that $\|y_0 - x_0\| \leq \eta$, $K\|A_0^{-1}\|\|y_0 - x_0\| \leq c_0$, and $\|A_0^{-1}\|\|B_0 - A_0\| \leq b_0$. Moreover, if $x_1 \in \Omega$, then

$$\|x_1 - y_0\| \leq b_0\|y_0 - x_0\| \leq b_0\eta,$$

$$\|x_1 - x_0\| \leq (1 + b_0)\|y_0 - x_0\| \leq (1 + b_0)\eta,$$

$$\|A_0^{-1}\|\|A_1 - A_0\| \leq a_0.$$

Now, from the Banach lemma on invertible operators, there exists A_1^{-1} and $\|A_1^{-1}\| \leq f(a_0)\beta$, since $\|I - A_0^{-1}A_1\| \leq a_0$.

Taking now into account Lemma 5.25 and (5.38), we have

$$\|F(x_1)\| \leq ((1 + b_0)\|A_1 - A_0\| + \|A_0 - B_0\|)\|y_0 - x_0\|,$$

$$\|y_1 - x_1\| \leq \|A_1^{-1}\|\|F(x_1)\| \leq f(a_0)g(a_0, b_0)\|y_0 - x_0\|,$$

$$\|y_1 - x_0\| \leq \|y_1 - x_1\| + \|x_1 - x_0\| \leq (1 + b_0 + f(a_0)g(a_0, b_0))\|y_0 - x_0\|,$$

$$K\|A_1^{-1}\|\|y_1 - x_1\| \leq f(a_0)^2 g(a_0, b_0)c_0,$$

$$\|A_1^{-1}\|\|B_1 - A_1\| \leq f(a_0)(1 + b_0)c_0 + f(a_0)^2 g(a_0, b_0)c_0.$$

Taking again into account Lemma 5.25 and (5.38) and denoting $b_1 = f(a_0)(1+b_0)c_0 + f(a_0)^2 g(a_0, b_0)c_0$, we obtain

$$\|F(y_1)\| \leq \|B_1 - A_1\|\|y_1 - x_1\|,$$

$$\|x_2 - y_1\| \leq \|A_1^{-1}\|\|F(y_1)\| \leq b_1\|y_1 - x_1\|,$$

$$\|x_2 - x_1\| \leq \|x_2 - y_1\| + \|y_1 - x_1\| \leq (1 + b_1)\|y_1 - x_1\|,$$

$$\|x_2 - x_0\| \leq \|x_2 - x_1\| + \|x_1 - x_0\| \leq (1 + b_0 + (1 + b_1)f(a_0)g(a_0, b_0))\eta,$$

provided that $x_2 \in \Omega$. Moreover,

$$\|A_1^{-1}\|\|A_2 - A_1\| \leq \|A_1^{-1}\|(\|x_1 - x_0\| + \|x_2 - x_1\|)$$

$$\leq f(a_0)(1 + b_0)c_0 + (1 + b_1)f(a_0)^2 g(a_0, b_0)c_0.$$

Then, we denote

$$c_0 f(a_0)^2 g(a_0, b_0) = c_1, \qquad f(a_0)(1 + b_0)c_0 + c_1 = b_1, \qquad b_1(1 + c_1) = a_1,$$

and define, for all $n \in \mathbb{N}$, the following scalar sequences:

5.5 Optimization 4

$$c_n = c_{n-1}f(a_{n-1})^2 g(a_{n-1}, b_{n-1}),$$
$$b_n = f(a_{n-1})(1 + b_{n-1})c_{n-1} + c_n, \qquad (5.43)$$
$$a_n = b_n(1 + c_n).$$

After that, we give two properties of the sequences defined in (5.43) that are needed later.

Lemma 5.26 *Let f and g be the two scalar functions defined in (5.42). If*

$$a_0 < \frac{3 - \sqrt{2}}{2}, \qquad b_0 < \frac{a_0^2 - 3a_0 + 1}{a_0 + 1} \qquad \text{and} \qquad c_0 < \frac{(a_0 - 1)^2 b_0}{1 + 2b_0}, \qquad (5.44)$$

then $\Delta = f(a_0)g(a_0, b_0) < 1$, and the sequences $\{a_n\}$, $\{b_n\}$, and $\{c_n\}$ are decreasing.

Next, we guarantee that the sequence (5.38) is well defined from the following lemma, where a system of recurrence relations, which the sequence (5.38) satisfies, is given. The proof of the lemma follows in a similar way to the abovementioned proof and using mathematical induction.

Lemma 5.27 *Let f and g be the two scalar functions defined in (5.42). Suppose (W1)–(W2)–(W3). If (5.44) is satisfied and $B(x_0, R\eta) \subseteq \Omega$, where $R = \frac{1+b_0}{1-\Delta}$ and $\Delta = f(a_0)g(a_0, b_0)$, the following recurrence relations are true for all $n \in \mathbb{N}$:*

- $(i_n) \quad \|A_{n-1}^{-1}\| \|A_n - A_{n-1}\| \leq a_{n-1}$
- $(ii_n) \quad \|A_n^{-1}\| \leq f(a_{n-1}) \|A_{n-1}^{-1}\|$
- $(iii_n) \quad \|y_n - x_n\| \leq f(a_n) g(a_n, b_n) \|y_{n-1} - x_{n-1}\|$
- $(iv_n) \quad \|y_n - x_0\| \leq (1 + b_0) \dfrac{1 - \Delta^{n+1}}{1 - \Delta} \|y_0 - x_0\| < R\eta$
- $(v_n) \quad K \|A_{n-1}^{-1}\| \|y_n - x_n\| \leq c_n$
- $(vi_n) \quad \|A_{n-1}^{-1}\| \|B_n - A_n\| \leq b_n$
- $(vii_n) \quad \|x_{n+1} - x_n\| \leq (1 + b_n) \|y_n - x_n\|$
- $(viii_n) \quad \|x_{n+1} - x_0\| \leq (1 + b_0) \dfrac{1 - \Delta^{n+1}}{1 - \Delta} \|y_0 - x_0\| < R\eta$

where $A_n = [x_{n-1}, x_n; F]$ and $B_n = [x_n, y_n; F]$, for all $n \in \mathbb{N}$.

Observe from (i_n) that $x_n \in \Omega$, for all $n \geq 0$, provided that the hypotheses of the last lemma are satisfied.

5.5.3.2 Semilocal Convergence

Once we have seen that the sequence $\{x_n\}$ defined in (5.38) is well defined, we can prove that (5.38) is convergent, since it is a Cauchy sequence. Moreover, we can also draw conclusions about the existence and uniqueness of solutions of the equation $F(x) = 0$.

Theorem 5.28 *Let $x_{-1}, x_0 \in \Omega$, and suppose that conditions (W1)–(W2)–(W3) are satisfied. If $c_0 = K\beta\eta$, $b_0 = K\beta(\alpha + \eta)$, and $a_0 = b_0(1 + c_0)$ satisfies (5.44) and $B(x_0, R\eta) \subseteq \Omega$, where $R = \frac{1+b_0}{1-\Delta}$, $\Delta = f(a_0)g(a_0, b_0)$, and f and g are the functions defined in (5.42), then the sequence (5.38) starting at x_{-1} and x_0 converges to a solution x^* of $F(x) = 0$, and the solution x^* and the iterations x_n belong to $\overline{B(x_0, R\eta)}$. Moreover, the solution x^* is unique in $\Omega_0 = B(x_0, r) \cap \Omega$, where $r = \frac{1}{K\beta} - \alpha - R\eta$, provided that $R < \frac{1+c_0-b_0}{c_0}$.*

Proof To prove that the sequence (5.38) is convergent, it is enough to see that (5.38) is a Cauchy sequence. For $m \in \mathbb{N}$, we have

$$\|x_{n+m} - x_n\| \leq \sum_{i=0}^{m-1} \|x_{n+i+1} - x_{n+i}\|$$

$$\leq \sum_{i=n}^{n+m-1} (1 + b_i) \left(\prod_{j=0}^{i-1} f(a_j)g(a_j, b_j) \right) \|y_0 - x_0\|$$

$$\leq \sum_{i=n}^{n+m-1} (1 + b_0) \Delta^i \|y_0 - x_0\|$$

$$\leq (1 + b_0) \Delta^n \frac{1 - \Delta^m}{1 - \Delta} \|y_0 - x_0\|,$$

since $f(a_j)g(a_j, b_j) \leq \Delta$, for $j = 1, 2, \ldots, m + n - 1$. As a consequence, the sequence (5.38) converges to a solution x^* of the equation $F(x) = 0$. Moreover, if $n = 0$ in the last, we obtain

$$\|x_m - x_0\| \leq (1 + b_0) \frac{1 - \Delta^m}{1 - \Delta} \|y_0 - x_0\| < R\eta$$

and $x_m \in B(x_0, R\eta)$, for all $m \in \mathbb{N}$.

On the other hand, from

$$\|F(x_n)\| \leq (\|A_n - A_{n-1}\|(1 + b_{n-1}) + \|B_{n-1} - A_{n-1}\|) \|y_{n-1} - x_{n-1}\|$$

$$\leq (\|A_n - A_{n-1}\|(1 + b_{n-1}) + \|B_{n-1} - A_{n-1}\|) \Delta^{n-1} \eta$$

and $\Delta < 1$, it follows $F(x^*) = 0$ by the continuity of F.

To show the uniqueness of x^*, we suppose that there exists another solution z^* of $F(x) = 0$ in Ω_0 such that $z^* \neq x^*$. From $[z^*, x^*; F](z^* - x^*) = F(z^*) - F(x^*)$, we obtain $z^* = x^*$, provided that the operator $J = [z^*, x^*; F]$ is invertible. For this, we do

5.5 Optimization 4

$$\|I - A_0^{-1}J\| \le \|A_0^{-1}\|\|A_0 - J\| \le K\beta \left(\|z^* - x_{-1}\| + \|x^* - x_0\|\right) < K\beta(\alpha + r + R\eta) = 1$$

and use the Banach lemma on invertible operators. ∎

5.5.3.3 Application

In this section, we present an application of the previous analysis to the nonlinear integral equation (5.35) given in Example 5.22

$$x(s) = 1 + \frac{1}{2}\int_0^1 G(s,t)x(t)^2\,dt, \quad s \in [0,1],$$

where $x \in \mathcal{C}([0,1])$ and the kernel G is the Green function in $[0,1] \times [0,1]$.

We have seen in Example 5.22 that Eq.(5.35) can be transformed into the nonlinear system (5.37)

$$F(\mathbf{x}) \equiv \mathbf{x} - \mathbf{1} - \frac{1}{2}D\mathbf{y} = 0, \quad F: \mathbb{R}^m \to \mathbb{R}^m,$$

where

$$\mathbf{x} = (x_1, x_2, \ldots, x_m)^T, \quad \mathbf{1} = (1, 1, \ldots, 1)^T, \quad D = (d_{jk})_{j,k=1}^m, \quad \mathbf{y} = \left(x_1^2, x_2^2, \ldots, x_m^2\right)^T$$

and

$$d_{jk} = \begin{cases} w_k t_k (1 - t_j) & \text{if } k \le j, \\ w_k t_j (1 - t_k) & \text{if } k < j, \end{cases}$$

with w_k the weights and t_j, t_k the nodes ($j, k = 1, 2, \ldots, m$) used in the Gauss-Legendre formula used to approximate the integral involved in the problem.

Moreover, for $\mathbf{u}, \mathbf{v} \in \mathbb{R}^m$, $[\mathbf{u}, \mathbf{v}; F] = ([\mathbf{u}, \mathbf{v}; F]_{ij})_{i,j=1}^m \in \mathcal{L}(\mathbb{R}^m, \mathbb{R}^m)$, where

$$[\mathbf{u}, \mathbf{v}; F]_{ij} = \frac{1}{u_j - v_j}\left(F_i(u_1, \ldots, u_j, v_{j+1}, \ldots, v_m) - F_i(u_1, \ldots, u_{j-1}, v_j, \ldots, v_m)\right),$$

$\mathbf{u} = (u_1, u_2, \ldots, u_m)^T$ and $\mathbf{v} = (v_1, v_2, \ldots, v_m)^T$, we have $[\mathbf{u}, \mathbf{v}; F] = I - S$, where $S = \frac{1}{2}\text{diag}\{\mathbf{u} + \mathbf{v}\}D$ and $S = (s_{ij})_{i,j=1}^m$ with $s_{ij} = \frac{1}{2}(u_j + v_j)d_{ij}$.

If we now choose $m = 8$ and the starting points $\mathbf{x}_{-1} = (0.9, 0.9, \ldots, 0.9)^T$ and $\mathbf{x}_0 = (1, 1, \ldots, 1)^T$, we obtain $\alpha = 0.1$, $\beta = 1.133\ldots$, $\eta = 0.06996\ldots$ and $K = 0.06178\ldots$ for Theorem 5.28, since

$$\|[\mathbf{u}, \mathbf{v}; F] - [\mathbf{w}, \mathbf{z}; F]\| = \|S - P\| \le \frac{1}{2}\|D\|(\|\mathbf{u} - \mathbf{w}\| + \|\mathbf{v} - \mathbf{z}\|),$$

where $\mathbf{u} = (u_1, u_2, \ldots, u_8)^T$, $\mathbf{v} = (v_1, v_2, \ldots, v_8)^T$, $\mathbf{w} = (w_1, w_2, \ldots, w_8)^T$, $\mathbf{z} = (z_1, z_2, \ldots, z_8)^T$, $S = (s_{ij})_{i,j=1}^8$ with $s_{ij} = \frac{1}{2}(u_j + v_j)d_{ij}$, $P = (p_{ij})_{i,j=1}^8$ with $p_{ij} = \frac{1}{2}(w_j + z_j)d_{ij}$, and $K = \|D\| = 0.06178\ldots$ In addition,

$$a_0 = 0.011950\ldots < \frac{3 - \sqrt{2}}{2} = 0.7930\ldots$$

$$b_0 = 0.011890\ldots < \frac{a_0^2 - 3a_0 + 1}{a_0 + 1} = 0.9530\ldots,$$

$$c_0 = 0.004894\ldots < \frac{(a_0 - 1)^2 b_0}{1 + 2b_0} = 0.0113\ldots$$

Therefore, the hypotheses of Theorem 5.28 are satisfied, and the method (5.38) converges to the solution $\mathbf{x}^* = (x_1^*, x_2^*, \ldots, x_8^*)^T$ given in Table 5.6 of Example 5.22 after eight iterations.

Bibliography

1. G. Alefeld, On the convergence of Halley's method. Am. Math. Monthly 88, 530–536 (1981)
2. D. Altman, Concerning the method of tangent hyperbolas for operator equations. Bull. Ac. Pol. Sci., Ser. Sci., Math., Ast. et Phys. **9**, 633–637 (1961)
3. S. Amat, S. Busquier, Geometry and convergence of some third-order methods. Southwest J. Pure Appl. Math. **2**, 61–72 (2001)
4. S. Amat, S. Busquier, J.M. Gutiérrez, Geometric constructions of iterative functions to solve nonlinear equations. J. Comput. Appl. Math. **157**(1), 197–205 (2003)
5. I.K. Argyros, Quadratic equations and applications to Chandrasekhar's and related equations. Bull. Aust. Math. Soc. **32**(2), 275–292 (1985)
6. I.K. Argyros, On a class of nonlinear integral equations arising in neutron transport. Aequationes Math. **36**(1), 99–111 (1988)
7. I.K. Argyros, The Secant method and fixed points of nonlinear operators. Monatsh. Math. **106**, 85–94 (1988)
8. I.K. Argyros, Remarks on the convergence of Newton's method under Hölder continuity conditions. Tamkang J. Math. **23**(4), 269–277 (1992)
9. I.K. Argyros, D. Chen, Results on the Chebyshev method in Banach spaces. Proyecciones **12**(2), 119–128 (1993)
10. I.K. Argyros, On the Secant method. Publ. Math. Debrecen **43**(3–4), 223–238 (1993)
11. I.K. Argyros, D. Chen, Q.S. Qian, A local convergence theorem for the Super-Halley method in a Banach space. Appl. Math. Lett. **7**(5), 49–52 (1994)
12. I.K. Argyros, On the method of tangent hyperbolas. Approx. Theory Appl. **12**(1), 78–96 (1996)
13. I.K. Argyros, Comparing the radii of some balls appearing in connection to three local convergence theorems for Newton's method. Southwest J. Pure Appl. Math. **2**, 24–28 (1998)
14. K.H. Becker, M. Dörfler, *Dynamical Systems and Fractals* (Cambridge University Press, Cambridge, 1991)
15. V. Berinde, *Iterative Approximation of Fixed Point* (Springer, New York, 2005)
16. D.D. Bruns, J.E. Bailey, Nonlinear feedback control for operating a nonisothermal CSTR near an unstable steady state. Chem. Eng. Sci. **32**, 257–264 (1977)
17. R. Campbell, *Les intégrales euleriennes et leurs aplications* (Dunod, Paris, 1966)
18. V. Candela, A. Marquina, Recurrence relations for rational cubic methods I: the Halley method. Computing **44**, 169–184 (1990)
19. V. Candela, A. Marquina, Recurrence relations for rational cubic methods II: the Chebyshev method. Computing **45**, 355–367 (1990)
20. D. Chandrasekhar, *Radiative Transfer* (Dover, New York, 1960)

21. D. Chen, On the convergence of the Halley method for nonlinear equation of one variable. Tamkang J. Math. **24**(4), 461–467 (1993)
22. D. Chen, I.K. Argyros, Q. S. Qian, A local convergence theorem for the Super-Halley method in a Banach space. Appl. Math. Lett. **7**(5), 49–52 (1994)
23. Z. Ciesielski, Some properties of convex functions on higher orders. Annales Polonici Mathematici **7**, 1–7 (1959)
24. J.B. Conway, *A Course in Functional Analysis* (Springer, Berlin, 1990).
25. A. Cordero, J.R. Torregrosa, A class of multi-point iterative methods for nonlinear equations. Appl. Math. Comput. **197**, 337–344 (2008)
26. H.T. Davis, *Introduction to Nonlinear Differential and Integral Equations* (Dover, New York, 1962).
27. J.E. Dennis, Toward a unified convergence theory for Newton-like methods, in *Nonlinear Functional Analysis and Applications*, ed. by L.B. Rall (Academic Press, New York, 1971)
28. J.E. Dennis, R.B. Schnabel, *Numerical Methods for Unconstrained Optimization and Nonlinear Equations* (SIAM, Philadelphia, 1996)
29. B. Döring, Einige sätze über das verfahren der tangierenden hyperbeln in Banach-räumen. Aplikace Mat. **15**, 418–464 (1970)
30. J.A. Ezquerro, *Construction of Iterative Processes Through Accelerations of Newton's Method* (Spanish), PhD (University La Rioja, Logroño, 1996)
31. J.A. Ezquerro, M.A. Hernández, Different acceleration procedures of Newton's method. Novi Sad J. Math. **27**(1), 1–17 (1997)
32. J.A. Ezquerro, J.M. Gutiérrez, M.A. Hernández, M. A. Salanova, The Application of an Inverse-free Jarratt-type approximation to Nonlinear Integral Equations of Hammerstein-type. Comput. Math. Appl. **36**(4), 9–20 (1998)
33. J.A. Ezquerro, M.A. Hernández, On a convex acceleration of Newton's method. J. Optim. Theory Appl. **100**(2), 311–326 (1999)
34. J.A. Ezquerro, J.M. Gutiérrez, M.A. Hernández, M.A. Salanova, Chebyshev-like methods and quadratic equations. Rev. Anal. Numér. Théor. Approx. **28**(1), 23–35 (1999)
35. J.A. Ezquerro, J.M. Gutiérrez, M.A. Hernández, M.A. Salanova, Solving nonlinear integral equations arising in radiative transfer. Numer. Funct. Anal. and Optimiz. **20**(7–8), 661–673 (1999)
36. J.A. Ezquerro, M.A. Hernández, Chebyshev's approximation algorithms for operators with ω-conditioned first derivative. J. Comput. Anal. Appl. **9**(1), 93–101 (2007)
37. J.A. Ezquerro, M.A. Hernández, Halley's method for operators with unbounded second derivative. Appl. Numer. Math. **57**(3), 354–360 (2007)
38. J.A. Ezquerro, M.A. Hernández, An optimization of Chebyshev's method. J. Complexity **25**(4), 343–361 (2009)
39. J.A. Ezquerro, M.A. Hernández, N. Romero, An extension of Gander's result for quadratic equations. J. Comput. Appl. Math. **234**(4), 960–971 (2010)
40. J.A. Ezquerro, A. Grau, M. Grau-Sánchez, M.A. Hernández, Construction of derivative-free iterative methods from Chebyshev's method. Anal. Appl. (Singap.) **11**(3), 1350009 (2013)
41. J.A. Ezquerro, M.A. Hernández-Verón, How to improve the domain of parameters for Newton's method. Appl. Math. Lett. **48**, 91–101 (2015)
42. J.A. Ezquerro, M.A. Hernández-Verón, *Newton's Method: An Updated Approach of Kantorovich's Theory*. Frontiers in Mathematics (Birkhäuser/Springer, Cham, 2017)
43. J.A. Ezquerro, M.A. Hernández-Verón, Domains of global convergence for Newton's method from auxiliary points. Appl. Math. Lett. **85**, 48–56 (2018)
44. J.A. Ezquerro, M.A. Hernández-Verón, How to obtain global convergence domains via Newton's method for nonlinear integral equations. Mathematics **7**, 553 (2019)

45. J.A. Ezquerro, M.A. Hernández-Verón, *Mild Differentiability Conditions for Newton's Method in Banach Spaces*. Frontiers in Mathematics (Birkhäuser/Springer, Cham, 2020)
46. J.A. Ezquerro, M.A. Hernández-Verón, The Newtonian operator and global convergence balls for Newton's method. Mathematics **8**, 1074 (2020)
47. J.A. Ezquerro, M.A. Hernández-Verón, *Restricted Global Convergence Domains for Integral Equations of the Fredholm-Hammerstein Type*. Studies in Systems, Decision and Control, vol. 340 (Springer, Cham, 2021), pp. 125–148
48. W. Gander, On Halley's iteration method. Am. Math. Monthly **92**, 131–134 (1985)
49. J. Garay, M.A. Hernández, *Degree of Logarithmic Convexity* (Spanish). Publicaciones del Seminario Matemático García de Galdeano, Serie II, Sección 1, no. 26 (Universidad de Zaragoza, 1988)
50. S. Gorn, Maximal convergence intervals and a Gibbs type phenomenon for Newton's approximation procedure. Ann. of Math. (2) **59**, 463–476 (1954)
51. M. Grau, M. Noguera, A variant of Cauchy's method with accelerated fifth-order convergence. Appl. Math. Lett. **17**(5), 509–517 (2004)
52. J. M. Gutiérrez, *Newton's method in Banach spaces* (Spanish), Ph.D., University La Rioja, Logroño (Spain), 1995.
53. J.M. Gutiérrez, M.A. Hernández, Recurrence relations for the Super-Halley method. Comput. Math. Appl. **36**(7), 1–8 (1998)
54. J.M. Gutiérrez, M.A. Hernández, An acceleration of Newton's method: Super-Halley method. J. Comput. Appl. Math. **117**(2–3), 223–239 (2001)
55. M.A. Hernández, A note on Halley's method. Numer. Math. **59**(3), 273–276 (1991)
56. M.A. Hernández, Newton-Raphson's method and convexity. Zb. Rad. Prirod.-Mat. Fak. Ser. Mat. **22**(1), 159–166 (1992)
57. M.A. Hernández, M.A. Salanova *Convexity in Solving Nonlinear Scalar Equations* (Spanish) (Servicio de Publicaciones de la Universidad de La Rioja, Logroño, 1996)
58. M.A. Hernández, M.A. Salanova, Chebyshev method and convexity. Appl. Math. Comput. **95**, 51–62 (1998)
59. M.A. Hernández, M.A. Salanova, Index of convexity and concavity: application to Halley method. Appl. Math. Comput. **103**, 27–49 (1999)
60. M.A. Hernández, Chebyshev's approximation algorithms and applications. Comput. Math. Appl. **41**(3–4), 433–445 (2001)
61. M.A. Hernández, The Newton method for operators with Hölder continuous first derivative. J. Optim. Theory Appl. **109**, 631–648 (2001)
62. M.A. Hernández, N. Romero, On a characterization of some Newton-like methods of R-order at least three. J. Comput. Appl. Math. **183**(1), 53–66 (2005)
63. M.A. Hernández, N. Romero, Toward a unified theory for third R-order iterative methods for operators with unbounded second derivative. Appl. Math. Comput. **215**(6), 2248–2261 (2009)
64. L.V. Kantorovich, On Newton's method for functional equations (Russian). Dokl. Akad. Nauk SSSR **59**, 1237–1240 (1948)
65. L.V. Kantorovich, The majorant principle and Newton's method (Russian). Dokl. Akad. Nauk SSSR **76**, 17–20 (1951)
66. L.V. Kantorovich, Some further applications of principle of majorants (Russian). Dokl. Akad. Nauk SSSR **80**, 849–852 (1951)
67. L.V. Kantorovich, G.P. Akilov, *Functional Analysis* (Pergamon Press, Oxford, 1982)
68. H. Keller, *Numerical Methods for Two-Point Boundary Value Problems* (Dover, New York, 1992)
69. K. Kneisl, Julia sets for the super-Newton method, Cauchy's method, and Halley's method. Chaos **11**(2), 359–370 (2001)
70. A.G. Kurosch, *Ecuaciones algebraicas de grados arbitrarios* (Mir, Moscú, 1976)

71. J.M. McNamee, *Numerical Methods for Roots of Polynomials* (Elsevier, Amsterdam, 2007)
72. M. Nadir, A. Khirani, Adapted Newton-Kantorovich method for nonlinear integral equations. J. Math. Stat. **12**(3), 176–181 (2016)
73. B. O'Neill, *Elementary Differential Geometry* (Academic Press, New York, 1966)
74. J.M. Ortega, The Newton-Kantorovich theorem. Am. Math. Monthly **75**, 658–660 (1968)
75. A.M. Ostrowski, *Solution of Equations in Euclidean and Banach Spaces* (Academic Press, New York, 1943)
76. A.M. Ostrowski, *Solution of Equations and Systems of Equations* (Academic Press, New York, 1966)
77. F.A. Potra, V. Pták, *Nondiscrete Induction and Iterative Processes* (Pitman, New York, 1984)
78. L.B. Rall, A note on the convergence of Newton's method. SIAM J. Numer. Anal. **11**, 34–36 (1974)
79. L.B. Rall, *Computational Solution of Nonlinear Operator Equations* (Robert E. Krieger Publishing Company, New York, 1979)
80. J. Rashidinia, A. Parsa, Analytical-numerical solution for nonlinear integral equations of Hammerstein type. Int. J. Math. Model. Comput. **2**(1), 61–69 (2012)
81. W.C. Rheinboldt, A unified convergence theory for a class of iterative processes. SIAM J. Numer. Anal. **5**, 42–63 (1968)
82. W.C. Rheinboldt, An adaptive continuation process for solving systems of nonlinear equations. Math. Models Numer. Methods, Banach Cent. Publ. **3**, 129–142 (1978)
83. A.W. Roberts, D.E. Varberg, *Convex Functions* (Academic Press, New York, 1973)
84. J. Rokne, Newton's method under mild differentiability conditions with error analysis. Numer. Math. **18**, 401–412 (1972)
85. R.A. Safiev, The method of tangent hyperbolas. Sov. Math. Dokl. **4**, 482–485 (1963)
86. E. Schröder, Über unendlich viele Algorithmen zur Auflösung der Gleichungen. Math. Ann. **2**, 317–365 (1870)
87. S.M. Shakhno, On an iterative algorithm with superquadratic convergence for solving nonlinear operator equations. J. Comput. Appl. Math. **231**, 222–235 (2009)
88. J.F. Traub, *Iterative Methods for the Solution of Equations* (Prentice Hall, New Jersey, 1964)
89. J.L. Varona, Graphic and numerical comparison between iterative methods. Math. Intelligencer **24**, 37–46 (2002)
90. S. Weerakoon, T.G.I. Fernando, A variant of Newton's method with accelerated third-order convergence. Appl. Math. Lett. **13**, 87–93 (2000)
91. S. Wolfram, *The Mathematica Book* 5th ed. (Wolfram Media/Cambridge University Press, 2003).
92. T. Yamamoto, On the method of tangent hyperbolas in Banach spaces. J. Comput. Appl. Math. **21**, 75–86 (1988)
93. T.J. Ypma, Historical development of the Newton-Raphson method. SIAM Rev. **37**(4), 531–551 (1995)
94. H. Zhengda, A note on the Kantorovich theorem for Newton method. J. Comput. Appl. Math. **47**, 211–217 (1993)

The manufacturer's authorised representative in the EU is Springer Nature Customer Service Centre GmbH, Europaplatz 3, 69115 Heidelberg, Germany. If you have any concerns regarding our products, please contact ProductSafety@springernature.com

Printed and bound by CPI Group (UK) Ltd, Croydon, CR0 4YY

26/03/2026

02078943-0016